D0821911

THE FROZEN EARTH
FUNDAMENTALS OF GEOCRYOLOGY

Studies in Polar Research

This series of publications reflects the growth of research activity in and about the polar regions, and provides a means of synthesising the results. Coverage is international and interdisciplinary: the books are relatively short and fully illustrated. Most are surveys of the present state of knowledge in a given subject rather than research reports, conference proceedings or collected papers. The scope of the series is wide and includes studies in all the biological, physical and social sciences.

Other titles in this series:

The Antarctic Circumpolar Ocean
Sir George Deacon

The Living Tundra*
Yu.I. Chernov, transl. D. Love

Arctic Air Pollution
edited by B. Stonehouse

The Antarctic Treaty Regime
edited by Gillian D. Triggs

Antarctica: The Next Decade
edited by Sir Anthony Parsons

Antarctic Mineral Exploitation
Francisco Orrego Vicuna

Transit Management in the Northwest Passage
edited by C. Lamson and D. Vanderzwaag

Canada's Arctic Waters in International Law
Donat Pharand

Vegetation of the Soviet Polar Deserts
V.D. Aleksandrova, translated by D. Love

Reindeer on South Georgia
Nigel Leader-Williams

*Also available in paperback

The Biology of Polar Bryophytes and Lichens
Royce Longton

Microbial Ecosystems
Warwick Vincent

The Frozen Earth

FUNDAMENTALS OF GEOCRYOLOGY

PETER J. WILLIAMS and
MICHAEL W. SMITH

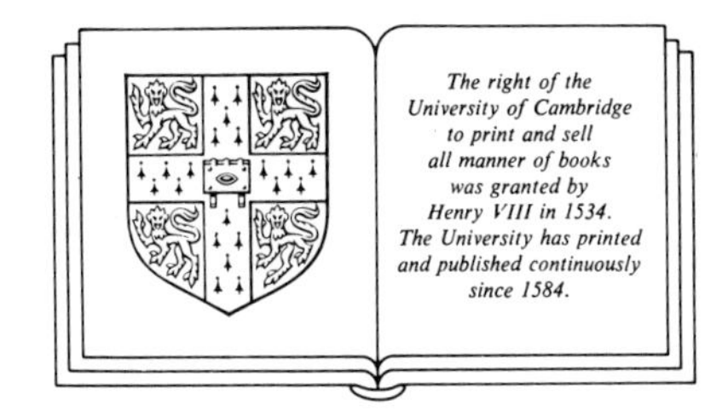

CAMBRIDGE UNIVERSITY PRESS
CAMBRIDGE
NEW YORK NEW ROCHELLE MELBOURNE SYDNEY

Published by the Press Syndicate of the University of Cambridge
The Pitt Building, Trumpington Street, Cambridge CB2 1RP
32 East 57th Street, New York NY 10022, USA
10 Stamford Road, Oakleigh, Melbourne 3166, Australia

First published 1989

Printed in Great Britain by The Alden Press, Oxford

British Library cataloguing in publication data

Williams, Peter J, (Peter John), 1932–
The frozen earth.
1. Periglacial processes
I. Title II. Smith, Michael W.
III. Series
551.3′8

Library of Congress cataloguing in publication data

Williams, Peter J. (Peter John), 1932–
The frozen earth: fundamentals of geocryology/Peter J. Williams and Michael W. Smith.
p. cm. — (Studies in polar research)
Bibliography: p. 23
Includes index.
ISBN 0-521-36534-1
1. Frozen ground. I. Smith, Michael W., 1944-. II. Title.
III. Series.
GB641.W55 1989
551.3′84—dc19 88-30458 CIP

ISBN 0 521 36534 1

AL

Contents

List of symbols

A	cross-sectional area
A_s, A_z	amplitude of temperature wave at surface and at depth z respectively
C	volumetric heat capacity; cohesion
$C_a(T)$	apparent volumetric heat capacity at temperature T
C_s, C_m, C_o, C_w	volumetric heat capacity, soil, soil minerals, organic material and water respectively
C_v	consolidation coefficient
c	mass heat capacity
c_p	mass heat capacity, air at constant pressure
D_v	diffusion coefficient for water vapour in soil
D_θ	soil moisture diffusivity
E	internal energy
e	saturation vapour pressure over ice
G	free energy (absolute)
Gg	geothermal gradient
ΔG	(relative) free energy
ΔG_l	free energy, liquid phase
ΔG_s	free energy, solid phase
ΔG_i	free energy, ice
ΔG_w	free energy, water
H	component of movement perpendicular to slope
ΔH	change of heat storage
h	vertical depth below water table; height of water column
K	thermal conductivity
K_H	turbulent transfer coefficient
K_v	turbulent transfer coefficient, water vapour
$K\downarrow$	short wave radiation, incoming
$K\uparrow$	short wave radiation, outgoing (reflected)
k	constant; hydraulic conductivity, permeability
k_f	permeability of frozen soil to water substance
L	downslope component of movement
L_f	latent heat of fusion
L_v	latent heat of vaporisation
$L\downarrow$	long wave radiation, incoming
$L\uparrow$	long wave radiation, outgoing
l	length
m_v	coefficient of volume compressibility

P	period of temperature wave; pressure; load (weight) of soil
P_a	gas pressure; air pressure
P_w	water pressure
P_i	ice pressure
Q^*	net radiation (radiative surplus or deficit)
Q_G	ground heat flux
Q_H	sensible heat flux (atmospheric)
Q_{LE}	evaporative–condensative heat flux
q	flow, volume per second; specific humidity; flux of water
q_l	flow, liquid phase
q_f	flux through frozen soil
q_u	flux through unfrozen soil
q_v	mass vapour flow; mass flux per unit area
q_s	specific humidity of surface
q^*	saturation specific humidity
R	thaw-consolidation ratio; gas constant for water vapour
r_a	aerodynamic resistance
r_{aw}	radius of meniscus, air/water
r_{iw}	radius of ice/water interface
S	shear strength; specific surface area
s	entropy
s_i	entropy of ice
s_w	entropy of water
T	temperature
ΔT	depression of freezing point; difference (change) in temperature
T_a	air temperature
T_0	freezing point, normal conditions
T_s	surface temperature
T_z	temperature at depth z
t	time
u	pore water pressure
V	volume; specific volume
V_a	specific volume, air
V_i	specific volume, ice
V_w	specific volume, water
w	soil moisture content (gravimetric)
X_m, X_o, X_w	volume fraction soil minerals, organic material, and water
Z_H	component of depth normal to slope
Z	distance; depth

z_p	depth at which geothermal gradient offsets negative temperature at surface
α	albedo
β	angle of slope
γ	unit weight of soil
γ_w	unit weight of water
ε	emissivity; strain
θ	volumetric moisture content
θ_u	water content of frozen soil (unfrozen water content)
θ_i	volumetric ice content
λ	a period of time
κ	thermal diffusivity
κ_a	apparent thermal diffusivity
ρ	density
ρ_d	dry density of soil
ρ_i	density of ice
ρ_s	bulk density of soil
ρ_w	density of water
σ	normal stress
σ'	effective normal stress
σ_n	neutral stress
σ_1	axial stress
σ_3	confining stress
σ_{aw}	interfacial energy (surface tension) air/water
σ_{iw}	interfacial energy (surface tension) ice/water
τ	shear stress
τ_f	shear stress for failure
ϕ	angle of internal friction
$\chi(\Omega)$	factor relating to stress partition
ψ	moisture potential

Preface

Activities in the cold regions of the earth have expanded greatly in recent decades, largely as a result of interest in natural resources (especially oil, gas and minerals), creating a demand for knowledge of the special natural conditions found there. Over this period there have been improvements in geotechnical engineering and a greater scientific understanding of the effects of freezing and thawing upon earth materials. The benefits of this have also extended to regions which experience significant seasonal freezing and thawing. While there are several texts which describe the terrain features of the cold regions, as well as a large technical literature in the form of research papers and conference proceedings, there appeared to be a need for a textbook which integrated the various approaches inherent in our contemporary understanding of the frozen earth.

Much of our knowledge of geocryology is recent, and has been built upon the extensive work of field scientists, as well as the work of geotechnical engineers. As practical needs have become more pressing, the work of laboratory and theoretical scientists–applying the principles of physical chemistry and thermodynamics–has been of fundamental importance. The challenge of this book was to combine and integrate these, and indeed other perspectives into an effective synthesis, one which could provide students and professionals alike with the basic knowledge appropriate to the study of the unique terrain and geotechnical conditions in cold climates.

We have tried to include sufficient description of the natural terrain in order to understand the significance and effects of the thermal, mechanical and hydrological processes which are characteristic of the cold regions. These processes we have dealt with at several levels: so far as possible basic introductory accounts are given, which outline the physical science involved and which provide a starting point for more detailed accounts given in other chapters. The reader may find some element of repetition, which we hope really represents a progression (not necessarily coincident with the order of the chapters). Many topics, of course, turn out to be related, increasing the need for cross-referencing. Depending on the reader's background, some chapters will inevitably seem more familiar than others. But with the exception of the first chapter, which introduces the range of issues covered, and the last, which summarises current themes in the subject, the order of chapters is not designed with any particular kind of reader in mind. Rather, we hope regardless of the order in which the chapters are first read, that we have presented a reasonably comprehensive picture of a rapidly developing subject, one which to a high degree builds upon many different branches of intellectual inquiry. While permafrost and related topics are normally the concern of earth scientists, we hope that this book will also be useful to engineers, ecologists, hydrologists and others.

Acknowledgements

The writing of this book has taken many years, and we are indebted to many colleagues and students for help of various kinds. Dr Branko Ladanyi read and commented upon Chapter 9. Drs Chris Burn and John Wood provided much stimulating discussion of several parts while they were graduate students. Dan Patterson and Dan Riseborough have also worked closely with us and given many helpful comments. We have enjoyed the hospitality over lengthy periods of two of the world's leading institutions for cold regions studies while preparing the book: the Scott Polar Research Institute in Cambridge, England, and the United States Army Cold Regions Research and Engineering Laboratories in Hanover, New Hampshire. We have much appreciated conversations with many of our graduate students through the years, and also with a number of the scientists world wide who have provided much of the recent knowledge of the subject. We hope that the content of the book reflects the importance of their work.

On a more personal note our thanks are due to Beatrice Williams for advice on thermodynamics and to Pauline Smith for advice on presentation–our families have also contributed by cheerfully accepting our preoccupation with this book.

We are grateful to those who have kindly lent photographs and they are acknowledged in the photo captions. We thank the following for permission to reproduce copyrighted materials: Department of Natural Resources, Government of Manitoba for Figure 5.10; Geological Survey of Canada for Figure 4.19, from Figure 14 (Smith 1976); Longman for Figure 7.5 from Figure 6.4 (Williams 1982); Hemisphere Publishing Corporation for the quotation, p. 271 from pp. 135–6 (Tsytovich, 1975); Norwegian Geotechnical Institute for Figures 1.9 and 7.3, from Figures 5 and 8 (Williams 1976).

1

Periglacial conditions

1.1 The significance of freezing in soils and rocks

Water is ubiquitous in the surface regions of the earth, and is unique in the extent of its occurrence in all three phases: solid, liquid and gaseous. Its formative influence on the nature of the earth's surface, on the behaviour of earth materials, and its role in the complex pattern of transfers of energy and mass in the ground and atmosphere mean that water is a central component of study in the earth sciences. However, the solid phase, and transitions between solid and liquid or vapour, have in this context received relatively little attention. While the subject of glaciology has concerned itself with snow and ice, the study of freezing and thawing within soils and rocks has been more limited and mainly of recent date.

Water freezes at 0 °C and in doing so expands by 9% of its volume. In freezing it may exert great expansive pressures. These statements serve for many everyday considerations. But they require qualification in that only pure water, under a pressure of one atmosphere, and in sufficient quantity, freezes at 0 °C. Those conditions are generally not met when freezing occurs within soil or rock materials. This is not a matter of mere scientific pedantry, of trivial deviations of freezing point arising from the fact that natural waters are never pure in the ultimate sense, or from the fact that they are commonly subject to some gravitationally produced pressure arising from weight. Freezing of water in most earth materials occurs over a range of temperature to many degrees below 0 °C (although never at temperatures significantly above 0 °C).

This circumstance should not cause surprise. The boiling point of water is 100 °C – but we do not assume that transitions of water to vapour are limited to that temperature. Meteorologists and climatologists accept as fundamental the role of transitions to and from the vapour phase at all naturally occurring temperatures in the earth's atmosphere, and this is essential to the understanding of the nature of the atmospheric climate.

Although the circumstances are different, phase transition, between solid and liquid, is the basic element in understanding frozen and freezing soil materials. We must consider thermodynamic and thermal properties, as well as the mechanical properties of the materials, in order to understand the behaviour of the ground.

Those parts of the earth's surface which experience freezing temperatures reveal this in a variety of ways. Where frost is an occasional or short-term occurrence the effect may be largely an ecological one, affecting the growth conditions for plant species, perhaps those species of economic significance, so that, for example, the study of night frosts is important in agronomy. But, for perhaps 35% of the earth's land area, mainly in the northern hemisphere, the effects of freezing, and freezing and thawing, are such as to radically affect the nature of the ground surface. This area is referred to as the periglacial regions. In comparison, only some 3% of the earth's surface is covered by perennial snow and ice.

The periglacial regions are characterised by unique displacements of soil material and migrations of water substance, the development of unique terrain features, and of distinctive vegetation, all related to freezing and thawing. These regions are by no means uniform in appearance, however. The cold northern forests, perhaps underlain by permafrost, are to the casual observer as distinct from the sub-arctic peatlands, or from the open tundra, as they are from the climatically quite different temperate regions and their varied terrains. Common to the diverse landscapes of the periglacial regions, however, is the particular behaviour and properties of the soils and rocks at freezing temperatures. The explanation of this involves the thermodynamics of phase transitions of the water in pores and other small openings within the granular mineral materials, the soils and rocks, at the earth's surface.

Perhaps we can imagine a different, and purely hypothetical situation, where freezing and thawing involved only the solidification and liquefaction of water contained in the soils or rocks, at a single temperature and without any of the attributes of the phenomenon as noted above. The effect would be to have two distinct mechanical states, one of rigidity and one of relative weakness. In addition, there would be two hydrological situations one of relative impermeability and the other of permeability. These characteristics are not at all uncommon in earth materials in general and, while they indeed have implications for the stability of slopes, for the moisture conditions of the ground and for the nature of the earth's surface, they are quite insufficient to explain the dominating peculiarities associated with periglacial conditions. Clearly, more is involved in the effects of freezing and thawing.

One of the most fundamental phenomena, well known to inhabitants in cold climates, is that of *frost heave* (Figure 1.1). Many soils, when frozen, are found to have a far greater moisture content than they could possibly have in the unfrozen state. A sample of such soil, thawed in a beaker, may show several centimetres of water lying above the mineral material. The excess moisture accumulates in the soil, as ice, at the time of freezing. The 'heave' refers to the volume increase of the soil which results. It may be revealed by an extension of 10 or 20% or more, of the height of a soil column. This can have great practical significance, through displacement of

0 2 cm

Figure 1.1 Frozen clay showing typical ice layers (or 'lenses'). This ice is mainly from water which is drawn to the freezing zone, and is the cause of the expansion constituting frost heave. See also Chapter 8, Figure 8.12. The ice layers vary greatly in size and form, from soil to soil. Photograph from Williams (1986).

building foundations or road surfaces. Equally important is the effect of thawing of such material (Figure 1.2); the large quantities of water released and the settlement that ensues as drainage occurs have wide-reaching implications. The frost heave process, and also the eventual consequences on thawing, are of the greatest significance in the explanation of many of the unique terrain features of cold regions.

1.2 Freezing and thawing in porous materials

The frost heave phenomenon relates to the occurrence of freezing at temperatures *below* 0 °C. A simple application of the principles of physical chemistry aids our understanding. A property of substances known as the Gibbs free energy is used in studies of phase transition (see, for example, Nash 1970; S. S. Penner 1968; Spanner 1964; also Chapter 7). Two phases coexist when the free energies of the phases are equal. This is a definition of freezing (or solidification, or melting) point. Strictly

Figure 1.2 Aklavik. Northwest Territories. Canada, 1957. Thawing of frost heaved material, with drainage impeded by permafrost, resulted in roads being impassable for wheeled vehicles every summer. Modern construction methods, based on an understanding of the freeze–thaw processes, overcome this problem. Photograph from Williams (1986).

speaking, it is the partial molar free energies which are equal, but we defer a more rigorous analysis until Chapter 7. At temperatures below the freezing point the free energy of the liquid, ΔG_L, is greater than that of the solid, ΔG_s:

At freezing point $\Delta G_s = \Delta G_L$

below freezing point $\Delta G_s < \Delta G_L$

The ΔG is used because we cannot actually measure free energies, but only differences of free energy, relative to some datum.

Two phases of differing free energy cannot coexist in a stable fashion. Instead, the quantity of the phase of lower free energy is augmented at the expense of that of higher free energy. Thus, for pure water under normal conditions, lowering the temperature through 0 °C results in all the water being transferred to ice, which has the lower free energy.

In soils the situation is different. The water commonly contains some dissolved salts, which lower the free energy. As is well known, pure ice forms on the freezing of a solution. But, because the solution has a lower free energy than pure water, for ice to form it too must have a different free energy from ice forming in pure water. Changing the temperature changes the free energy of substances, and in the case of ice at a slower rate than that of the solution. Accordingly, as illustrated in Figure 1.3, at some temperature below 0 °C the free energy of ice is lower than that of the solution, such that ice forms from the solution. This is referred to as a *depression of the freezing point*, and the phenomenon is not limited to solution.

Generally, the concentration of dissolved salts in the soil water is so weak that the freezing point is only 0.1 °C or so below 0 °C on this account. As ice forms in the pores of soils or rocks, however, a far more important effect occurs. The decreasing amount of water is increasingly affected by *capillarity* and *adsorption*. These phenomena cause the free energy of the water to fall further. Consequently for freezing to continue a still lower temperature is required.

Capillarity is an effect associated with molecular forces at the interface between phases, when the interface is confined. The rise of water in capillary tubes follows from the confinement, in the capillary tube, of the interface (the meniscus) between the air and water. The rise is greater as the tube diameter is smaller. The rise is often referred to as the effect of a suction generated in the water at the meniscus. This suction can be equated with a decrease in free energy relative to normal, or 'free' water at the same temperature.

In the freezing soil, formation of ice results in the water being confined progressively in smaller spaces. The free energy of the water falls on this

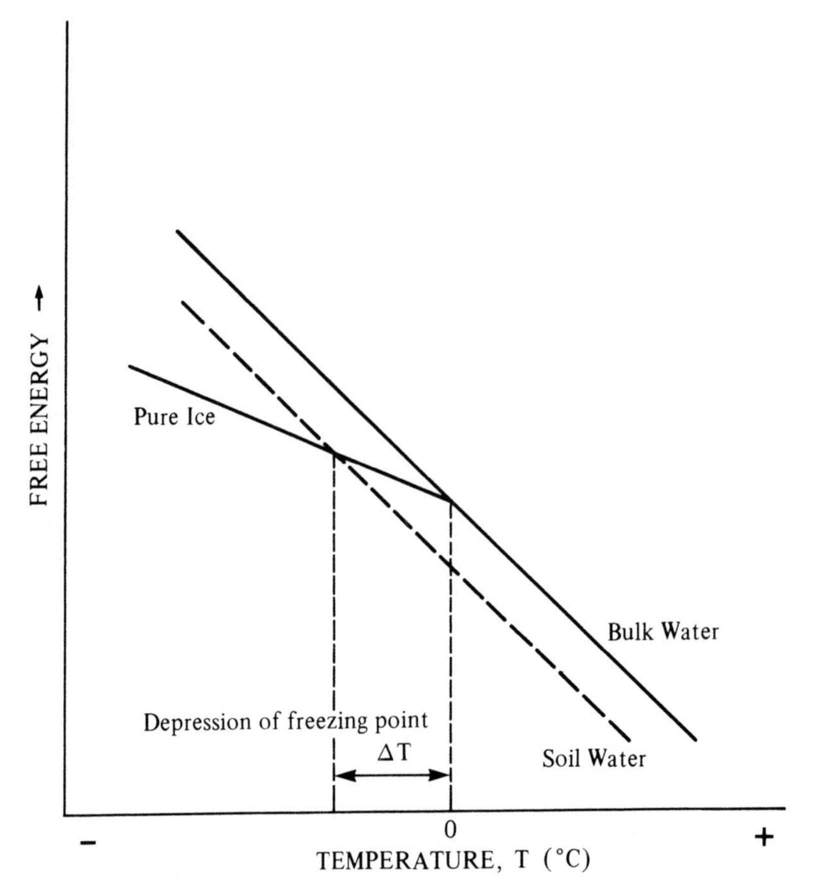

Figure 1.3 Ice and water normally coexist (have equal free energies) at 0 °C. But if the free energy of the water is reduced (dashed line) the freezing point is changed by ΔT (diagram modified after Everett 1961).

account. This effect can be referred to as the *suction*, or *cryosuction*, generated by soil freezing, and it is the cause of migration of water to the freezing zone.

Adsorption has a similar effect. It refers to the influence of forces emanating from the mineral particle surfaces, which reduce the free energy in a thin layer, the 'adsorbed layer', of water on the particles. For freezing to continue, it is increasingly this water which must be converted to ice, and consequently lower and lower temperatures are required.

Figure 1.4 shows the amounts of water remaining unfrozen at temperatures below 0 °C, in different soils. Below about − 1.5 °C this is almost exclusively 'adsorbed'. At temperatures between − 1.5 °C and 0 °C this capillarity is responsible for much of the freezing point depression. The term freezing point is somewhat awkward in the case of a soil since freezing occurs progressively over a range of temperature, any particular tem-

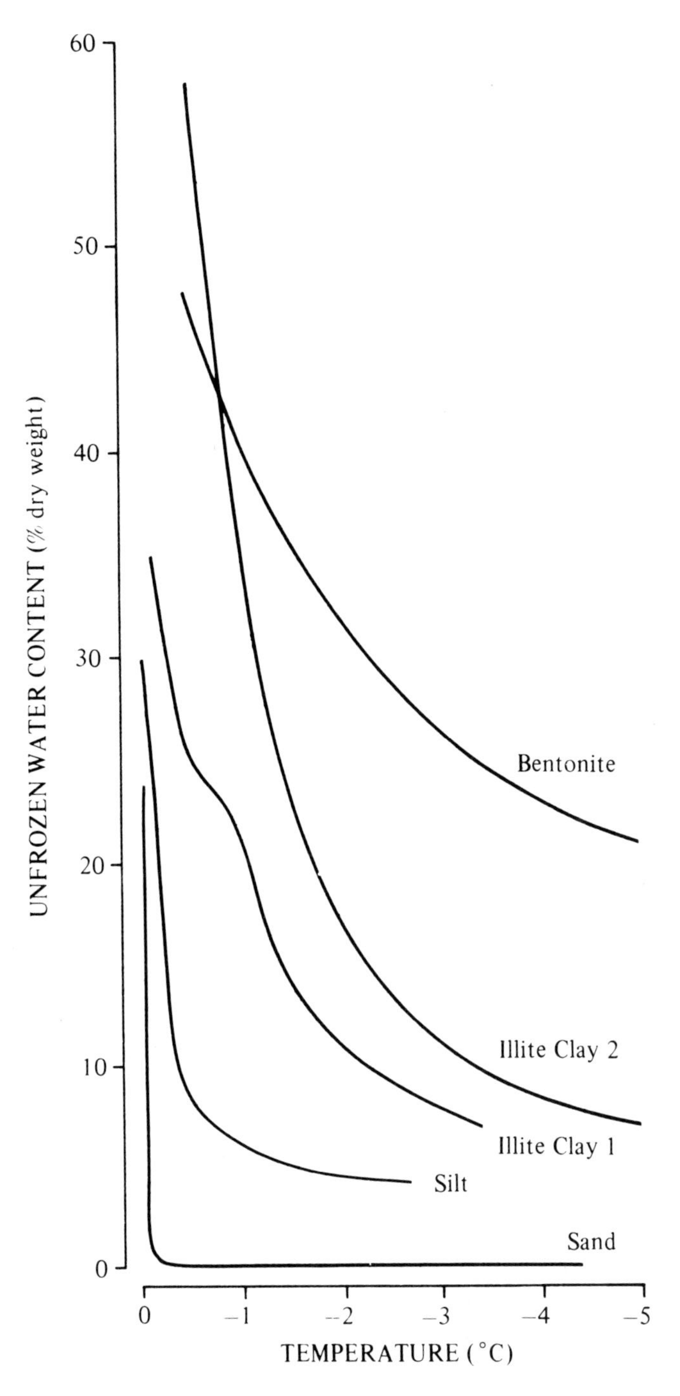

Figure 1.4 Amount of water remaining unfrozen at temperatures below 0 °C, various soils.

perature being associated with a particular quantity of water remaining. Regardless of what temperature is considered, the free energy of the ice and water are equal, or at least are in the process of immediately becoming so. Furthermore, the free energy depends on the temperature and not on the type of soil.

As soon, therefore, as freezing commences and the temperature falls below 0 °C, the free energy of the remaining water is less than that of 'normal' water. Now, it is an additional attribute of the free energy property that a transfer of mass tends to occur along a free energy gradient. If the water in the soil adjacent to ice has a lower free energy than the water further away from the frozen zone, then a migration of water towards the frozen zone will occur. It is this migration which is ultimately the cause of frost heave. The water accumulates as ice, commonly in layers or 'lenses' (Figure 1.1).

We may refer to potential of the water, in this context the term being essentially synonymous with free energy. Reference has also been made to suction as being approximately equivalent to free energy, and all the terms have implications for water movement. Suction suggests a hydrostatic pressure effect. Suction, and also negative pore water pressure, are frequently referred to in studies of soil–water relations, but it should be remembered that the changes of free energy discussed are not limited to hydrostatic pressure effects in a strict sense. With this qualification, the terms are useful because they permit ready analysis of the pressures produced by the heaving soil under different ground water conditions.

The amount of heave occurring depends greatly on soil type, on ground water conditions, rates of freezing and other factors. It is absent when the pores are large, that is, in purely coarse-grained materials. A considerably more detailed analysis is necessary before the nature of frost heave can be properly understood, and this is undertaken in Chapters 7 and 8. We will merely note that the suctions leading to frost heave are often very great, and may be thought of as tensile stresses of many atmospheres (1 atmosphere is about $10^5\,\mathrm{N\,m^{-2}}$ or 100 kPa).

1.3 Climate and ground freezing

Frozen materials are limited to a fairly thin surface layer of the earth. The maximum depth of frozen ground exceeds 1000 m in rather limited areas. Over much larger areas, frozen ground is only tens or hundreds of metres thick, while probably the greater part of the earth's land surface experiences repeated short-term freezing of a much thinner layer.

Ultimately, all specific knowledge of the extent of frozen ground has its origin in direct observation involving sampling procedures or measurements of ground temperature. Such procedures are costly and give information only on the immediate vicinity of the observations. Prediction of the extent of frozen ground at different times and places is one of the most pressing geotechnical problems and the subject of much current research. Problems arise because the temperatures in the ground are related in a complex fashion to atmospheric climate, and consequently, are not immediately predictable by reference to commonly measured climatic parameters (see Chapter 3).

The temperature of the surface layers of the earth's crust follows from heat flows in upward and downward directions. Firstly, the high temperatures of the earth's interior give rise to a flow of heat outwards, the geothermal flux. It varies somewhat from place to place but varies little with time, and is in any case relatively small, some $0.05\,\mathrm{W\,m^{-2}}$. Secondly, there is a continually changing but usually orders of magnitude greater, flux to or from the ground surface, affecting the upper metres or tens of metres of the ground. The origin of this ground heat flux is a variety of heat transfer processes at the ground surface, which are ultimately of extra-terrestial origin. The radiative energy from the sun arriving at the surface is dissipated in various ways, by reflection, absorption, reradiation, involvement in evaporation and in some part in warming the air and the ground. The latter process reverses during night time, with flow of heat out of the ground and consequent cooling. The energy exchange at the earth's surface is illustrated in Figure 1.5 where the physical processes involved and their relative importance as energy paths are shown.

The ground heat flux changes continually with time but the sums of ingoing and outgoing values through one year commonly approach zero. During the summer months there is a net cumulative intake of heat by the surface layers, which is more or less completely lost through the surface during the winter season. The annual passage of the seasons is a cyclic change occurring in a regular and repeated manner. Consequently, the mean annual ground temperature at the surface of a particular site remains fairly constant from one year to the next, the day by day and annual cycles of temperature change occupying a predictable range about this mean. The mean annual temperature will change, however, if there is a significant change of the nature of the surface. The situation where there is, year after year, a repeated small net intake or repeated loss of heat through the surface is not rare and is extremely important. But it should preferably be regarded as a deviation about the normal quasi-equilibrium situation since it is necessarily preceded by a change in environmental conditions and is always a transient event. If it were not, the earth would be either steadily

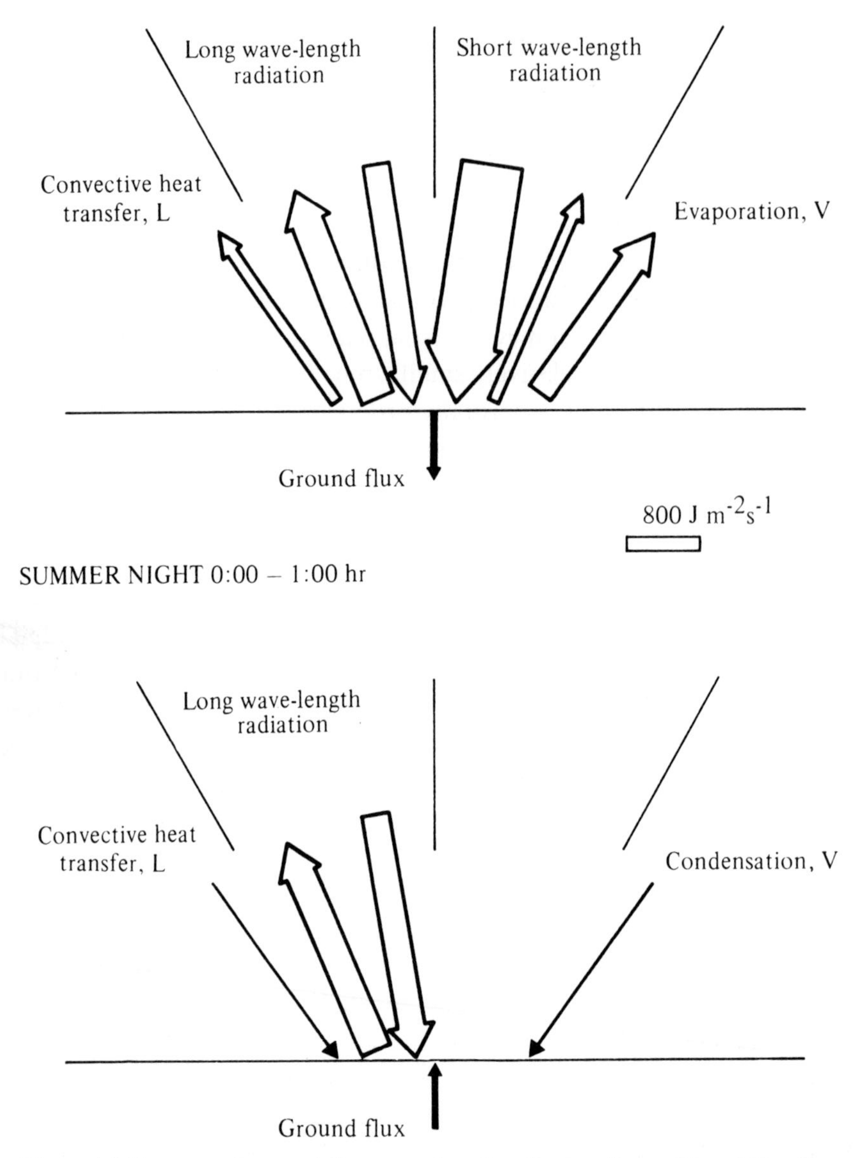

Figure 1.5 Energy exchange at the ground surface for two times. The width of the arrows corresponds to the rate of energy transfer (energy flux) by each path. This example is for Potsdam, Germany and modified from Geiger (1965).

cooling or steadily warming, with catastrophic implications. Such is found not to be the case (we ignore the slow loss of heat from the earth due to internal radioactivity).

The various flows of heat energy to and from the surface of the ground can be expressed by an equation for the heat balance, which in its simplest form is:

$$Q^* = Q_{LE} + Q_H + Q_G \tag{1.1}$$

Q^* refers to the net radiation, which is that radiative energy largely direct solar radiation, arriving at the earth's surface, less the components reflected back and emitted from the surface. This difference between incoming and outgoing radiative energy (positive during the daytime) gives rise to other forms of heat transfer at the earth's surface. The processes of evaporation or condensation involve the absorption or release of latent heat by water, and the energy used in these processes is represented by Q_{LE}. The term Q_H refers to energy transfer which is manifested by temperature, the so-called sensible heat. In the equation it refers to the flux of heat by convection and advection (wind), from the ground surface and to the air. Heat is moved upwards in warmed air in turbulent motion. Q_G refers to the ground heat flux which occurs largely by conduction, through the soil or rock, to or from the surface. According to the conservation principle, energy cannot simply disappear, and the component fluxes at the ground surface must balance at all times as indicated by the equation. The magnitude of the different component fluxes varies greatly from place to place and over short periods of time. If one flux is modified there must be balancing modifications in another one or more. More details are given in Chapter 3.

The temperature of the ground surface and thus of the ground below depends on the fluxes. All the mathematical expressions describing the fluxes include the surface temperature and thus this temperature is functionally related to the fluxes. Any change in the environmental conditions represents a disturbance which modifies the fluxes, and necessarily causes some change in the temperature of the surface, and in due course, of the ground below.

As a result of these energy exchanges, a continuing characteristic temperature, the mean annual ground temperature, is maintained at any place. Temperature variations around this mean are experienced in the surface layers within each year, following the passage of the seasons. They extend in a progressively dampened manner to a depth of some 10 or 20 metres. Below this depth, effectively, no change in temperature is detectable from one year to the next (Figure 1.6). If we consider several adjacent sites, for example, one bared of vegetation and snow, one with thick moss cover, and one with trees, it is clear that at all times the values of the fluxes will be different at each site and accordingly at most times, so will the surface temperature. It is not then surprising to find that with these different microclimatic environments, the mean annual ground temperature is also somewhat different in each case.

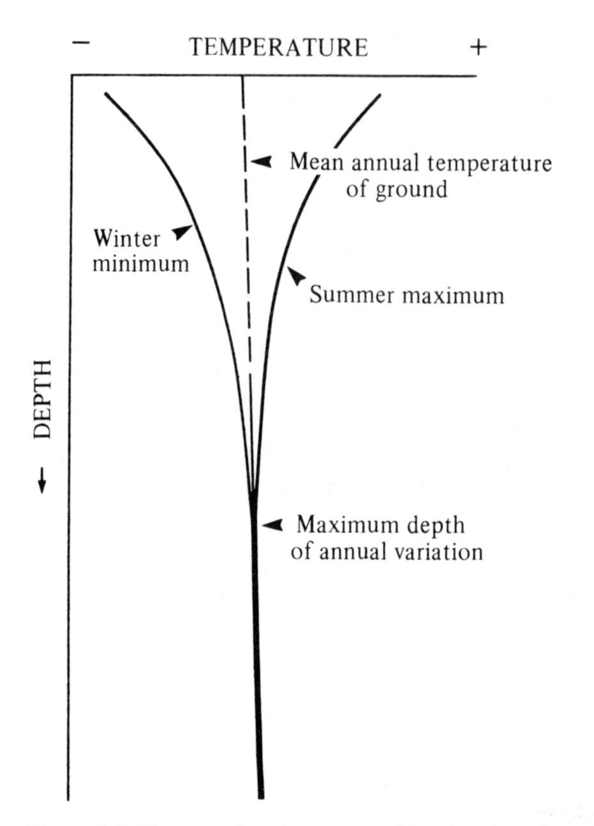

Figure 1.6 Diagram showing range of temperature (summer maxima and winter minima) as a function of depth. The dashed line shows the mean annual temperature of the ground.

1.3.1 Permafrost

Permafrost is defined as ground remaining frozen for more than a year. Most permafrost is thousands of years old but permafrost of recent origin or shortly to disappear is usually of greater importance to man. If the mean annual ground temperature is lower than the freezing point, then permafrost exists at depths where the seasonal variation is insufficient to raise the temperature above freezing point. The distribution of permafrost in the northern hemisphere is shown in Figure 1.7. Examples of the variations in thickness are illustrated in the diagram Figure 1.8.

Figure 1.7 Map of permafrost in the northern hemisphere. The intensity of the shading increases according to the extent, laterally and with depth, of the permafrost. The outlying regions (the isolated occurences on the map) represent mountainous regions where high altitude causes low temperatures. Modified from Péwé (1983), with additional information from Zhou Youwou & Guo Dongxin, 1983; Kudriavtsev, 1978.

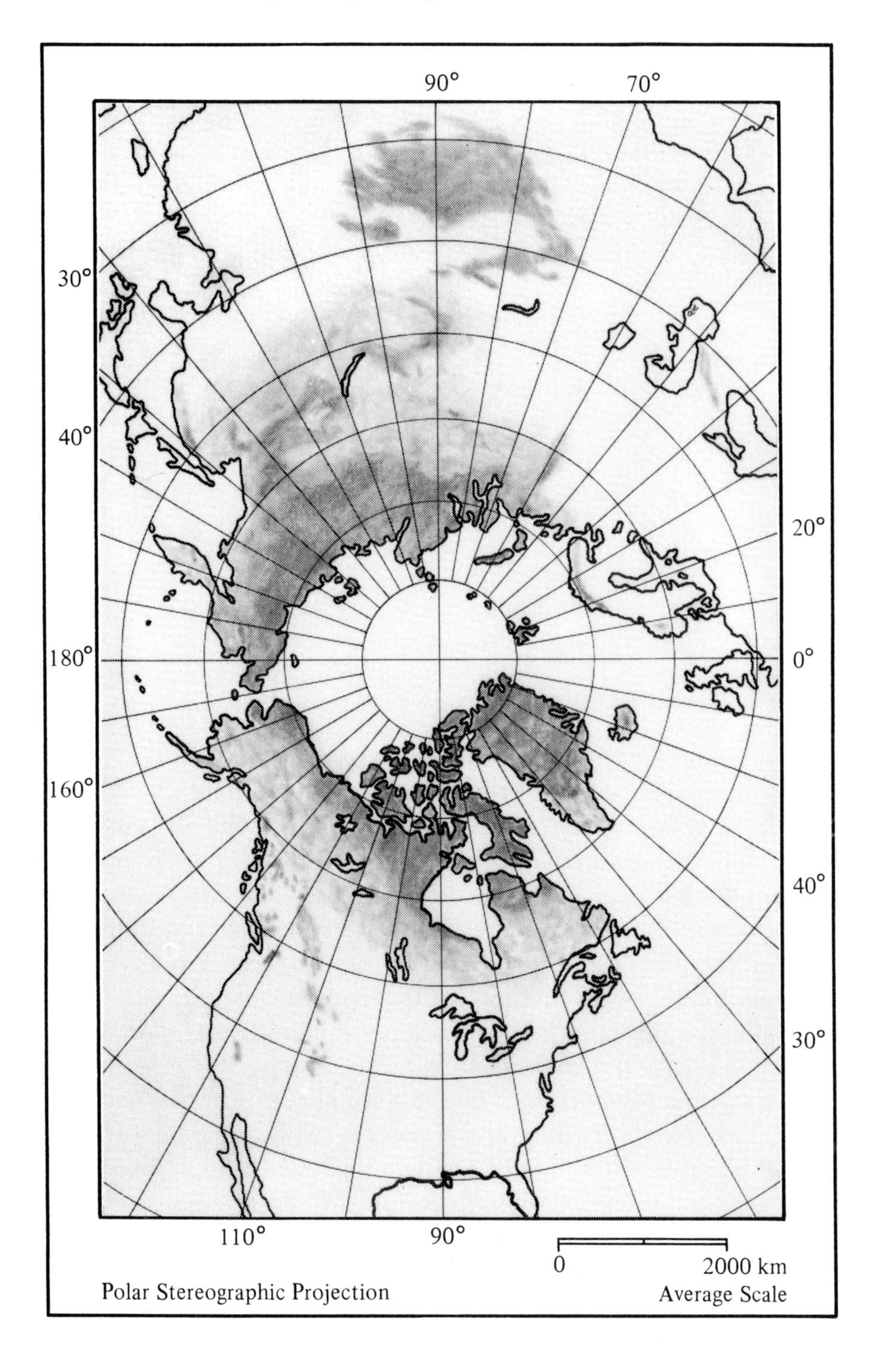
90°
70°
30°
40°
20°
180°
0°
160°
40°
30°
110°
90°
0
2000 km
Average Scale
Polar Stereographic Projection

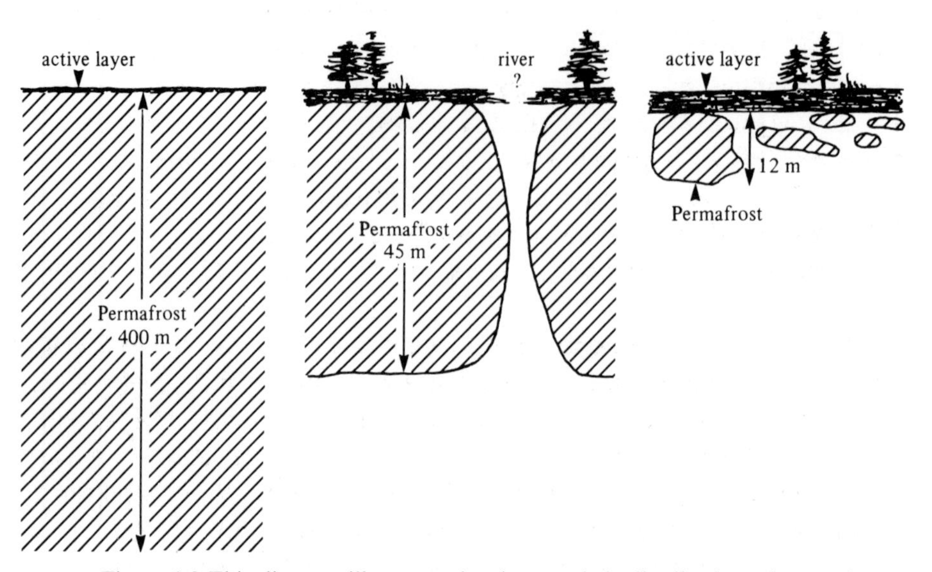

Figure 1.8 This diagram illustrates the characteristic distribution of permafrost as exemplified by three places: (a) very cold (mean ground temperatures many degrees below 0 °C) (b) fairly cold (mean ground temperatures usually below 0 °C, often by several degrees) (c) mean ground temperatures around 0 °C.

Mean annual temperatures as commonly reported are those of the air. Unfortunately, mean annual ground temperatures commonly differ by several degrees from mean annual air temperatures and the difference between them is very variable. Consequently, air temperature records are of limited practical use in predicting the occurrence of permafrost. Near the poles, of course, where the land climate is extremely cold, there will be permafrost throughout. Elsewhere, even where the mean annual air temperature is several degrees below 0 °C, uncertainty prevails. The mean annual ground temperature itself varies very significantly over short distances laterally, even within a few metres.

The mean ground temperature is determined primarily by heat and mass exchange processes external to the soil or rock. In particular the nature of the ground surface (vegetation, snow, bare soil, etc) has a profound controlling influence being interposed between the gross, atmospheric and extra-terrestrial climatic elements and the ground. Consequently in regions where mean ground temperatures are within a few degrees of 0 °C, permafrost will occur in a scattered or discontinuous fashion, depending on the nature of the ground surface.

Thus the existence of permafrost, which requires mean ground tempertures below 0 °C, frequently depends on an appropriate combination of atmospheric climate and ground surface, or microclimatic, conditions. Over many thousands of square kilometres of the earth's surface there is a local and discontinuous distribution of permafrost which can only be explained through an understanding of the energy exchange processes. Even in regions where mean air temperatures are so low that 'common sense' would suggest permafrost to be uniformly present we find this is not necessarily the case. There is no permafrost below the Mackenzie River nor below many lakes in the Mackenzie Delta, such is the special thermal situation of large water bodies, even though mean annual air temperatures of −10 °C occur.

Once the surface temperature of a uniform solid body has been defined, and in this context we may regard the ground as such, then the temperature variations within that body can be analysed by means of simple heat conduction theory. Soil is not in fact a homogenous, solid body, and freezing or thawing also complicates the situation somewhat. But, once the surface temperatures are defined, the temperatures and heat flows below the surface can be analysed without further references to the diverse processes above the surface. This is considered further in Chapter 4.

1.3.2 Ephemeral freezing

Seasonal freezing and thawing may extend two or three metres into the ground. Its geomorphological effects are as significant in many respects as those of permafrost, and the economic effects have to date been far greater, particularly because of the associated costs of foundations for buildings and highways. Many of the more populous areas in Russia, Middle North America, Northern Europe and parts of the southern hemisphere have winter frost which penetrates a metre or more into the ground. If permafrost occurs, this does not preclude an annual freezing and thawing of the uppermost layers of ground, this constituting the *active layer*.

The importance of annual freezing depends in large measure on the depth to which it penetrates. Its effects also depend on the nature of the materials which are frozen, and to a lesser degree on the manner in which the frost leaves the ground, whether by thawing from the surface (always the case when the active layer is underlain directly by permafrost), or, in part, from the bottom of the frozen layer (as occurs in significant degree in certain climates, particularly those where summers are hot).

Winter frost penetration cannot be very satisfactorily correlated with winter cooling as evidenced by atmospheric climate parameters. Even over distances laterally of some tens of metres (and less), the depth of a frost penetration may vary by a factor of two or more due to relatively inconspicuous differences in the nature of the ground surface. Maps of frost penetration, at best, can indicate average depths of penetration for some assumed uniform ground surface and soil characteristics. The absence of snow from ploughed roads or wind-exposed areas gives frost penetration often two or three times that in adjacent areas. Similar considerations apply to the depth of ground thawing, in the summer, above permafrost.

The following characteristics determine annual frost penetration in a general sense:

1. The mean annual ground temperature;
2. The amplitude of the annual temperature cycle at the ground surface (reflecting the 'continentality' of the climate);
3. The nature of the ground surface cover, particularly snow cover;
4. The thermal properties of the soil material.

If the annual passage of temperature for the surface of the soil is known, a prediction of frost penetration can be made by evaluating the thermal properties of the soil, the heat capacity and thermal conductivity, and utilising these in relatively simple equations for determining the passage downwards of the 0 °C isotherm. The temperature at which freezing of soil commences is normally very slightly below 0 °C, so that strictly speaking calculations should relate to the isotherm for −0.1 °C or similar. The quantity of water that freezes in the soil is a most important factor, because of the unusually high latent heat of fusion of water (the quantity of energy that must be removed to freeze it). This is 333 $J\,g^{-1}$, and must be regarded as a component of the soil's heat capacity.

By contrast, lowering the temperature of water by one degree involves removing only 4.19 $J\,g^{-1}$ – its specific heat capacity in the strict sense. In practice, precise calculations of frost penetration are difficult because of uncertainty as to the amount of water migrating to the freezing layer, and frequently because of uncertainty as to the temperatures or heat flows at the ground surface. As outlined above, ground surface temperatures must be seen as an element in the system of mass and energy fluxes at the ground–air interface.

The process of winter freezing constitutes a part of the cycle of heat exchange occurring annually and it is a mistake to consider only the effect of winter temperatures. It has been common engineering practice to predict frost penetration on the basis of *frost index* or freezing-degree days (the sum of days times their negative temperatures, and having units: −°C days) but this leads to anomalous results. The frost penetration at Oslo, Norway (mean annual air temperature +5 °C), is often greater than that at Ottawa, Canada (mean annual air temperature +6 °C), although the frost index is about one-half. The explanation lies in the fact that heat lost during the winter is largely heat added to the ground during the previous summer. The much warmer summers of Ottawa mean that a much higher percentage of the heat leaving the ground in the winter is stored summer heat, which has raised the temperature of the ground, rather than heat released in the freezing process as latent heat of fusion.

Inland or continental type climates are characterised by a large annual exchange of heat in the near surface layers. Clearly, for any freezing to occur, the mean ground temperatures must be near enough to 0 °C that the annual cycle of temperature around the mean reaches below 0 °C. The presence of vegetation, and particularly of snow commonly prevents ground surface temperatures falling as low as air temperatures. In maritime climates the surface of the ground may experience an annual range of temperature of only some 10 °C. There will then be no significant freezing of the ground unless the mean ground temperature is less than about +5 °C. In such climates, for example, coastal districts of Iceland, the frost penetration will be small (usually < 1 m) even if the mean temperature is within a degree or two of 0 °C. In markedly continental climates, winters may be very cold, but the counteracting effect of warm summers, means that annual frost penetration will be deep only when the mean temperature is within a few degrees of 0 °C.

The climatic conditions associated with the deepest annual freezing and thawing are those where the mean ground temperature is around 0 °C, and the annual cycle is characterised by extremes (high continentality). Some general implications are important. We associate the absence of trees (which require summer temperatures exceeding about 11 °C) with 'cold' climates. Yet, if the climate is very maritime (small difference of temperature summer to winter) there may be no trees, relatively mild winters and little frost penetration as in much of Iceland. Conversely, in continental areas, warm summers result in forest overlying permafrost. Where permafrost occurs, the depth reached by the annual thaw (the active layer) is likewise greatest for mean temperatures near 0 °C. Clearly the depth decreases as mean temperatures become lower. For example, in the Canadian Arctic islands an active layer of some 10 cm or less is common.

Although these general considerations help us to avoid pitfalls in too rapid conclusions concerning the relationship of climate and geographical situation to frost penetration, they are by no means sufficient to enable us to characterise the frost penetration at specific sites. This is because of the great importance of surface cover and soil conditions. A relatively modest difference, for example, in vegetative cover, so modifies the heat exchange processes as to give a difference in frost penetration, that could correspond to the effect of a climate difference experienced only on travelling tens or hundreds of kilometres. W. G. Brown (1964) showed that with soil conditions (lithology and water content) the only variable, a range of frost penetration involving a factor of two occurs. Tsytovich (1975) extends this to three.

Recently, fairly accurate methods of predicting frost penetration have been developed involving consideration of the component heat exchange processes and their relationship to the specific surface conditions and soil thermal properties of the site (see Chapter 3). These have served to further illustrate the inadequacy of atmospheric climatic considerations alone.

In many climates there are shorter periods, even less than 24 hours, during which a very shallow surface layer freezes. Diurnal freezing and thawing is frequent in middle latitudes, high altitude regions such as the Alps, the Rockies, and Scandinavia. The winters are not particularly cold, nor the summers warm, while the diurnal cycle is marked. Consequently there are long periods each year when there can be fluctuations around 0 °C. In the high arctic by contrast, the diurnal variation is less marked, indeed daylight or darkness may last 24 hours, while the annual cycle is very marked with long winters during which there is no thawing.

The protective or damping effect of ground cover is important. Only rarely will the soil experience as many freeze–thaw events per year as air temperatures indicate. The depth reached by diurnal freezing is small indeed, at most a centimetre or two. Cold periods of several days duration will only penetrate a few centimetres.

Mathematical analysis of the penetration of a temperature change into a body (in this case the ground) shows the rate of penetration to be proportional to the square root of the time from onset, or of the time period of a cycle. Consequently, if a yearly temperature cycle is experienced to 12 metres into the ground (Figure 1.6), the diurnal cycle extends $1/\sqrt{365}$ or 1/19 of this, or 0.63 metres. In the case of penetration of frost, however, the situation is made complex by the heterogeneous thermal properties of the materials. In particular, the very large quantities of heat exchanged in freezing or thawing are important. The general relationship still serves as a guide in comparing depths of freezing and thawing over different time periods.

1.4 Characteristics of permafrost

Contrary to the implication in the name, permafrost is characterised by lack of permanency, and by its inherent instability. Most of the world's permafrost is at temperatures warmer than −10 °C, the exceptions mainly being found in the Canadian Arctic islands, Northern Siberia and scattered Antarctic localities. Most permafrost is warmer than −5 °C. Consequently there is usually a significant quantity of water coexisting with the ice.

Any material, a significant component of which is at its melting point, is not 'stable', and, as a major stratum in the vicinity of the earth's surface permafrost is quite unique in this respect. The full implications are only currently being realised. The strength and deformation properties of frozen ground are peculiar, and of great importance in geotechnical considerations. The mechanical properties are highly temperature dependent as well as varying with particle size, ice content and other characteristics. The geomorphological and geological significance of the properties is also important. The most fundamental rheological property is that of creep, and the associated phenomenon whereby the strength of frozen ground is observed to fall as the duration of stress application is extended. The strength of frozen ground, if measured by application of load to cause rupture in a period of minutes, may be ten or more times that if a load is applied continuously for a period of days or weeks. In other words, smaller loads cause deformation if applied for a longer time. In this respect much frozen ground shows creep properties, and is not unlike ice in its behaviour. But there are many differences that require careful consideration (Chapter 9).

Because transitions between ice and water occur over a range of many degrees of temperature, it follows that volume changes (ice occupying 9% more volume than water) are a continual occurrence under changing temperatures (Figure 1.9). The water component can also move, and, because of the relationship between the state of suction and the temperature, there are strong hydraulic gradients associated with temperature gradients in frozen soil. Thus, the volume changes can be augmented by movements of water within the frozen ground. Ultimately, these give additional accumulations of ice, at the expense of water or ice elsewhere in the ground. Little direct evidence is available as to these effects in the ground; certainly the translocation of moisture occurs only slowly, although over years or hundreds of years there may be major redistributions. Phase transitions and very local 'microscopic' migrations of water are also important in the deformation of frozen ground.

There is an additional instability inherent in permafrost of more obvious significance: it may thaw. In any year in which summers are warm, that is, in which more heat enters the ground than in the previous year, the active

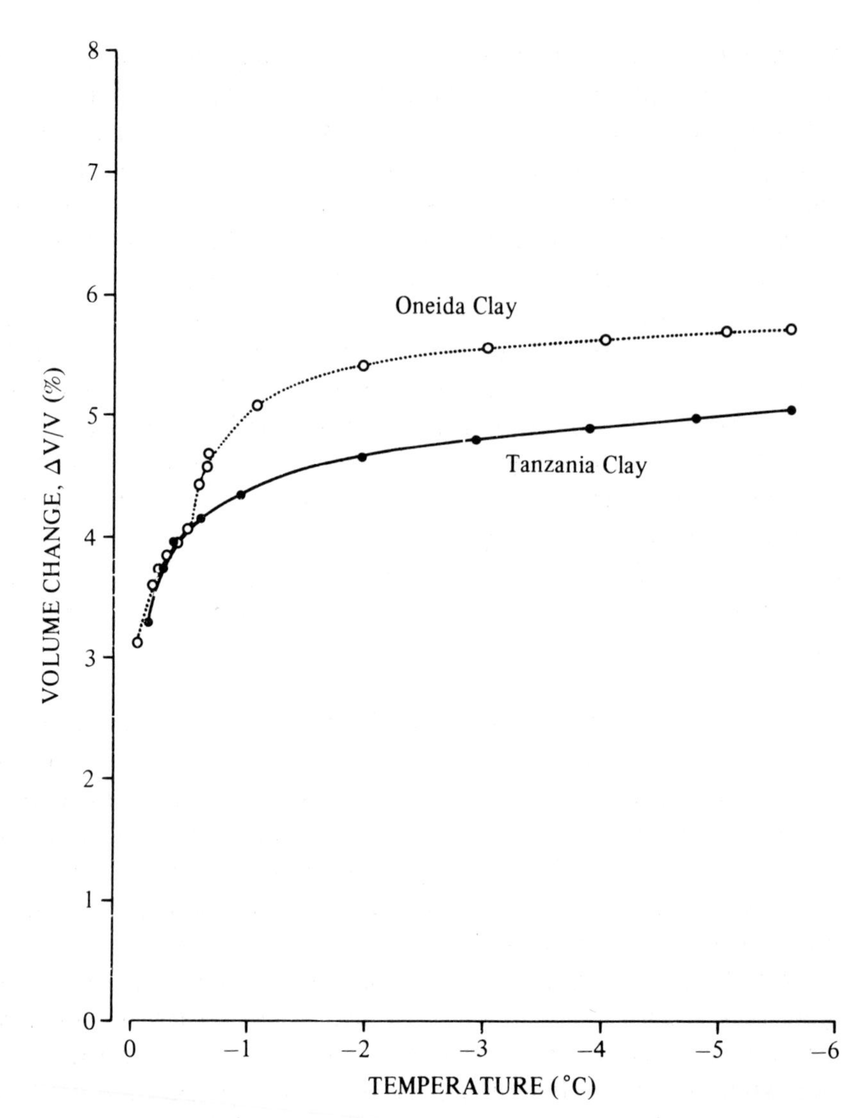

Figure 1.9 Volume changes in two soils on cooling below 0 °C. The expansion was due to formation of ice from water already present in the samples. At still lower temperatures the effects of thermal contraction will overcome those of freezing and the volumes will decrease (from Williams 1976).

layer will be deeper, although perhaps only by a small amount. A following colder summer will cause a shallowing.

More important is the *aggradation* or *degradation* of the permafrost over several or more years – these terms referring to an increase or decrease in the extent of permafrost. Such occurs in association with a change of mean ground temperature, because over a period of years there is then a net flux of heat into or out of the ground. The causes may be microclimatic, that is, a change of ground surface conditions modifying the component heat fluxes at the surface, or climatic, that is, following from a change in atmospheric climate. If the changes are minor, or of short duration, then significant aggradation or degradation may occur only in areas where the permafrost is already very close to 0 °C. There, the lateral extent of the permafrost bodies may change substantially. If the changes are greater, perhaps a climatic change lasting for tens or hundreds of years, much larger areas may become free of permafrost, or conversely develop a significant thickness. A change of mean ground temperature affects the depth reached by the permafrost. On geological time-scales, it is the changing climate which has been responsible for major shifts in the worldwide distribution of permafrost. There is evidence of permafrost during the Pleistocene glaciation, in middle Europe, southern England, the northern United States, much of Russia, and elsewhere. Quite large areas currently underlain by permafrost, however, may have been free of permafrost when overlain by the great continental glaciers – such is the 'microclimatic' role of such an ice cover.

The aggradation or degradation of permafrost, being the result of a balance of heat entering or leaving the ground over years, is in contrast to the mere existence of permafrost. The element of permanency in permafrost arises precisely because the cold ground temperatures are, broadly, in equilibrium with the climatic and other conditions. This point is often confused in elementary 'explanations' of permafrost, where the *existence* of permafrost is often erroneously ascribed to a 'negative heat balance'.

Aggradation or degradation is not limited to circumstances where the mean ground temperature changes. An increase in the annual amplitude of temperature at the ground surface produces a deepening of the active layer. The active layer is rarely deeper than some 3–4 m, however, and is usually much less, so that the lowering of the permafrost is limited, unless there is a concurrent change in mean annual temperature. If the permafrost layer is relatively thin, not extending deeper than a few metres, then it may totally disappear due solely to a change in surface temperature extremes.

The thermal regime of permafrost, and the change in distribution of permafrost in time and space, is a problem no less complex, if not entirely analogous, to that of the mass balance and distribution of glaciers – the glacier regimen. Glacier regimen is the key to understanding glacial landforms. Similarly, the ground thermal regime is the fundamental element in the formation of the characteristic periglacial landforms soils and terrains.

1.5 Manifestations of freezing and thawing of the ground

Solifluction refers to particular forms of downslope movement of a shallow soil layer, maybe no deeper than is frozen and thawed annually, which give a variety of characteristic surface relief features – small terraces lobate wave-like flow forms and similar (Figure 1.10). The term solifluction is best used for such features exclusively found in periglacial situations and in which it can be established that processes associated with freezing and subsequent thawing have an essential role. This latter qualification is useful because downslope movements of material not specifically the result of freezing and thawing, such as mudslides, landslides, and surface wash, occur as much in the periglacial regions as elsewhere. Indeed, careful studies of rates of downslope movement in an arctic–alpine tundra area (Rapp, 1960) established that movement processes totally *unique* to the periglacial situation have a minor role in the overall rate of denudation.

Figure 1.10 These characteristic solifluction lobes, in the Trollheim mountains, Norway, move a few millimetres or centimetres downslope, each year.

In field studies it is not always easy to make such a distinction. But it is certain that there are forms of downslope movement which can only be explained by involving soil behaviour arising directly from freezing of the soil. The frost heave process results in a disruption of soil structure, and a consequent loss of strength upon thawing, which is unique. Nothing similar is found elsewhere in the study of soil mechanics – unless some parallel can be drawn with the disruptive effects of salt crystallisation in certain shale materials. The deformation properties of ground while frozen are also unique and striking in their dependence on temperature.

This can hardly be said of the loss of strength associated with high water pressures occurring frequently in the thawed soil in the spring. High pore water pressures are promoted by impeded drainage, due perhaps to underlying frozen ground, but also to other impermeable layers. Frictional strength, which is proportional to the stress acting across particle to particle contacts, is always reduced by pressure in the pore fluid, so the effect is not exclusive to the periglacial environment.

Together with solifluction, the development of *patterned ground* is frequently regarded as the most obvious characteristic of the periglacial regions (Figure 1.11). Patterned ground refers to a surface sorting of material, and a systematic, patterned arrangement of microrelief, variously described as polygons, nets, circles, hummocks and similar. These forms are, in most cases, the direct result of the properties and behaviour of freezing and thawing soils. Explanation of their formation commonly involves frost-heave oriented in certain directions. The direction of the heave is influenced by the direction of heat flow, which in turn depends on the thermal properties of conductivity and heat capacity. It is also influenced by, and itself largely determines, stress patterns in the ground. Of the many types of patterned ground, the formation of only a few has been studied in detail with modern scientific methods. Contemporary understanding of the mechanical principles, however, is sufficient to demonstrate that no great mystery attaches, even, or perhaps least, to the impressive *ice wedge polygons* (Figure 6.7) which characterise the terrain over millions of square kilometres. They are initiated by cracking associated with intense cold and thermal contraction. The cracks fill progressively with ice. The vertical wedges of ice demarcate these polygons and occur below the depth of seasonal freeze–thaw. The surface expression is merely a slumping of the soil above the ice wedges.

Other types of patterned ground involve just the depth disturbed by seasonal freezing and thawing. Patterned ground and solifluction do not develop in forests, except on a local and much-reduced scale. This is mainly

Figure 1.11 Examples of patterned ground. The larger forms (*a*) are 2–3 metres in diameter. These are incompletely developed stone circles (Photo B. Van Vliet Lanoe). (b) Unsorted polygons with raised centres and vegetation demarcating each unit.

because the roots of trees and undergrowth stabilise the soil, and also modify the ground surface in such a way as to exclude the required pattern of heat flows.

Of much wider significance, though until recent years little investigated, is the presence of many forms of ice masses in permafrost which occur in addition to ice wedges. Ice lensing, well known in seasonally freezing ground (Figure 1.1), is abundant in permafrost although the concentration of lenses is reduced with depth. It appears that below some 6 m (doubtless the figure varies) such lenses as occur have been formed largely from water already in the soil (Williams, 1968) rather than by moisture migration from more distant groundwater.

Under conditions of very slow freezing, normally the case in formation of permafrost, very large lenses or masses of ice may be formed by the frost heave process. Pingos are conical hills, tens of metres high and resembling volcanic cones (see section 6.2.1), which have ice cores many metres thick. The manner of their formation (see Chapter 6) often involves permafrost aggradation in old lake beds and a complex thermal and hydrological situation.

The ice bodies in permafrost have diverse origins. They may be merely buried snow or lake ice; it is also possible that sometimes they develop within the permafrost many years after its formation. Much of this *ground ice* is not revealed by any characteristic surface relief. However, when degradation of permafrost occurs the disappearance of the ground ice can, if abundant, give rise to a terrain known as *thermokarst*. There are many pits and depressions, and a sequence of striking surface changes occur. Ultimately, the surface may have abundant water bodies, around the margins of which (in forested areas) are leaning trees and innumerable slumps of soil, mud flows and landslides.

Such developments may be totally destructive of engineering structures, and consequently great care is necessary to avoid even minor changes to the surface cover of a kind that may initiate degradation (Pollard & French 1980). Disruption of the ground due to thawing and settlement occur widely in association with climatic change, and, in future millenia, would be recognised by conspicuous 'flow' structures in the sedimentary profile.

Thus the role of climatic change, and the associated succession of environments, is, through the thermodynamic processes, added to the geological record. In many present-day temperate areas of the world, there is evidence of former periglacial conditions. The interpretation of this 'fossil' evidence requires understanding of the periglacial processes. The relationship between specific features and particular climate and microclimate, soil type and hydrological conditions, must be understood before the geological record can be correctly interpreted – see Washburn (1979). From the more immediate point of view of the plant and animal ecologist, just as of the engineer, it is important to understand the effects of even small changes in surface conditions on the behaviour of the freezing and thawing soils.

2

Morphology of permafrost and seasonally frozen ground

2.1 Frozen ground as a geological material

Geologists classify the materials of the earth according to their lithological nature. Whether granite or sand, quartzite or shale the materials of the surface region are described according to their mineral composition, the arrangement of components and their shapes. Because water is fluid, and at the surface of the earth the contained amount is constantly changing the water component tends to be ignored in this sort of classification. However, when frozen it is not nearly so ephemeral in amount and disposition. Indeed the ice in frozen ground, which takes many forms, can be very usefully described in much the same way as we describe the form and orientation of other mineral components. In geotechnical work the appearance of the ice in core samples (or its form and extent in the ground estimated by other means) is given greater prominence than the description of the other mineral components of the ground. Geotechnical properties are of course dominated by the presence of the ice (see especially Chapter 9). This is because the ice is close to its melting point (while all other minerals are hundreds of degrees below theirs) and it is consequently of similar importance to know where the ground, the soils and rocks, are actually frozen.

Thus this chapter considers, firstly, the structure of frozen ground with regard to its characteristic component, ice, and, secondly, the extent of frozen ground, that is, the extent of sub-freezing temperatures in the lithosphere.

2.2 Ice in the ground

It has been emphasised in the previous chapter that the unique features of the terrain and of soil behaviour in cold climates arise because of the processes of freezing and thawing. In concentrating upon the significance of freezing within confined spaces (the soil pores), and the associated stresses and temperature conditions leading to ice segregation and frost heave, it must be remembered that many bodies of ice also occur, especially in permafrost, whose origin is different. These include large bodies of ice which form annually, some at the ground surface, and others whose origin is far back in geological history. Some are the product of slow accretion. In so far as such bodies are in contact with soil, the thermodynamic conditions appropriate to ice in a porous medium will apply. The presence of all this ice, of course, greatly modifies the properties and behaviour of the ground. It is important to know the possible forms, extent, and origins of the ice included in frozen ground. The properties of ice in general are also important and are reviewed in, for example, Hobbs (1974).

2.2.1 Excess ice

This is defined as ice present in excess of the volume of the soil pores had the soil been unfrozen. On thawing, the soil volume will decrease as water drains away. Bands of ice, often interspersed with sediments, can be many metres thick in permafrost and are known as *massive ice* (Figure 2.1). On the other hand, the ice may be more uniformly dispersed in thin layers or flakes and the amount of excess ice may still be large. Ground which is apparently uniform, lacking any obvious concentrations of ice, may also contain excess ice in expanded pores, and thus become supersaturated on thaw. Ice concentrations in seasonally frozen ground are, of course, limited in size and less diverse in origin. Ice concentrations larger than pore size are the last to thaw. Indeed soil which has thawed except for such discontinuous pieces of ice, larger than pore size, may require special study notably in submarine engineering where it may occur extensively.

Ice in permafrost, being so widespread, has great significance in geotechnical engineering and the thousands of boreholes made for oil exploration or in connection with pipelines and other structures have greatly increased the information available. Areas tens or even hundreds of square kilometres in extent are underlain by sediments containing 50% or more by volume of excess ice. In near-vertical exposures, massive ice layers may tower over the observer; they may constitute the greater part of exposed surfaces.

The amount of excess ice decreases with depth, and although information is somewhat limited, most massive ice occurs within 50 or 60 metres of the ground surface, its concentration being greater towards the surface. Characteristic involuted hills underlain by a high concentration of massive ice (Rampton, 1973) suggests that the ice in permafrost has an important role in formation of topography, or at least, in the origin of minor relief.

Figure 2.1 Exposures of massive ice, near Fairbanks, Alaska.

This may be fairly general and is not restricted to particular features such as pingos (section 6.2.1). On the other hand, in some situations there is little topographic expression. Quite extensive areas occur where there is, in fact, little excess ice present. Normally, these are areas of well-drained relatively coarse sediments.

2.2.2 *Frost heave ice: segregation ice*

The origin of excess ice in permafrost is as varied as the forms of the ice. Most of the ice inclusions, larger than pore size, in seasonally frozen ground are a result of water migration due to gradients of potential associated with freezing at temperatures below 0 °C. The suctions developed and the presence of unfrozen water for freezing at such temperatures is a consequence of the porous structure and particle surface area of the soil. Accordingly, in the terminology of soil physics or hydrology, the gradients can be designated as arising from matric potential (section 7.2). Because of the importance of temperature, however, they may be more specifically referred to as thermal-matric, or simply thermodynamic, potentials. Ice so formed is often referred to as *segregated* ice; other ice bodies could also be said to be segregated, but the term can usefully be restricted to ice formed by attraction of water from, or through, the adjacent soil pores. Although most such 'frost-heave' ice is in layers or lenses, there is enormous variation. In seasonally frozen ground, as in permafrost, the ice may appear as thin closely spaced flakes, as layers centimetres or more in thickness, or as 'hairlines'. Usually the segregations are aligned more or less horizontally, but there may also be vertical layers (Figure 1.1) such that the soil appears separated into roughly cubic pieces each a centimetre or more across. It is usually assumed the ice layers tend to be aligned perpendicular to the heat flow, but other factors may dominate. The heave normally occurs in the direction of least resistance and perpendicular to the layers. For these reasons, the layers are usually more or less parallel to the surface of the ground. When vertical layers occur as well as horizontal ones this is normally the result of lateral shrinkage of a compressible soil as water is drawn into the segregations (Mackay, 1974b). Vertical cracks occur and ice lenses tend to form within them as freezing proceeds further downwards. Beskow's (1935) classic work has striking illustrations of the forms the ice may take. Ice segregations in compressible soils can be the result of migration of water only from adjacent soil pores, with a corresponding consolidation of the soil matrix (section 7.2.2). In such cases, the ice is not, strictly speaking, excess ice (in contrast to other ice concentrations) even though part of the water may not be reabsorbed on thawing.

In conditions of deep seasonal freezing the ice layers can be a half metre or more thick but this is rare. In permafrost, massive ice layers up to several metres thick occur, due to this kind of freezing. The ice core of a pingo is a particular example. The development of massive ice in association with heave is aided in the pingo by high pore water pressures which allow the uplift and deformation of a considerable thickness of frozen overburden (see section 6.2.1). High pore water pressures, which promote frost-heave, can originate because of the hydrological situation – perhaps drainage restricted by frozen ground, and elevation potentials in hilly terrain. Artesian pressures are not uncommon in association with permafrost water being trapped sometimes at tens of metres depth where there is a *talik* (an unfrozen soil layer). Of importance, too, is water under high pressure which has been pushed away from pores of saturated coarse-grained material by the expansion of water on freezing, the frost heave suction effect being absent in such soils.

It has been observed that massive ice is often largely surrounded by relatively coarse-grained and, therefore, permeable materials, while in contact with fine-grained frost susceptible (frost heaving) sediments (Mackay, 1971). This situation aids growth of the massive ice, with water being supplied by the porous material yet with the frost heave process active on the face of the ice adjacent to the fine-grained material.

While many observations have shown that frost heave and ice segregation occur during seasonal freezing only in suitably fine-grained sediments, it seems that some frost heave and ice segregation occur in permafrost in relatively coarse-grained materials, over long periods. The thermodynamic potentials are present in coarse-grained as in fine-grained materials, but the low unfrozen water contents of the former result in low permeability (section 7.5) and, presumably, only very slow heave.

Segregation ice can often be distinguished by the alignment of the ice crystals perpendicular to the layer and by the elongation of bubbles in the same direction. However, recrystallisation can be expected over long periods, possibly with a loss of these orientations.

Large areas of the warmer permafrost regions have organic soils, which range from more or less pure plant remains to mineral soils with a substantial organic component. Segregation of ice by the frost heave process does not occur in pure peat. Indeed this material is used as a non-heaving material in railway foundations (Skaven Haug 1959). However, massive ice as well as smaller bodies are found in natural peat soils and this can be segregation ice formed in association with inclusions, perhaps sedimented layers, of silt or clay. Even without such excess ice, frozen peat soils often show settlement on thawing because of their high ice content – peat soils often contain several times as much water by volume as solid matter. Following thaw, decomposition may occur and in any case the material is highly compressible.

2.2.3 *Intrusive ice: ice wedge ice, and other forms*

Intrusive ice, Figure 2.2, which includes that found in veins, sills or other forms (which may sometimes be massive), results essentially because of an elevated hydrostatic pressure in the water of the adjacent soil. Such a situation is in contrast to the frost-heave or segregation ice process where it is the freezing-induced *lower* pressure (potential) of water at the ice–water interface which causes the accumulation of ice. Thus intrusive ice may be formed in soil that is too coarse for segregation ice. More frequently it seems that, as in pingo ice, the high pore water pressures will merely be a factor promoting ice segregation and frost heave, in which case the distinction between the categories becomes uncertain. Indeed, in some cases, both effects may be essential for the occurrence of the ice mass and an 'intrusive-segregation' category seems appropriate.

Figure 2.2 Intrusive ice which has caused rupture of surface (the ground surface also shows abundant effects of seasonal frost heave). Photo: Bruce Brockett.

Downward percolating meltwater is the source of much of the ice in ice wedges (section 6.4.1) and these might be thought a further example of intrusive ice. In fact, the downward movement of the water due to gravity is distinct from that involving elevated hydrostatic pressures. Ice wedge ice is to be regarded as a further particular category of excess ice and, indeed, an abundant one. Other subsurface flows, downwards, can give ice accumulations at various depths.

Genetic classification schemes such as that of Mackay (1972) shown in Figure 2.3 are valuable for the understanding of the behaviour and complexity of ground ice. More categories can be defined to take into account for example, ice accumulations following from diffusion of vapour, or from a combination of processes. The processes considered so far have concerned only ice formed within the ground.

2.2.4 *Ice of external origin: buried glacier, snow, lake and river ice; icings*

There is uncertainty about the amount and nature of that ice in the ground which has an external origin. During the glaciations much ground below glaciers was probably unfrozen, just as today ground is unfrozen below 'temperate' glaciers (those with the base at the pressure-melting point). Elsewhere the ground can have been frozen to great depths, but glacier ice would be buried only very locally, perhaps trapped within

ORIGIN OF WATER PRIOR TO FREEZING	PRINCIPAL TRANSFER PROCESS	GROUND ICE FORMS
Atmospheric water	vapour diffusion	open-cavity ice
Surface water	gravity transfer	single-vein ice, ice wedge } thermal contraction tension-crack ice – tension rupture
Subsurface water	vapour diffusion	closed-cavity ice
	thermodynamic and pressure potentials	epigenetic ice, aggradational ice } segregated ice ↕
– including pore water expelled on freezing	pressure potential	sill ice, pingo ice } intrusive ice
	freezing in pore (may expel water)	pore ice

Figure 2.3 Diagram showing sources of water and transfer processes giving rise to the various forms of ice in the ground (modified from Mackay 1972).

moraines or where covered by landslides. Østrem (1963a, b) discusses entrapped ice of recent origin near glaciers and demonstrates that buried snow patches may be one source. Buried lake or river ice is probably more common than buried glacier ice – the inherent instability of shore lines results in sediments collapsing upon, or being deposited on, ice masses, thus protecting them.

ICINGS

These are accumulations of ice, on the surface of the ground or on or within rivers. They are also known by the Russian name, *Naled* or the German, *Aufeis*. Icings vary greatly in size and appearance. They may for example be a flattish dome or sheet of ice ten or more metres across, on more or less level ground. They frequently occur on slopes, or rock faces, especially near the foot and fed by springs, where they appear as 'ice-falls' (resembling a frozen water fall); they may be conspicuous as mounds of ice in rivers, causing spring flooding and subsequent late-lying ice masses raised above the river bed. Icings are conveniently divided into several categories according to origin (Carey, 1973). Icings may be important for hydrological and geotechnical considerations. They may also become incorporated in sediments to be an occasional origin for ice masses in permafrost.

Ground icings (sometimes known as *seepage icings*) occur where ground is saturated in summer but where there is little relief and no permanent surface drainage path. *Spring icings* are those formed at sites where springs normally run, the water moving in definite channels. As winter freezing occurs the spring water will gradually accumulate with successive layers of ice which may continue through much of the winter if there is ground water available at temperatures slightly above freezing point. The water producing ground icings may well be forced up from a body of unfrozen soil progressively encroached upon by seasonally freezing ground (Figure 2.4), if the soil is sufficiently coarse for ice formation in the pores to lead to expulsion of the pore water. The ice may form just below the ground surface causing a 'frost warp', or blister – a temporary mound. Icings, especially spring icings, are common in cold areas without permafrost. But even in areas having relatively thick permafrost (extending tens of metres downwards) there are usually underground water sources which reach the surface to give icings. Not only is water trapped between the seasonally freezing ground and the permafrost, but there are 'holes' right through the permafrost below larger lakes and rivers. Occasionally, sufficiently great artesian pressure builds up that an icing explodes and blocks of ice can be hurled tens of metres. The hydrology of icings is discussed in Chapter 8.

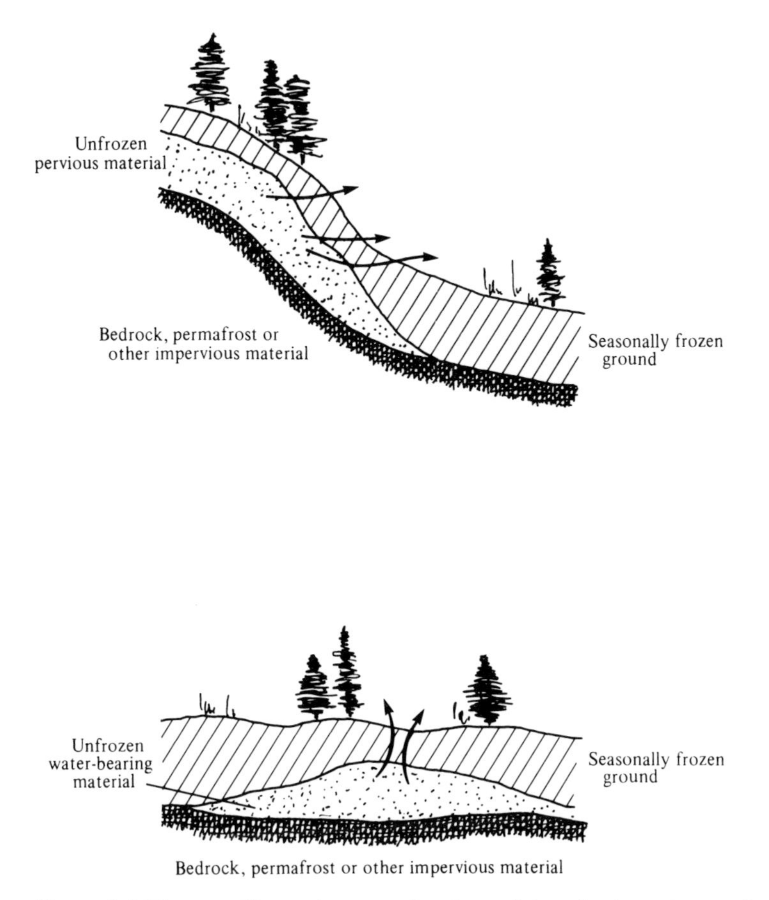

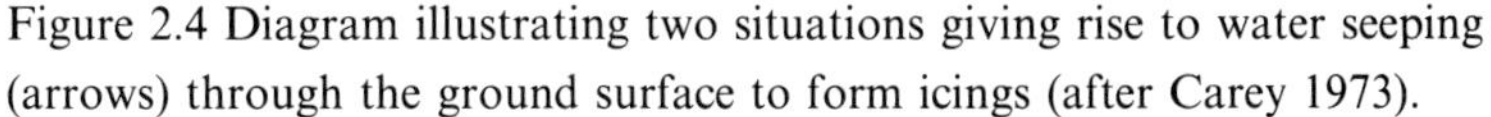

Figure 2.4 Diagram illustrating two situations giving rise to water seeping (arrows) through the ground surface to form icings (after Carey 1973).

River icings occur by accumulation of overflowing water, which becomes layers of ice, on top of the original surface ice during the winter. The water may originate in springs adjacent to the river bank, or, sometimes, rising in the river bed itself. Freezing from all sides of a small lake may also result in uplift of the ice surface, according to Shumskii (1964). Large icings may also form in rivers as a result of partial damming of the rivers by moving ice masses or frazil ice – a loose accumulation of small crystals (Gold & Williams, 1963). The icings may persist through the summer when river flow is low, or sometimes perennially, on islands or gravel bars.

Icings have geotechnical significance, especially for highways which may be blocked or rendered dangerous. Highway culverts blocked by ice are condusive to icing formation. Icings may also build up in association with restriction of drainage by pipeline foundations or other extended structures.

2.2.5 Age of ice in permafrost

That some ice in the ground is of great age is demonstrated by the ice masses investigated by Mackay *et al.* (1972) which were deformed more than 40 000 years ago along with the surrounding frozen sediments. The deformation was caused by glacier movement (although there is no evidence of glacial origin for the ice).

Ice wedges are found at considerable depths and examples are reported (Mackay *et al.*, 1979a) which formed more than 50 000 years ago according to the age of the surrounding sediments, and of the discontinuity representing the ground surface when the wedges formed (see also Black, 1983).

These demonstrably old, preserved ice forms are in contrast to the instability of the ice discussed in Chapters 7 and 8. This instability is demonstrated by the formation or loss of segregation ice layers occurring often more or less continuously, near the upper surface of contemporary permafrost where there is often a concentration of segregation ice (Mackay *et al.*, 1979b and see section 8.6.2). These authors review Russian experimental studies of changes of 'cryo-lithological texture' that is, of the forms and distribution of the ice in frozen ground. These changes would presumably be substantial over geological time, although perhaps only taking place when the temperature was within a few degrees of 0 °C.

Ice which was buried or formed *in situ* simultaneously with the deposition of sediments is called *syngenetic ice*. Ice that formed after the ground itself is called *epigenetic ice*, or *aggradational ice*, the latter usually referring to ice formed in association with the extension of permafrost (but see section 8.6.2). A distinction is thus made between the age of the permafrost, that is how long the ground has been frozen, and the age of particular ice inclusions. For example, ice wedges are formed in cracks in permafrost so the wedge is commonly younger than the permafrost (although in the case of syngenetic permafrost, cracking and the initiation of the wedges may occur almost immediately). Ice masses older than the permafrost itself cannot, of course, be segregation ice, but could be buried glacial ice of much greater age. The age of permafrost and thus of segregation ice, can vary abruptly over small distances. Such is the case where an isolated discontinuity in the permafrost, for example ground below a transient lake, refreezes. Even in areas where the general climate is overall cold enough for permafrost and has been so for a long time some ice in frozen ground may be of quite recent origin. In addition to the lateral variations in mean ground temperatures, there are short-term variations of the depth of the active layer (that is to say, of the top of the permafrost) also due to micro-climatic or climate fluctuations.

ISOTOPE VARIATIONS

Although there has been a great deal of research on oxygen isotope variations in glacier ice, similar studies of ground ice are not numerous. Urey (1947) first proposed the use of the ratio ($^{18}O/^{16}O$) in glacier ice for indicating past climatic conditions.

The interpretation of the isotope record from permafrost is more complicated than for glacier ice cores, because the accumulation of ice is not necessarily continuous. A portion of the isotope record may be lost and replaced by younger water, following periods of degradation. Nonetheless, oxygen isotope variations have been used to estimate paleoclimatic conditions during formation of ground ice and have assisted in the interpretation of permafrost history (e.g. Mackay & Lavkulich, 1974; Stuiver *et al.*, 1976; Michel & Fritz, 1978, 1982). More recently, Lorrain & Demeur (1985) used data on both hydrogen (deuterium) and oxygen isotopes from ground ice masses on Victoria Island, NWT, to infer a buried glacial origin.

In very young ground ice, the amount of tritium present can also be analysed. This tritium originated from the atmospheric testing of nuclear weapons in the 1950s and 1960s, becoming incorporated into the ground water/ice system via the natural Processes of the hydrologic cycle. Tritium is one of the few viable tracers for studying the process of water movement in permafrost, for whereas dissolved solutes alter the behaviour of water below 0 °C, the isotopes of oxygen and hydrogen are directly incorporated into the water molecule. Van Everdingen (1978) used natural isotope data to study the origin of water and the freezing process of frost blisters, and Michel & Fritz (1978, 1982), Chizov *et al.* (1983) and Burn & Michel (1988 in press) have examined the tritium content of natural permafrost.

Since there is no other method for differentiating the relative age of ground ice masses, the use of isotope data is clearly worth further development. However, it should be seen, ultimately, as another tool providing information to be fitted into a composite picture of the stratigraphic record (e.g. Burn *et al.* 1986).

2.2.6 *Distribution of ground ice*

Although several modes of formation of ice masses in permafrost are well understood, it is often difficult to predict the nature and extent of ice in the ground. Ice inclusions observed in cores obtained by drilling are often impossible to describe according to origin. Indeed, even in large exposures it is difficult to judge the shape of the whole mass, let alone its origin. It is rare that an ice wedge, for example, is exposed such as to reveal a wedge shape. Melting on exposure in any case disturbs the form.

Many detailed and systematic investigations have been made along proposed pipeline routes in the Canadian north and a certain pattern of distribution can often be recognised in a particular region. For example, massive ice sometimes occurs predominantly in low-lying valley bottom situations where water flows down through the active layer or through bedrock to feed the growth of ice. Elsewhere, lithology, or geological history may give a quite different distribution. Ice masses and high excess ice contents are more frequent in the proximity of water bodies. Such ice masses are for the most part, presumably, segregation ice, developing in fine-grained material, and an important requirement often seems to be that interspersed coarse-grained sediments provide a passage for water supply. These general correlations between ice distribution and soil type and other environmental conditions can be obscured by strictly local factors.

Frost-heave (segregation) ice frequently varies in amount and form even within a metre of ground. The reason for such variations is often not clear. The effects make prediction of amounts of ground ice or of possible frost heave extremely difficult and this is a major problem in geotechnical engineering. It is worth emphasising that ice masses occur in gravel-rich or other coarse-grained permafrost (Kudriavtsev, 1978) in contrast to seasonally frozen materials of this kind. The origin for the ice in the coarse-grained permafrost is often uncertain. As noted in section 2.2.2 some may be segregation ice.

The formation of segregation ice will normally give topographic expression with pingos and palsa (see Chapter 6) being extreme examples. The humpy, hummocky nature of the ground surface in cold regions is characteristic, while even shallow seasonal freezing can give visible surface irregularity. Ice wedge ice is, of course, revealed by polygonal patterning. In the Tuktoyaktuk area of the Mackenzie Delta ground ice is abundant but there is little relief. Rampton (1973) and Mackay (1983a, p. 71) have proposed that the ice was formed by freezing of glacial meltwater. As the glaciers retreated exposure of the ground occurred, the mean ground temperature fell (ground being frequently unfrozen below glaciers) and permafrost formed with freezing of ponded, or perhaps subterranean, water. The ice segregation process would also be expected with the presence of so much water, at least wherever the sediments were fine grained.

An interesting but little studied matter is that of ice in the ground with respect to the hydrological cycle and in particular, relative to 'desert' conditions as normally understood. The latter arise where evaporation at the ground surface exceeds precipitation (and other sources of water supply) on a continuing basis. The continuing water deficit results in a substantial depth of ground becoming dry. In very cold regions, could ice be lost in this way over a sufficiently long period?

Precipitation in the north of Alaska (for example) is, at some 24 cm per year, almost as low as in the Tunisian desert. Yet the ground conditions are wet in summer because of the poor conditions for evaporation and also because of poor drainage due to permafrost and abundant ground ice. In parts of the Yukon, which is substantially warmer, but with low precipitation (in the vicinity of Whitehorse some 26 cm), there is a moisture deficit in summer and salt accumulation may be observed on the surface following from dominantly upward moisture flow. Nevertheless, there is abundant ground ice.

The persistence of masses of ice fairly near the ground surface can be ascribed to the low vapour pressure of ice (see Figure 2.5), and the small dependence on temperature, compared with water at higher temperatures. Under temperate conditions more evaporation occurs because of temperature-induced vapour pressure gradients, near the soil surface, and

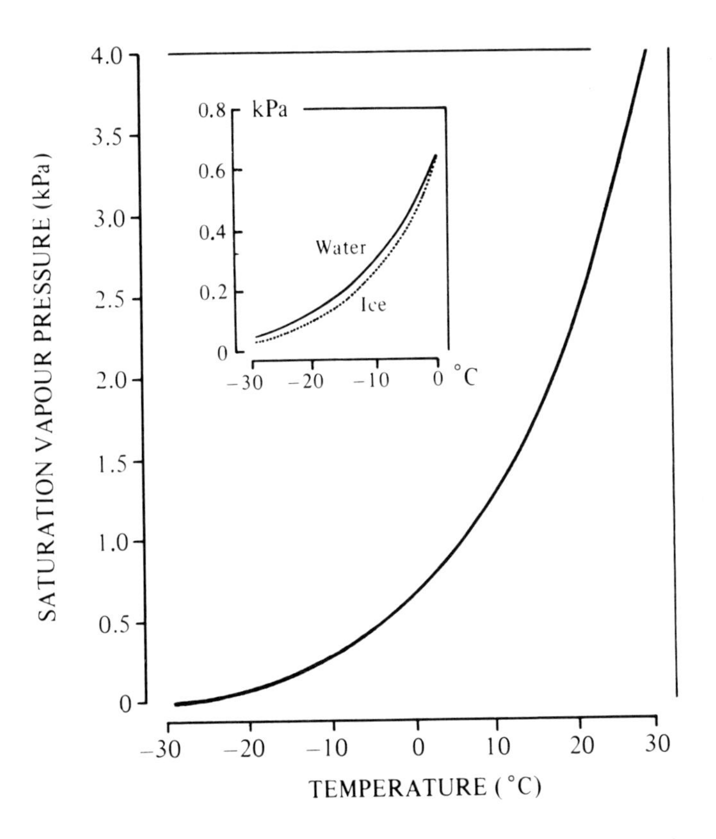

Figure 2.5 Saturation vapour pressure of water as a function of temperature and compared to that of ice (insert). After Byers (1965).

between the surface and the air. Such pressure gradients will be smaller at low temperatures.

Nummedal (1983) has pointed out how the low vapour pressures of ice at low temperatures could mean that ice persists in the ground even at the low atmospheric vapour pressures of Mars. Thus we appear to have a general explanation of why the moisture (that is, ice content) remains high even in regions of extremely small precipitation.

Another important factor in the cold regions may be the persistent snow cover of the long winters. There is unlikely to be much evaporative loss to the atmosphere from the ground through the intervening snow. Nevertheless, while certain characteristics of the hydrological cycle in cold regions are receiving attention (Ryden, 1981; Woo, 1986; see also Chapter 8), there is little known about 'dryness' of the cold regions (Bovis & Barry in Smiley & Zumberge 1974).

In spite of the circumstances noted, there are indeed regions of 'dry' permafrost around McMurdo in the Antarctic and in northern Greenland (Smiley & Zumberge, 1974) where ice is apparently absent in the ground. Cracks of the kind which would have been elsewhere filled with ice wedges, are filled with apparently windblown sediments (Péwé in Smiley & Zumberge, 1974). Both ground and air temperatures are persistently very low in these areas and it is then more likely that the very low relative humidities of the air and of the generally dry snow-free surface lead, over long periods, to disappearance of any ice inclusions in the ground. The likelihood of the (saturation) vapour pressure of ice being greater than the vapour pressure of the very dry atmosphere (low relative humidity) means sublimation may be occuring even when the atmosphere is somewhat warmer than the ice. 'Polar desert' (Smiley & Zumberge, 1974) is a term that may properly be applied to these areas. To extend it, as some do, to include any rather barren polar area, including, for example, the North Slope of the Brooks Range, Alaska, seems mistaken. Most people would not feel it appropriate for that landscape of swamps and ponds.

2.2.7 Submarine frozen ground

Seawater has a concentration of salt of $35\,\mathrm{g\,l^{-1}}$ and thus has a freezing point of $-1.9\,°\mathrm{C}$. This requires special consideration when the freezing of marine sediments is considered. Depending on the depth of sediments in the seabed and their history, the concentration of salt in the pore water varies from as low as that found on land to in excess of that in seawater – the latter circumstance being due to concentration by freezing.

Only in near-shore regions is there ice extending down to the sea bed. Elsewhere, sea water (at temperatures down to −1.9 °C) determines the temperature of the sea bed.

Much confusion occurs because of salt-rich submarine sediments which are unfrozen at, say, −1.5 °C. According to a widely accepted definition any material below 0 °C may be called permafrost (Washburn, 1979). To apply the term to a material with none of the properties of frozen sediment (except that it might induce freezing in non-saline water in a pipe within the sediment) seems to us mistaken. A similar problem arises elsewhere, with very dry material or indeed 'ordinary' soil where the normal small quantity of salts prevents freezing precisely at 0 °C. But neither situation is of such practical importance as the extensive submarine sediments whose temperature lies below but within a degree or two of 0 °C.

Such sediments may be free of ice. There may also be found strong, thoroughly-frozen submarine sediments even at these 'warm' permafrost temperatures – such materials have originated as terrestrial sediments into which salt has not yet penetrated (see below, section 2.4.2). Alternatively, the sediments may contain ice in segregations (small or larger layers or more irregular forms) but no pore ice. The material will then not have much greater strength, possibly even less, than completely ice-free material, and (as discussed below) is essentially relict permafrost in the protracted final stage of a thawing which is controlled by the slow supply of the large quantity of heat required. If sufficient salt is present such melting will be complete at some temperature significantly below 0 °C.

Accordingly, terms such as 'ice-bearing', 'ice-bonded' (or 'unbonded'), and 'partially frozen' appear adjectivally with 'permafrost' in recent literature. Rates of thaw will generally be less than in terrestrial seasonally frozen materials. The rate of consolidation of submarine sediments on thawing will be slower than on land because of the high water pressures. Consequently, a loose condition but with ice still present ('ice-bearing unbonded') may be more persistent and widespread than occurs in terrestrial situations. It is important that the terms are used with precision and based on direct observation rather than, for example, temperature measurements, seismic soundings or determination of electromagnetic properties which can be open to various interpretations (National Research Council Canada, 1988). The uncertain findings of a geophysical remote-sending procedure do not justify being reported as 'acoustic permafrost'.

Generally speaking permafrost does not form at the sea bed, the concentration of salt being similar to that in the sea itself (but see section 2.4.2).

In submarine permafrost of terrestrial origin, the various forms of excess ice described earlier in this chapter may be present according to the terrestrial conditions that occurred earlier.

2.2.8 *Micromorphology of freezing soils*

Freezing affects the arrangement of soil pores and particles, that is, the structure of the soil, and these effects persist to some degree after the soil is thawed. Thus they are cumulative under repeated freezing.

The effects of freezing on soil structure are important in distinct ways. The structure is significant with respect to the soil as the medium for plant growth. Micromorphology is important to geologists and engineers concerned with the recent or past history of the earth materials and with past environmental conditions; the soil structures may eventually give valuable geotechnical information on processes of deformation.

The expansion of ice and the contractions resulting from the suctions in the unfrozen water can have opposite effects. While pores and other voids are often enlarged by formation of ice, aggregates of soil particles (often millimetres or centimetres across) are compressed (consolidated). In this compression, the suctions of the water are as important as the elevated pressures of the ice in defining the effective stress responsible (sections 7.2.2, 7.3). The aggregates are more dense the lower the temperature reached.

Aggregates persist after thaw because of the partial irreversibility of the consolidation process. Their size and shape depend on the frequency and orientation of ice lenses (Van Vliet-Lanoe *et al.* 1984), and consequently there are changes with depth. A sorting process during repeated freezing and thawing leads to small caps of finer-grained materials (Figure 2.6) on the upward surface of aggregates (Van Vliet-Lanoe 1982). At the same time a 'skeleton' of coarser particles expelled from the aggregates may be developed.

The structures formed are observed to be modified by creep on slopes, and the processes may themselves cause creep (sections 5.3.1, 9.8). Microscopic studies by Van Vliet-Lanoe *et al.* (1984) showed minute discrete slip planes, apparently the sites of ice lenses. C. Harris (1981) describes material in a solifluction lobe which showed orientation downslope of grains, coatings on larger grains sometimes on one surface only and local translocation at sites of ice lenses.

Somewhat faster movements have larger effects: aggregates may be rotated and in *cryoturbations* stones and particles within the soil may be arranged vertically and there are various 'injection', or flow-like features with tongues of materials often tens of centimetres in length visible in the

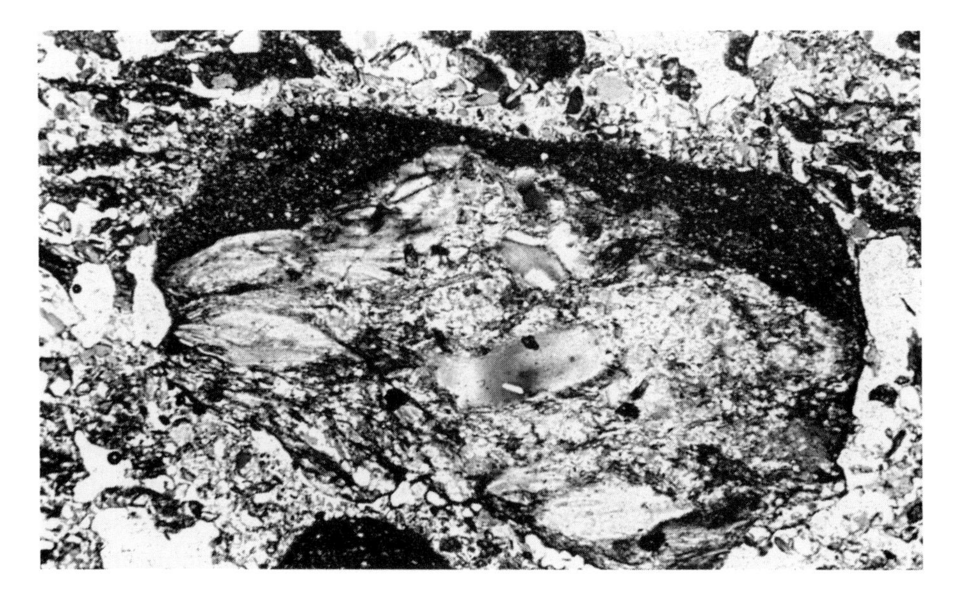

Figure 2.6 Example of microstructure developed by freezing and thawing of a soil. A capping of silt (dark, fine-grained material) over soil aggregate. Sample taken from a solifluction feature (× 25, Photo: C. Harris).

profile. Another important effect is the formation of voids or vesicules caused by air expelled from advancing ice. The voids are often much in excess of pore size. Air bubbling out of the ground surface during thaw probably comes from collapse of such vesicules. C. Harris (1983), however, believes vesicules are developed immediately following thaw, as a consequence of a thixotropic liquefaction.

The mineral composition in relation to particle size is changed by repeated freezing. Quartz is particularly unstable under freezing conditions and thus is represented by more small particles than in soils not exposed to freezing (Konischev, 1982).

In nature, ice is commonly a polycrystalline material although the size and orientation of the crystals varies greatly. Recrystallisation occurs with time and in association with changes of temperature and stress. Ice in the ground reflects these circumstances, and the various forms show diversity in crystal structure (Östrem 1963b). It might seem, therefore, that studies of the crystallography would be an extremely valuable tool in studying the origin of such ice but they have not been widely used.

E. Penner (1961) found that ice segregations in frost-heaved laboratory samples were composed of small crystals orientated with their long axes in the direction of the heat flow during the segregation. Östrem (1963b) examined ice from ice-cored moraines and found that it generally had one single preferred orientation, and the crystals were smaller than in glacier ice. The size of crystals tends to increase with age as a consequence of thermodynamic relations.

A detailed study of ice-wedge ice (Gell, 1978) showed that crystal size increased from the centre outwards, which would be expected as the wedge expands progressively by addition of ice to a central crack (see section 6.4). The preferred orientation of the crystals changes progressively outwards, being normal to the line of the crack at the centre of the wedge, and becoming parallel to the sides of the wedge. These effects are ascribed to the orientation of stresses.

2.2.9 Pedology and soil freezing

The surface layers of the ground are exposed to exchanges of heat and moisture that result in many chemical and physical changes of the original parent material. These weathering or soil-forming processes are most rapid in regions of high temperature and abundant moisture, the former speeding chemical change and the latter aiding it through the processes of hydrolysis and through the dispersion of the reaction products. Freezing and thawing, and low temperatures generally, would be expected to have significant effects on these processes, which in any case, vary with local conditions.

Pedologists regard the layers affected by these processes as the 'soil', in contrast to the usage adopted generally in this book where the term applies to any particulate earth material. The affected layers, one or two metres thick in temperate climates, characteristically show horizontal stratification, *soil horizons*, resulting from particle sorting, and accumulation or removal of reaction products. The development of the soil, so defined, provides for fertility; the nature of the soil is especially important in studies of vegetation. It is of practical importance in agriculture, which is carried out on areas underlain by permafrost in continental Asia, parts of Alaska and elsewhere, where summer temperatures are sufficiently high. The soil may be important in other practical considerations. It may have corrosive effects, and it may exert some control on the hydrologic regime or on the mechanical bearing capacity of ground.

The diverse soils of the world have been classified under the United States system into 10 orders, six of which are represented in the cold regions (Rieger 1983). Other systems are in use in Canada (Canada Soil Survey, 1978), the USSR and elsewhere (Linnell & Tedrow, 1981) but brief reference to the US system will illustrate classification for the cold regions. It differs from the Canadian mainly in terminology. *Entisols* (Canadian:

Regosols) are soils little altered from the parent geological material. It might be thought that this was the common characteristic of the soils of cold regions. In fact, the entisols are largely restricted to recently deposited fluvial sediments (which are then known as *Cryofluvents*), or extremely cold and dry situations (where chemical weathering is virtually absent) or where disturbance of the soil (as by downslope movement or cryoturbation) is so rapid as to exceed the rate of soil formation.

Soils of cold regions often show development of horizons due to the translocation of iron, aluminium or organic material downwards and they are then classified as *Spodosols*. These are also often known as 'podzols'. *Alfisols* are those where there is characteristically a downwards translocation of fine clay particles. The nature of the parent material is particularly important in the *Mollisols* which develop on calcareous, basaltic or other basic rocks. The associated vegetation and bacterial activity results in a characteristic soil profile which develops only slowly in cold regions because of the low biological activity.

Yet another group of soils derives from accumulations of vegetation residues. This is the *Histosols* – often known as organic soils. Such soils cover enormous areas in those parts of the cold regions where summer temperatures are, of have been, sufficient to produce the plant growth necessary. These are particularly the areas of peat bog (muskeg). Although the pedologist is concerned with the changes or the organic or mineral material within the first metre or two, it is significant that the peat (plant residue material) extends downwards often tens of metres. Thus it is also a parent material to be compared with those geological materials occurring below the depths affected by weathering and the pedological soil-forming processes.

The particular near-surface processes and the resulting soil profiles and horizons for cold regions are described in detail by Rieger (1983), with particular regard to the classification of the many subtypes of those groups briefly outlined above. The effect of permafrost is to greatly reduce downward moisture movement. So great is the effect on the translocation processes, a class of *Pergelic* (Canadian: Cryosolic) has been defined, as having a mean annual soil temperature of 0 °C or lower. This, of course, means that permafrost is present at depth. Changes in the thickness of the active layer modify the processes occurring at a particular location.

The *Inceptisols* (Canadian: Brunisols) include wet soils with restricted drainage, and thus often occur above permafrost. Under such conditions *gleying* occurs. This is the formation of ferrous iron, giving a greyish colour, under the reducing conditions of saturated, oxygen-depleted soil. These soils may show an *apparent thixotropy*, a tendency to lose strength and flow when disturbed.

In the USSR and in Canada temperature indices, the yearly sum of mean daily degrees above some datum (usually +10 °C) giving a degree day (°day) value, are used to define soil regions. Thus there is the 'tundra and arctic regions', with less than 600 °days, so defined, the 'frozen taiga', 600–800 °days, or the 'taiga forest boreal regions' with 600–2400 °days (Rieger, 1983, p.5).

The disturbances of the soil including differential frost heave, cryoturbation and sorting processes, and the creep of material on slopes, described in the previous section, frequently destroys such horizons as may develop. A few disorganised remnants of a profile is often all that can be seen. Some special situations occur, such as where material moving downslope overruns the surface to give a persistent and distinct buried organic layer.

The sorting and translocation of fine particles is promoted by freezing and thawing. The formation of aggregates, or peds, of a few millimetres or sometimes centimetres across, is basic to the development of soil structure. Similarly, the development of a hard dense layer, a fragipan, is characteristic of many types of soil. In the case of soils exposed to freezing the mechanical thermodynamic effects (outlined in the previous section and described in detail in later chapters) commonly produce such structures.

As the study of soil profile formation in cold regions continues, it is likely that the effects, physical rather than chemical, of freezing on granular materials will still be seen to have a dominant role. Nevertheless, many of the processes important in warmer climates can also occur in the permafrost regions – though largely restricted to the active layer.

2.3 Gas hydrates (clathrates)

A gas hydrate is a solid in which molecules of gas are combined with molecules of water. Of particular interest is the hydrate of methane (natural gas, $CH_4 6O H_2O$ – the amount of CH may vary) which is described as an ice-like substance (Davidson *et al.* 1978). This is common in continental margin sediments in all the major oceans under 500 m or more of water as the material is stable at such pressures, at the temperatures of the ocean bottom. At lower temperatures it is stable under lower pressures, hence the particular importance and relative abundance of gas hydrates in polar regions. Small quantities of hydrates of other gases than methane may be present as well (Kvenvolden, 1982). Sufficient loss of pressure or rise of temperature will cause the hydrate to become a mixture of gas and ice or water. In drilling operations this may occur explosively, the gas occupying a much greater volume than the hydrate.

The existence of gas hydrates in nature was only verified in 1968; they have been found in permafrost, for example, in the Mackenzie Delta, the Canadian Arctic Islands and elsewhere in the Arctic (Judge 1982; Weaver & Stewart 1982 and Makogon 1982). The extent of their occurrence in permafrost and below it, depends on the mean temperature of the surface of the ground and the geothermal gradient in a manner which can be understood from the pressure–temperature relationship shown in Figure 2.7. The maximum temperature at which the material is stable falls rapidly for pressures less than about 2000 kPa. At this pressure the material breaks down at −10 °C while at 1000 kPa it breaks down at −30 °C. Two thousand kPa is the pressure which, it is usually assumed, would occur under the weight of about 100 metres of soil. With a geothermal temperature gradient of about 0.025 °C per metre, for the temperature to be −10 °C at 100 metres the near-surface temperature would have to be −12.5 °C. In fact, only relatively small areas of the earth's land surface, for example the high latitude Canadian arctic islands, have mean near-surface (ground) temperatures as low as this. Consequently, gas hydrates would not be expected to occur at depths much less than 100 metres below the land surface, even in these cold areas.

Below the oceans the pressure due to the water may be great enough for hydrates to occur at the seabed. For much of the world seabed temperatures are some 5 °C. In the high arctic they may be one or two degrees below 0 °C. Consequently, hydrates occur at lesser depths of water (Figure 2.7). Low geothermal gradients increase the chance of hydrates occurring. They may also occur well below the permafrost. In the northern Canadian arctic islands the layer capable of sustaining hydrates may be more than 1000 metres in thickness. Below that depth the temperatures are likely to be too high (Figure 2.7).

Although the equilibrium conditions for the hydrate are understood in general, the location within the pores of the soil presumably means that thermodynamic conditions for equilibrium similar to those for water freezing within the pores, apply. Makagon (1974 in Weaver & Stewart, 1982) reports that higher pressures and lower temperatures are required for the hydrate to be stable in the pores of fine sand than when not so confined. Ultimately, the understanding of the extent and the behaviour of the hydrates may be as complex as that of the water–ice–pore system itself.

The amount of gas in the form of hydrates may be such that they will become a significant natural resource. On the other hand, the hydrates

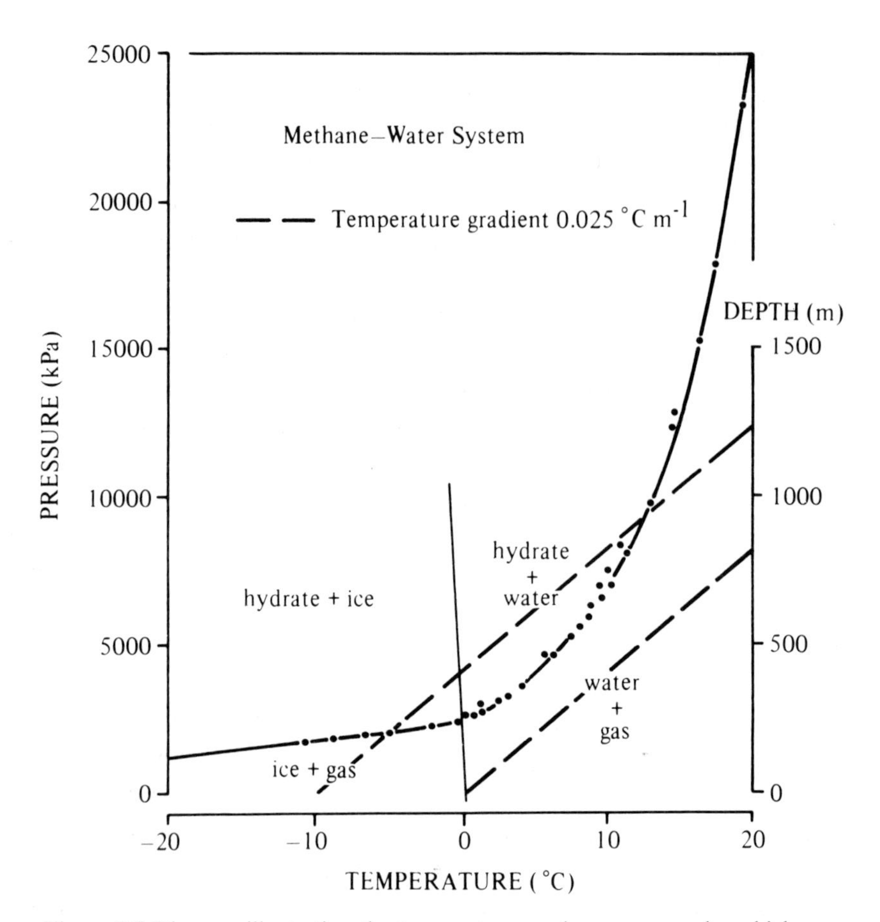

Figure 2.7 Diagram illustrating the temperatures and pressures under which gas hydrates occur. On the right hand axis the pressures are shown as equivalent depths of sea water. The two dashed lines represent the geothermal gradients for surface temperatures of 0 °C and – 10 °C. (After Davidson *et al.* 1978).

besides presenting a danger to drilling operations, including blow-outs due to gasification (Bily & Dick, 1974), can lead to misinterpretation of geophysical investigations of sediments and their stratigraphy (Judge, 1982). Furthermore, the presence of hydrate in the soil will, of course, change the geotechnical and other properties in various ways.

2.4 World distribution of permafrost

Some 26% of the world's land surface (Figure 1.7) is underlain by permafrost according to estimates reviewed by Washburn (1979). The extent of submarine permafrost is relatively small because of the higher temperatures of the seas relative to adjacent land surfaces, and the salinity which lowers the freezing point (to – 1.9 °C for a salinity of 35 g l^{-1}).

2.4.1 Land permafrost

The greatest known depth of permafrost (Washburn, 1979) is 1450 metres (in Siberia) and with further investigations a greater depth will surely be found. There is really no minimum value for the thickness of the permafrost layer, even a cm thick layer extending a few centimetres laterally, which remains frozen for more than one year is, by definition, permafrost.

The world map of permafrost distribution (Figure 1.7) shows that it occurs at high latitudes and high altitudes (where it is sometimes known as 'alpine permafrost', Péwé, 1983). These are characteristically cold locations considered by the normal criterion of air temperature. In so far as permafrost is climatically determined we can expect it to occur in such regions but it is emphasised that air temperatures do not 'cause' permafrost. Both air temperatures and ground temperatures are functions of the energy exchange acting in the vicinity of the ground surface (Chapter 3). Permafrost occurs where the mean annual ground temperature is below freezing. The mean annual ground temperatures and those of the air diverge by as much as 8–10 °C, usually the ground temperature being the warmer. The coldness of winters is not a good guide to the distribution of permafrost and such conventionally 'cold' places as Moscow or Winnipeg have mean annual temperatures well above 0 °C because of their warm summers. Permafrost extends south towards the interior of the North American continent. This is to be related to the lower mean annual temperatures in the continental interior, and not to the continentality (temperature amplitude and low winter temperatures) of the climate *per se* (Figure 1.7).

Although lying at latitudes where it is elsewhere widespread, Scandinavia is largely free of permafrost except at high altitudes (King, 1984). The Gulf Stream moving up the coast of Norway has a strong warming effect on the climate. By contrast, in the highlands of Tibet and Northwestern China permafrost is widespread at the same latitude as Palm Springs, Florida. Nearly two million square kilometres are underlain by permafrost in Mongolia, northwest China, the Himalayas and Tibet (Zhou & Guo Dongxin, 1983). Permafrost is often thought of as a northern, or polar phenomenon yet in parts of the Soviet Union, one travels south to come to these vast areas underlain by permafrost.

Altitude also accounts for occasional permafrost in the Rocky Mountains in Canada and the United States, even south of latitude 40° (Péwé, 1983, S. A. Harris, 1986). Likewise, permafrost has been reported on Mount Washington in the eastern United States, and in the Chic-Choc Mountains, Gaspé, Quebec (Gray & Brown, 1982).

If latitude and altitude are regarded as the two most general factors, there are large but more local features of a size represented on atlas maps or globes which also affect the distribution: large rivers or lakes are underlain by warmer ground than nearby and thus tend to be free of or have thinner underlying permafrost.

Certain geological structures lead to 'geothermally induced' higher ground temperatures where the flow of heat upwards to the ground surface assumes more significance than elsewhere. Heat carried in water to hot springs, as, for example, in Iceland obviously prevents permafrost. The geothermal heat flow by conduction varies from place to place, and the larger it is, the steeper will be the geothermal gradient (other conditions being equal). As a result the depth reached by permafrost should be less (see section 4.4.1).

While this is true in principle, the temperature gradient through the permafrost is unlikely to be linear. Different rocks have different thermal conductivities. If the geothermal heat flow is unvarying, the temperature gradients adjust themselves to become inversely proportional to the conductivities of the rocks in the profile. High conductivity rocks will, therefore, result in permafrost extending deeper for a particular surface temperature (consider Figure 4.3).

Most complex of all and presenting the most problems for the engineer are the myriad local variations of the vegetation, topographic form and soil conditions, that are responsible for the variation of the mean annual ground temperature by several degrees even over distances of only tens of metres (Chapter 3). Wherever the average temperature is within a few degrees of 0 °C the variability of the mean annual ground temperatures means that permafrost occurs in patches or in a discontinuous fashion. Moving into generally colder regions, bodies of permafrost become more frequent and cover larger areas. Even in very cold regions, however, there may be gaps through the permafrost. These circumstances coupled with the scattered nature of direct observations make precise mapping difficult.

Faced with the importance of permafrost to construction procedures and the evidently large areas where its presence is uncertain, early workers attempted to zone the cold regions as those of sporadic, discontinuous and continuous permafrost. However, there is little rational basis for delineation of boundaries. The map of permafrost in Canada (R.J.E. Brown 1967) shows a division into 'continuous' and 'discontinuous' permafrost, based on a curious criterion (ascribed to Russian workers) of observed mean annual ground temperatures above, or below, −5 °C. Frozen ground has properties depending on its composition and situation, but there is no special change at $-5\,°C$. Nor has this temperature any particular importance for the distance, whether 500 metres or 50 kilometres, to the nearest 'opening' or permafrost-free ground, and thus whether the region is one of 'continuous' or of 'discontinuous' permafrost. To draw the line representing the 'boundary' based on observations of this temperature is arbitrary and misleading.

The concept of 'continuous' and 'discontinuous' zones so illustrative when little was known about permafrost, has tended to suggest, unfortunately, that there are two always-distinct kinds of material or terrain. Rather, what is required is an understanding of the mechanisms controlling the temperatures in the ground (Chapters 3 and 4) and their spatial variations. In field studies, richer verbal descriptions of permafrost and its distribution are preferable: 'restricted to low-lying areas', 'patchy', 'often widely scattered', 'absent except in special situations', 'present under high ground', 'normally present' are examples which, with due regard for the scale, provide more meaningful information. The map (Figure 1.7) has been prepared so as to illustrate the uncertainty over the limits of permafrost.

2.4.2 Submarine distribution

In polar regions (even under the North Pole) the sea is commonly not frozen at depth so that sediments are also unfrozen at the seabed. Permafrost may occur, nevertheless, below the seabed, if the sediments are less saline and therefore frozen. They may be frozen even at temperatures higher than the seawater. Temperatures are, however, often lower at depth in the sediments (Figure 4.4) as a result of past colder climates and recent submergence due to isostatic adjustment (Sellman & Hopkins 1984). Consequently the distribution of the permafrost with depth will often be complicated. As with terrestrial permafrost the downwards extent of the permafrost will, ultimately, be limited by the rising temperatures due to the geothermal gradient.

Most of the permafrost off the northern coasts of North America and the Soviet Union probably originated as terrestrial, that is, land permafrost, during the last glaciation when sea levels were lower. Submergence often raises the temperature by ten or more degrees (the amount by which the seawater is warmer than the land surfaces) and this will result in a steady thawing from below of the now 'relict' permafrost. Molochuskin (1973) describes the temperatures of newly submerged permafrost in a region of very rapid coastal erosion by thermal abrasion. The near-surface temperatures rise rapidly but a new equilibrium may not be reached for a thousand years or more, the warming being retarded by the latent heat of fusion. Thus even though the sea in some cases had a mean temperature above 0 °C, permafrost extended a kilometre or more from the shore.

Terrestrial permafrost has low concentrations of salt and, once submerged, thaws slowly as a result of the progressive infiltration of salt water. The sea (water) and seabed temperature can even be a degree or two below 0 °C and thawing of the upper surface of permafrost still occurs, due to the infiltration of the salt. Accordingly, much submarine permafrost appears to be relict, the term being extended to include the delayed melting due to infiltration by salt. It seems unlikely that permafrost occurs beyond the continental shelf. Its presence has been established north of much of the Soviet Union probably extending hundreds of kilometres from the coasts in the Laptev and East Siberian seas (Are, 1984). Permafrost is apparently absent on the shelf of the Barentz and Kara seas, because the shelf has not been exposed by emergence. It is widely present in the Beaufort Sea off the north coast of Alaska and probably elsewhere off the northern mainland coast of Canada. A detailed geological study by Vigdorchik (1980) suggests that permafrost is absent below the Arctic Ocean depths.

Permafrost can develop at the seabed and below if sea ice extends to the bed for a sufficient part of the year. The presence of the ice implies freezing temperatures and the ice provides a conductive bridge transferring heat from the seabed to the cold sub-aerial ice surface. Such permafrost will be limited to places where the sea is shallow, perhaps 10 m, or somewhat deeper in the case of grounded icebergs. Accordingly it is to be expected along coasts.

2.5 Seasonal freezing

During seasonal (winter) freezing, frost may penetrate a metre or two and occasionally more. In milder climates, periods of freezing weather may result in only a few centimetres of frozen ground. But, as explained in Chapters 1, 3 and 4, the nature of the ground cover and the nature of the earth materials are responsible for similar variations within small distances. There are sharp, local differences, too, in the length of time that frozen ground may persist in the spring, and in the proportions of the thawing during the spring that occur from above and from below (Figure 2.8).

There are also great variations in the amount and distribution of ice in seasonally frozen ground. Excess ice is frequently present and is normally segregation ice, although water may enter cracks to give vein ice. Indeed, migration of water to give segregation ice often results in dessication cracks in the underlying soil. These cracks may subsequently fill to give more or

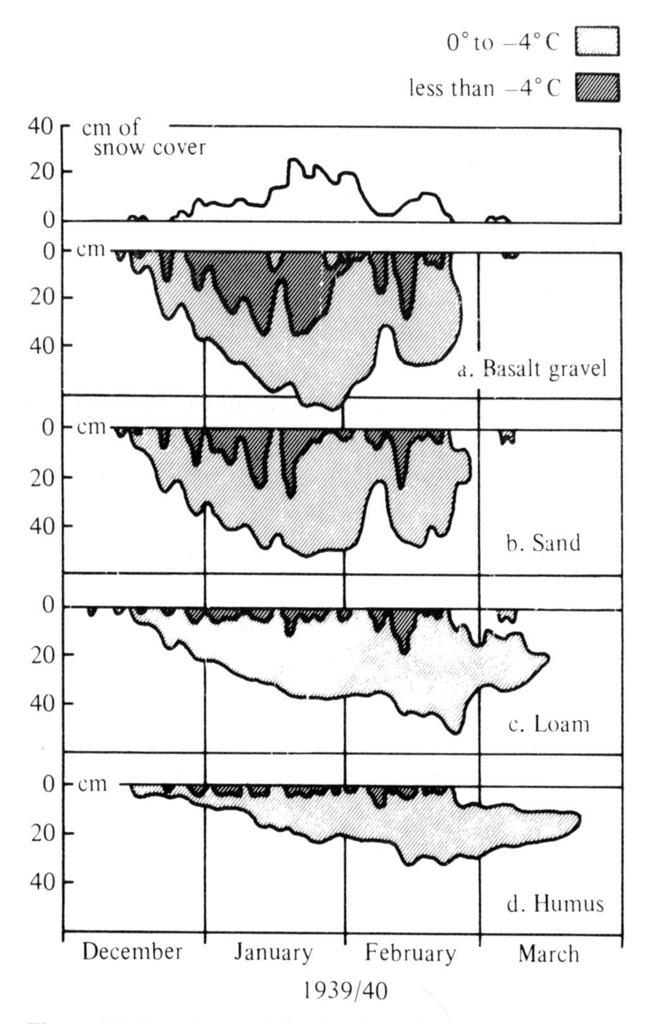

Figure 2.8 Duration and depth of ground frost in four different soils in winter of 1939–40 at Giessen (taken from Geiger 1965 after W. Kreutz). Climate has similar effects and the situation represented in (*d*) occurs more commonly in maritime climates, in many soils, while that in (*a*) is characteristic of climates with hot summers and cold winters (see section 1.3.2).

less vertical ice layers. Especially during midwinter thaws, such as occur especially in continental climates where there are sharp weather changes, some ice segregation occurs near the ground surface by water coming from the surface.

The thickness and concentration of ice segregations depends on the type of soil and the availability of moisture, as well as on the amount of freezing. It is unusual to see ice layers more than a few centimetres in thickness in seasonally frozen ground. In most situations (outside of permafrost conditions) there is insufficient removal of heat at freezing temperatures, during a winter, for thick layers to form – the total thickness of all layers in a profile is suggested by the thickness of ice on waterbodies in the region. Beskow (1935) in his classic study noted that frost heave of seasonally freezing soil may exceed 40% by volume (in fact, if only a small sample is considered, it could be much more). This implies a corresponding quantity of excess ice. However, the nature of the soil is a major control. Sands and other coarse-grained materials do not have segregation ice in seasonally frozen ground. The orientation and form of the segregation ice, whether a few thick lenses or numerous hair-line layers, is influenced by surface microclimate as well as layering, cracks, arrangement of particles and stones, and other features of the near-surface layers.

The relatively high permeability of silts in the unfrozen state as well as when frozen, with the occurrence of at least small quantities of unfrozen water, explains why silts are the soils generally showing the greatest heave during seasonal freezing. Although clay-rich soils are prone to ice segregation, the low permeability of such soils prevents large amounts of heave during the limited time of seasonal freezing.

The importance of avoiding ice segregation and heave in foundations of highways, airports and other constructions has led to extensive investigations into tests to measure the susceptibility of soils to frost heave (Chamberlain *et al.*, 1985). Because the amount of frost heave depends not only on the material but also on the environmental conditions (especially the availability of water, and the temperature gradients) tests on soil samples can never be fully satisfactory.

Narrow needles of ice crystals, called *pipkrake*, *needle-ice*, or *mushfrost* often develop on the surface of the ground during night frosts. They may reach several centimetres in length and often a gravel-sized particle is found on the top of the vertical crystal.

3

Climate and frozen ground

3.1 Introduction

The thermal regime in the upper layers of the earth is controlled by exchanges of heat and moisture between the atmosphere and the ground surface, together with the influence of ground thermal properties. The temperature at the ground surface undergoes fluctuations in response to changes in energy transfers, whilst the propagation of these fluctuations downwards depends on the thermal properties of the ground. The major surface temperature variation has a period of one year, corresponding to the annual cycle of solar radiation; there is also a diurnal variation corresponding to the daily cycle of radiation. Superimposed upon these periodic variations are other fluctuations with durations from seconds to years; these have a variety of causes, such as sporadic cloudiness, variations in the weather, and changes in climate.

The mechanisms of energy exchange at the earth's surface in cold regions are the same as those occurring elsewhere on the earth. Their significance springs from the fact that together they determine the surface temperature regime, and whether frozen ground will exist or not. The processes involved in the energy balance comprise the net *exchange of radiation* (represented by Q^*) between the surface and the atmosphere, the *transfer of sensible* (Q_H) and *latent heat* (Q_{LE}) by the *turbulent motion* of the air, and the *conduction of heat into the ground* (Q_G). The exact partitioning of the radiative surplus (or deficit) between Q_H, Q_{LE} and Q_G is governed by the nature of the surface and the relative abilities of the ground and the atmosphere to transport heat. Each of the energy transfer terms has an influence on the surface temperature, and the way in which the energy balance is achieved establishes the surface temperature regime (e.g. see Outcalt, 1972).

While the occurrence of ground freezing and permafrost depends basically upon climate, the exact relationship is not simple, since the surface temperature regime does not depend solely on geographic location. Local surface conditions, for example, type of vegetation or depth of snow cover, have a profound controlling influence on the surface energy regime, being interposed between the atmosphere and the ground. The influence of these conditions on energy exchange, i.e. the interaction of site-specific factors and the atmospheric climate, produces the *microclimate* of a particular locale. It is the microclimate which is ultimately of importance to ground thermal conditions, since a difference in the surface characteristics will cause a change in the exchange of heat energy between the atmosphere and the ground surface, with significant effects for the ground thermal regime (Figure 3.1). Permafrost forms where the mean annual ground surface temperature is maintained below 0 °C, but differences in microclimate affect the near-surface temperature regime, which influences permafrost thickness and the depth of the active layer (or seasonal frost).

3.2 Seasonal frost and permafrost

Freezing conditions in the surface layers of the earth occur not only in the polar regions, but also in temperate zones, and at altitude even in the tropics. The depth of ground freezing depends in some part on the duration and severity of freezing air temperatures. It also depends on the composition and water content of the earth materials, as well as heat flow conditions in the ground. Periods of freezing, and the existence of frozen layers, vary from a few hours in low latitudes to thousands (and even 100 000s) of years in high latitudes. Freezing in tropical and sub-tropical deserts is of relatively little interest since moisture contents of the materials are so low.

For our purposes, we can divide frozen ground into two basic forms: that which thaws annually (i.e. seasonal frost), and that which does not, which we call permafrost. Permafrost terrain itself is characterised by a seasonally 'active' surface layer that thaws each summer, underlain by perennially frozen ground. Throughout this book, however, the term 'seasonal frost' applies to non-permafrost areas. In the northern hemisphere, the areas of seasonally frozen ground and permafrost encompass large parts of Eurasia and North America.

Aspects of seasonal freezing were discussed in Chapters 1 and 2. Let us imagine some change in climatic conditions which causes the mean annual surface temperature to fall to below 0 °C, so that the depth of winter freezing will exceed the depth of the summer thaw. A layer of permafrost would grow downward from the base of the seasonal frost, thickening progressively with each succeeding winter. Were it not for the effect of the

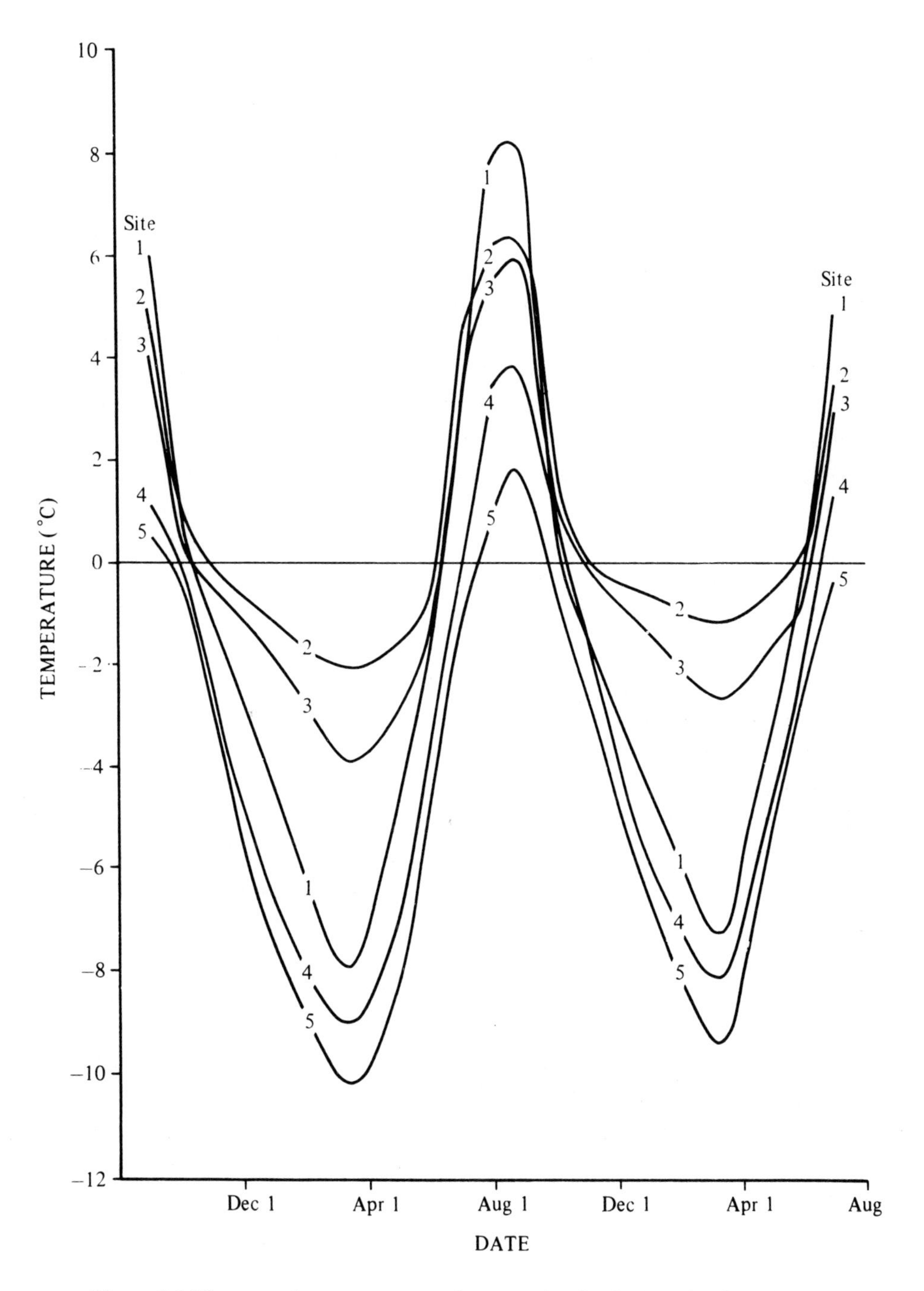

Figure 3.1 The annual temperature regime at a depth of 50 cm for five sites in the Mackenzie Delta, NWT (from Smith, 1975).

heat escaping from the earth's interior, the permafrost would grow to great depths in response to surface temperatures only slightly below 0 °C. However, this heat flow from the earth – along the geothermal gradient – results in a temperature increase with depth of about 30 K km^{-1} (the figure varies with regional geological conditions). Thus the base of permafrost approaches an equilibrium depth where the temperature increase due to the geothermal gradient just offsets the amount by which the surface temperature is below 0 °C (see section 4.4.1). However, such an equilibrium configuration might not be attained for thousands of years, and during this time the surface temperature will undoubtedly have changed.

Whereas the bottom of permafrost is determined by the mean 'surface' temperature and geothermal heat flow, through processes that act over long periods of time, the top of the permafrost is controlled by the seasonal fluctuations of temperature about the mean. Temperature variations are experienced with the passage of the seasons at the surface, and extend in a progressively dampened manner to a depth of 10 or 20 metres. Within the layer of annual variation, the maximum and minimum temperatures form an envelope about the mean, and the top of permafrost is that depth where the maximum annual temperature is 0 °C (Figures 1.6 and 4.7c). The thickness of seasonal frost corresponds to the depth where the minimum annual temperature is 0 °C.

In Figure 1.6, the mean annual temperature profile is shown as extending in a linear fashion right to the ground surface. In reality, this may not be exactly the case. Goodrich (1978) demonstrated that, where there is a marked variation of ground thermal properties with temperature, a deviation characteristically occurs in the equilibrium profile, with mean annual temperatures *increasing* in close proximity to the surface (see Figure 4.15). Burn & Smith (1988) report such offsets in the surface temperature of up to 2 °C at field sites near Mayo, Yukon. They noted that the effect terminates near the base of the active layer, where there is relatively little seasonal change in thermal conductivity. Thus it appears that the mean annual temperature at the base of the active layer constitutes a more appropriate thermal boundary condition for permafrost than the ground surface temperature *per se*. However, for the sake of convenience, the term 'surface temperature' will be used throughout the rest of this book, as though the circumstances in Figure 1.6 prevailed. In most practical cases, the surface temperature is determined by simple linear extrapolation of the ground temperature profile at depth.

The analysis of ground temperature conditions is dealt with in Chapter 4; for now, we wish to examine those largely atmospheric (climatic) processes that influence the surface temperature regime.

3.3 Climate and permafrost distribution

Except for a relatively few areas where adequate ground temperature or other observational data exist, maps which show the distribution of permafrost depend on broad extrapolations and some assumed relationship between mean annual air and ground temperatures. For example, in Canada, the 'southern limit' of permafrost is often represented, somewhat misleadingly, as coinciding with the -1 °C mean annual air isotherm (R.J.E. Brown, 1978*a*), although localised occurrences of permafrost are known south of this (e.g. Zoltai, 1971). While attempts have been made to regionalise permafrost occurrence on the basis of air temperature, little effort has been made to incorporate the influence of snow cover, other surface characteristics and ground properties into mapping permafrost over large areas in the absence of primary data (but see Nelson & Outcalt, 1983).

The general climatic conditions of a place depend on its position (latitude, altitude, aspect and slope angle), since this influences the amount of incoming solar radiation and the air temperature. Since the receipt of solar radiation at the earth's surface decreases consistently with latitude, there is a general corresponding decrease in the mean annual air temperature. However, as a result of the varied physical geographic conditions that can affect the energy exchange between the ground and the atmosphere, the otherwise simple zonal influence of solar radiation upon the temperature regime is much modified. Local factors (i.e. microclimate) commonly override the influence of larger scale macroclimatic factors in the severity of seasonal freezing and the occurrence and distribution of permafrost.

The permafrost map of Canada (R. J. E. Brown, 1978*a*) demarcates three geographical zones – continuous, widespread, and scattered – based on a broad extrapolation using air temperature. In addition, the general decrease in air temperature with altitude results in the occurrence of permafrost at mountain elevations in more temperate regions. This so-called alpine permafrost has been described by Ives (1973) and Harris & Brown (1982), for example. They identify a broad vertical zonation comparable to the general effects of latitude, but stress the significance of microenvironmental factors, such as aspect and snow cover.

Of course, sudden changes in ground thermal conditions do not occur at the borders of permafrost zones. Instead, the zonation really represents the gradual transition from seasonally frozen ground in the temperate regions to the perennially frozen ground of the Arctic. As one moves progressively southward from the far north, where permafrost is present almost everywhere (continuous), the climate ameliorates and permafrost gradually becomes thinner and less widespread (discontinuous). Where mean annual ground temperatures are close to 0 °C, permafrost is present at some sites, where specific local conditions are favourable, while it is absent at other nearby sites. Eventually, permafrost exists only in scattered islands, as in the peatlands of northern Alberta (Lindsay & Odysnky, 1965; Zoltai, 1971) and the Kenai Lowland of Alaska (Hopkins *et al.*, 1955), but it ultimately disappears altogether.

Whilst they may be of some value in portraying the broad geographical features of permafrost distribution, large-scale maps are of little value to specific studies of a scientific or applied nature. In such cases, original surveys must be carried out, since local environmental factors will always be of some, perhaps great importance.

3.4 The importance of microclimates

While ground thermal (and hydrologic) conditions are linked to the atmospheric climate, this influence is moderated by processes occurring in the boundary layer of vegetation, surface organic material and snow. Because of this, mean annual ground temperatures usually differ by several degrees from mean air temperatures, and the difference is not constant from place-to-place, but varies locally with specific conditions. Under the same boundary conditions, soils of different composition, structure and water content freeze to different depths. On the other hand, the same soil under different boundary conditions also freezes to different depths. Thus even in regions where the mean air temperature is low enough to suggest that permafrost should be present everywhere, this is not necessarily the case since its existence depends on an appropriate combination of climate and local/surface/subsurface conditions. Where ground temperatures are close to 0 °C, local factors, such as variations in snow depth, availability of water for evaporation, surface and subsurface materials, etc, can determine whether permafrost is present or not. Consequently, it may occur in a scattered or discontinuous fashion.

Wide variations in ground thermal conditions are known to occur within a small area of uniform climate due to local factors (e.g. R. J. E. Brown, 1973, 1978b; Smith, 1975). Figure 3.1, for example, shows substantial differences in the annual ground temperature regime between neighbouring sites in the Mackenzie Delta, where the mean annual air temperature is close to −10 °C (for further details see Smith, 1975). Soil conditions are fairly uniform among the sites, and the temperature differences are due principally to local variations in surface conditions, particularly snow cover, with vegetation shading of secondary importance. The following points can be noted:

1. Site 1 (no vegetation) has the greatest annual range; it is the warmest in summer (high radiation), but cools off the most in winter (least snow cover).
2. Site 4 (open vegetation) is warmer than site 5 (dense vegetation canopy) all year round.
3. Sites 2 and 3 show distinctly different regimes from the others, with a greatly reduced winter cooling wave because of deep snow cover.
4. Ground temperatures during the second winter were everywhere higher than in the previous year, even though monthly air temperatures were 2° to 9 °C lower. This was because snow depths were greater at all sites, and, further, snow was on the ground by the end of September, some weeks earlier than normal.

The range in ground thermal conditions observed is equivalent climatically to several degrees of latitude.

Figure 3.2 illustrates the effects of largely lithologic variations on ground temperatures; bedrock has a high thermal diffusivity, because of high density and low water content, and this leads to the deep penetration of summer warming (see also Figure 4.7). R. J. E. Brown (1965) reported large differences in ground temperatures at different sites in the Norman Wells

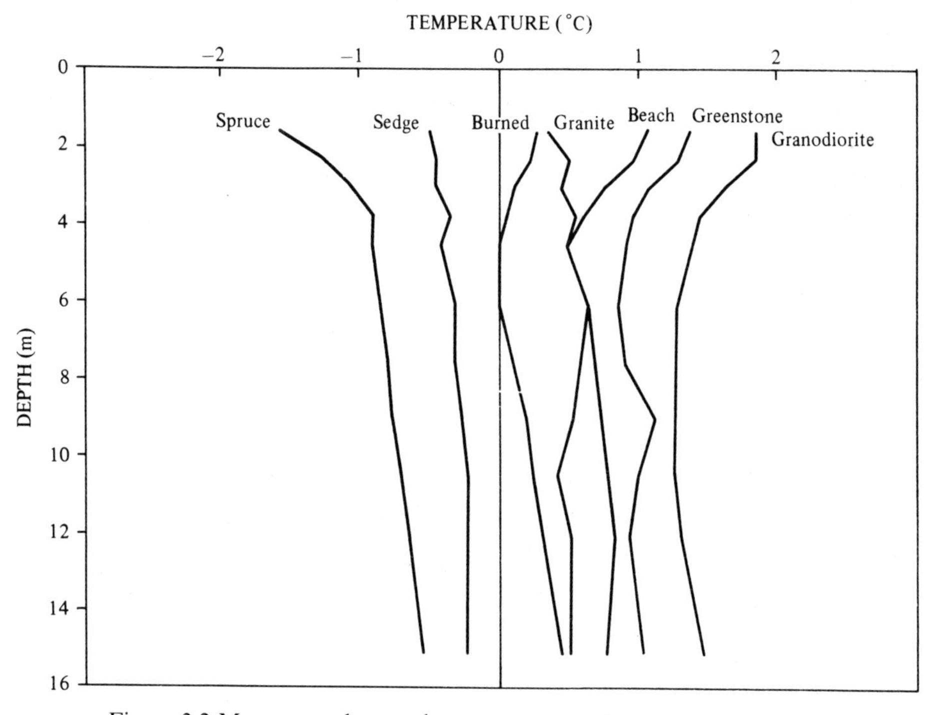

Figure 3.2 Mean annual ground temperature profiles at seven sites near Yellowknife, NWT (from R.J.E. Brown, 1973).

area as a result of the thermal effects of organic matter (see section 4.4.3). There appeared to be a general decrease in temperature with increased moss cover and peat thickness, and temperature amplitudes also decreased in the same order.

Other examples of microclimatic variability can be found in Smith & Riseborough (1983). Together, these observations illustrate the intricacies of the climate–ground temperature relationship, and the need to understand the interaction of climate and local conditions which influences the ground thermal regime. Wright (1981) has commented on two approaches which have been taken in climate–permafrost studies: those based on the spatial association (correlation) of permafrost with vegetation and terrain features, and those relying on energy balance models. In both approaches, the aim is to determine which characteristics of the environment control the ground thermal regime.

There have been numerous studies concerned with the relationship of permafrost to differences in vegetation, topography, hydrology, snow cover, soil conditions, etc. However, interpretations of the effects of site conditions on the details of permafrost distribution are often confused by the high degree of interdependence among these factors. For example, Smith (1975) studying the variation of permafrost along a successional sequence of vegetation in the Mackenzie Delta, found a complex interaction between topography, vegetation and snow cover (see Figure 3.9; also Viereck 1970). Price (1971) found that ground temperatures beneath a north-facing slope were *warmer* than an adjacent southeast-facing slope, which at first seems counter-intuitive. However, the presence of mosses and a thick plant cover on the south-facing slope accounted for the difference. Kudriavtsev (1965, p.17) discusses another example, in which the effect of slope aspect is moderated by the uneven distribution of snow cover. The prevailing winds of the region in question remove snow from south- and west-facing slopes and deposit it on north- and east-facing slopes, which, as a result, remain warmer in winter. In summer, on the other hand, the south- and west-facing slopes receive more solar radiation and are warmer. The net effect is that the two factors compensate, and mean annual soil temperatures are the same for all aspects. Dingman & Koutz, (1974) compared the distribution of permafrost near Fairbanks with the average annual solar radiation, but they found that the relationship was confounded by the effects of vegetation.

Many other such examples could be cited, but it is clear that ultimately we must unravel the interactions of climate, microclimate and lithologic conditions in order to understand the details of the ground thermal regime.

The spatial correlation approach may be a necessary first step, but investigation of the climate–permafrost relationship using a physically based approach allows for more critical examination of which factors are of ultimate significance in determining the ground thermal regime.

3.5 The surface energy balance

The basis for a rational understanding of the variations in ground thermal conditions can be found in terms of the surface energy balance. This may be written:

$$Q^* = Q_H + Q_{LE} + Q_G \tag{3.1}$$

the terms being previously defined. Equation (3.1) can be applied on any time scale, so that we may speak of an hourly, daily, monthly, yearly (or longer) energy budget. In subsequent illustrations, the daily regime is frequently used, although as indicated earlier in this chapter, it is the annual energy regime which is of ultimate importance to us.

Whatever the time-scale chosen, according to the conservation principle, energy cannot simply disappear, and the component fluxes in equation (3.1) must balance at all times:

$$Q^* \pm Q_H \pm Q_{LE} \pm Q_G = 0 \tag{3.2}$$

The chosen sign convention is such that a heat flux towards the surface is deemed positive (an energy input) and a flux away from the surface negative (energy loss). Some difference between the surface temperature and the air temperature is usually inevitable in establishing the balance in equation (3.2), and the difference depends on such variable factors as the amount of radiation absorbed, the degree of air turbulence, the availability of water for evaporation, and the thermal conductivity of the surface ground layer. In winter, the ground will usually be snow covered, and this insulating layer will also contribute to the difference.

The surface energy balance is shown diagrammatically in Figure 3.3(*a*). In the daytime, Q^* is positive (i.e. more radiation is absorbed than reflected and emitted), and energy is transferred away from the surface by convection of sensible heat (Q_H) and by evaporation (Q_{LE}) into the atmosphere, and by conduction of heat (Q_G) into the ground. The terms K and L refer to fluxes of shortwave and terrestrial (longwave) radiation respectively, the arrows indicating direction. At night Q^* is negative – since no solar radiation is received, but there is a loss of longwave radiation from the earth's surface – and there is a net transfer of energy towards the surface by Q_H, Q_{LE} and Q_G.

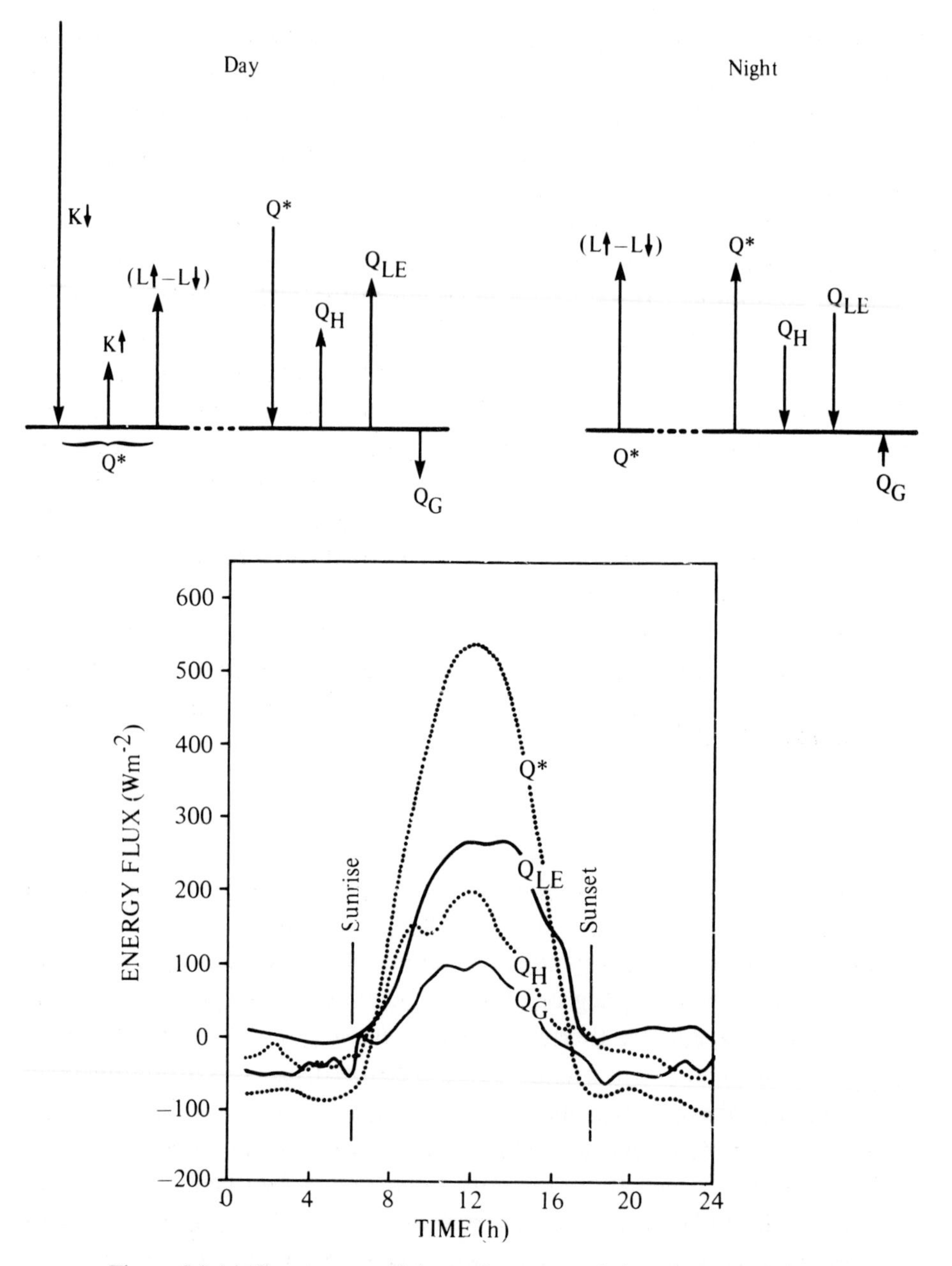

Figure 3.3 (*a*) Components of the surface energy balance. (*b*) Diurnal variation of energy balance components (see text for definitions. From Sellers, 1965).

Figure 3.3(*b*) shows a typical example of the *daily surface energy regime*. (In Figures 3.3(*b*) and 3.5, although the sign of Q_H, Q_{LE} and Q_G is opposite to Q^*, all values are plotted on the same axis, for convenience.) The net radiation varies systematically through the day, taking on small negative values from about an hour before sunset until an hour after sunrise. At these times, the small amount of incoming solar radiation just balances the net longwave radiation from the surface. Daytime values of Q^* increase rapidly to a maximum near noon; in northern latitudes in winter, Q^* may be positive for only a few hours, or even not at all. As stated previously, the sensible heat flux, Q_H, is normally directed upward during the day, when the surface is warmer than the air. At night, the surface is likely to be cooler and Q_H is directed downwards. The turbulent transfer of heat is assisted by a decrease of temperature with height (which leads to buoyancy of rising air), and therefore daytime values of sensible heat transfer are much larger than night-time values. The conduction of heat into the ground, Q_G, is often considered small, and this is probably true for daily totals, since the energy gained during the day may be largely lost at night. The heat flow is downward from a little after sunrise until the late afternoon. Over the 24-hour period, Q_G rarely exceeds 10 to 20% of the net radiation, although for bare ground it can be up to 30%.

The term Q_G may be neglected in the *annual energy balance*, since heat stored in the ground during spring and summer is all released in the fall and winter, except insofar as the earth may be warming or cooling on a longer time-scale (see sections 3.5.3 and 3.6). Thus equation (3.1) becomes:

$$Q^* = Q_H + Q_{LE}$$

The net radiation increases more-or-less uniformly from a minimum in winter to a maximum in summer, with the winter minimum being negative poleward of 40° (Sellers, 1965, p.105). At high latitudes in summer, the net radiation is favoured by long daylength, but limited by the low sun angle and the high reflectivity of snow- and ice-covered surfaces. In the more humid regions Q_{LE} is mainly a function of the available radiative energy, and the annual course of evaporation is similar to that of the net radiation, with a summer maximum and winter minimum (Sellers, 1965, p.109).

The energy exchanges mentioned above can be described in terms of standard microclimatological relationships. For a discussion of these and the methods and instrumentation used to measure energy balance terms the reader is referred to Oke (1978) or Rosenberg, Blad & Verma (1983).

The relative magnitude of the different energy balance components, then, varies greatly from place to place, and from one time period to another. For example, the net radiation depends on the albedo and cloud conditions; evaporation depends upon the net radiation and the availability of water; the transfer of sensible heat depends on the windspeed and the surface roughness, which increases air turbulence; and the ground heat flux depends on the conductivity of the ground. This interplay between climatic factors and the site-specific aerodynamic, radiative, thermal and hydrologic properties that affect energy exchange, is known as the microclimate.

Figure 3.4 provides an overview of the interactions occurring in the surface boundary layer that affect ground temperatures. The interplay between climatic conditions, site-specific factors and soil conditions determines the surface temperature, which, together with the influence of the soil thermal properties, drives the ground thermal regime. The soil properties will vary with the moisture content and the temperature of the soil (see section 4.3).

3.5.1 The nature of surface interactions

Lockwood (1979) provides a useful scheme for exploring the nature and significance of these surface interactions, and this is used as a basis for the following brief discussion.

DRY SURFACE WITH NO ATMOSPHERE

While not actually occurring on earth, a surface of this nature would assume a very simple energy balance:

$$Q^* = K\downarrow(1 - \alpha) - L\uparrow = Q_G$$

or,

$$K\downarrow(1 - \alpha) - \varepsilon k T_s^4 = Q_G \qquad (3.3)$$

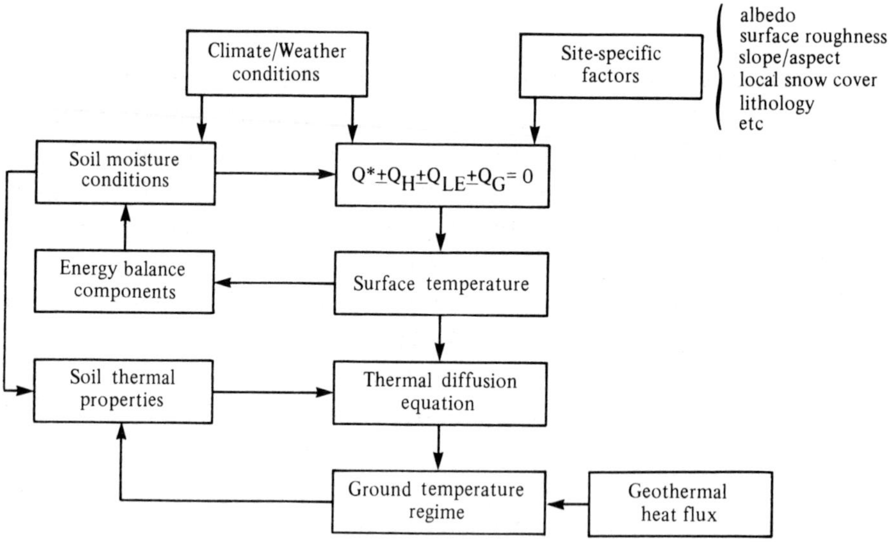

Figure 3.4 A flow diagram of climate–ground thermal interaction.

As mentioned, the net radiation consists of fluxes of *shortwave* (K), and *longwave/infrared* (L) radiation ($\mathrm{W\,m^{-2}}$), whose sources are *solar* and *terrestrial* respectively. α is the *albedo* (reflectivity) of the surface, T_s is the surface temperature, ε is the surface *emissivity* and k is a constant ($5.67 \times 10^{-8}\,\mathrm{W\,m^{-2}\,K^{-4}}$).

As mentioned previously, Q_G is generally very small compared to the other terms, and the surface temperature will change in close accordance with the variations in incoming radiation – from day to night, summer to winter. From equation (3.3) we can see that the surface temperature for a given value of incoming radiation will depend on the albedo, the infrared emissivity and the thermal conductivity of the surface material (i.e. site-specific factors). The higher the albedo, the more radiation is reflected and the lower the surface temperature will be. A surface with a relatively low emissivity will lose heat by longwave radiation more slowly; to compensate for this, the surface temperature will have to rise. Finally, if the thermal conductivity is high (as in most rocks), heat is transferred into and out of the ground more easily, and surface temperature variations are moderated. Conversely, if the thermal conductivity is low (e.g. dry soil), surface temperatures will be higher by day (and in summer) and lower by night (and in winter). Thus, the radiative and conductive properties of the surface materials of the ground can impart significant modulations to the surface temperature regime.

DRY SURFACE WITH ATMOSPHERE PRESENT

The surface energy balance equation now becomes:

$$Q^* = K\downarrow(1-\alpha) - \varepsilon k T_s^4 + L\downarrow(1-\varepsilon) = Q_H + Q_G \quad (3.4)$$

In addition to convecting sensible heat, the atmosphere will also modify the incoming radiation by reflection, scattering and absorption (climatic effects). Further, since the atmosphere also absorbs and radiates infrared (longwave) radiation, there is an infrared flux towards the surface ($L\downarrow$) as well as away from it. The magnitude of this flux depends largely on air mass and cloud cover conditions (climatic factors).

Equation (3.4) reveals the influence of both atmospheric (climatic) and site-specific (microclimatic) factors on the net radiation:

$K\downarrow$ is primarily a climatic factor, depending on latitude, time of year (and day), and weather conditions (cloudiness). However, the receipt of solar radiation at the earth's surface is also affected by the local topography (slope angle and aspect – e.g. see Oke, 1978). In addition, $K\downarrow$ at the ground surface may be reduced by the shading effects of vegetation (a site-specific property).

Table 3.1. Albedo values for various surfaces

Water/lake	<0.10
Wet tundra	0.11
Recent burn	0.07–0.12
Tundra	0.17–0.20
Lichen heath	0.19
Lichen woodland	0.20
Stoney plain	0.24
Dry sandy plain	0.36
Snow cover	0.80

Compiled from various sources

α (the albedo) is a site-specific property, and can vary widely (some typical values are given in Table 3.1).

$L\downarrow$ is a climatic factor, depending largely on cloud conditions and atmospheric humidity.

ε (the infrared emissivity) is a site-specific property, but does not vary as much as the albedo.

The convective transfer of sensible heat into the air may be expressed by:

$$Q_H = -\rho c_p K_H (dT/dz) \qquad (3.5)$$

or, in finite form:

$$Q_H = -\rho c_p K_H (T_2 - T_1)/(z_2 - z_1)$$

where T is measured at two levels in the air. ρ is the density and c_p the mass heat capacity of air. The turbulent transfer coefficient, K_H, depends on the windspeed (a largely climatic factor) and the *surface roughness* (a site-specific property). Alternatively, Q_H may be expressed in the height-integrated form:

$$Q_H = \frac{\rho c_p}{r_a}(T_s - T_a) \qquad (3.6)$$

where T_a is the temperature at some height in the air and r_a is an aerodynamic resistance term (an inverse form of K_H). Since the flux of sensible heat into the air depends on the windspeed and the aerodynamic roughness of the surface, rougher surfaces, such as forests, will tend to be cooler, other things being equal (e.g. Rouse, 1984).

WET SURFACE WITH ATMOSPHERE PRESENT

A wet surface is one from which evaporation can take place, and it is the most common type of surface found in nature. The energy balance becomes that in equation (3.1). Evaporation is the change of water or ice to vapour, and it proceeds continuously from water, soil, snow and ice surfaces. In addition, soil water is extracted by plant roots, passed upward through the plant and discharged as vapour into the air. This is known as transpiration, and the combined process of evaporation and transpiration is called evapotranspiration. Globally, evaporation accounts for about 80% of the energy transfer from the earth's surface (Sellers, 1965, p.103), and variations in evaporation are a principal cause of microclimatic differences.

The rate of evaporation depends on prevailing weather and climatic conditions (radiation, temperature, windspeed, humidity) and on the availability of water at the surface. If profile measurements of humidity and windspeed are available, the evaporative energy flux may be determined from:

$$Q_{LE} = -\rho L_v K_V (q_2 - q_1)/(z_2 - z_1) \quad (3.7)$$

where q is specific humidity and K_V is the turbulent transfer coefficient for water vapour (cf. equation (3.5)) and L_v is the latent heat of vaporisation. When the surface is wet, the surface specific humidity may be assumed to be the saturation value (q^*) for the surface temperature, i.e.

$$q_s = q^*(T_s) \quad (3.8)$$

This allows a functional approximation, first introduced by Penman (1948), which eliminates the need for measurements of humidity, temperature and windspeed at two levels, as in the profile and energy balance methods (e.g. see Oke, 1978). Measurements at one level suffice. A discussion of Penman's method, and various derivations from it, can be found in Brutsaert (1982).

When a surface is saturated, evaporation proceeds at the potential rate, the maximum rate at which water vapour can be added to the atmosphere under the given climatic conditions. As the surface or ground dries, however, remaining moisture is held under an increasing suction and eventually the evaporation will fall below the potential value. Whatever the relationship chosen for this, it is included in Figure 3.4 under '*soil moisture conditions*', which, as shown, serve to influence the disposition of energy at the surface.

Figure 3.5 illustrates the diurnal energy regimes at two adjacent sites which differ significantly only in terms of moisture availability – site 1 has saturated conditions at the surface whereas site 2 is drier. With the greater moisture availability, evaporation is much higher at site 1, and as a result of this cooling it experiences lower daily surface (and subsurface) temperatures. Figure 3.6 shows how this difference increases with higher daily

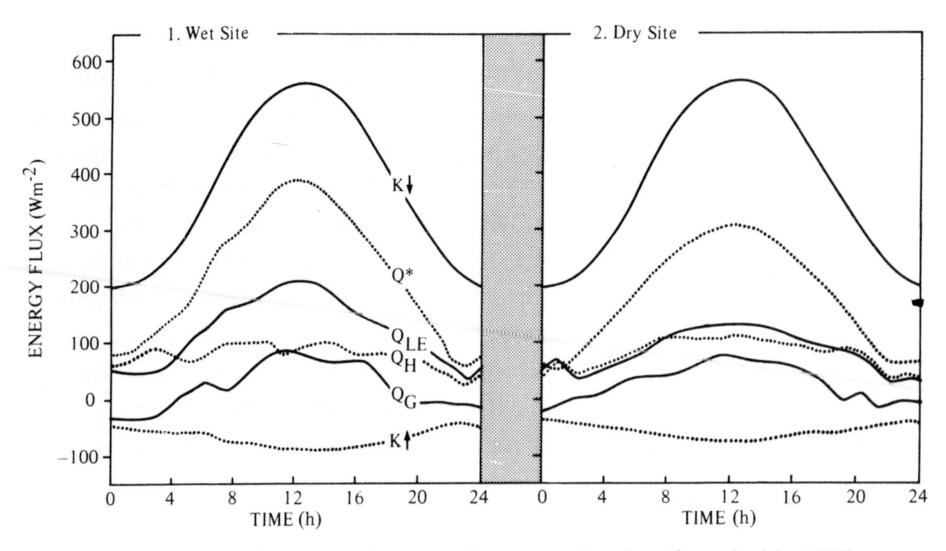

Figure 3.5 Diurnal energy regimes at adjacent tundra sites (from Smith, 1977).

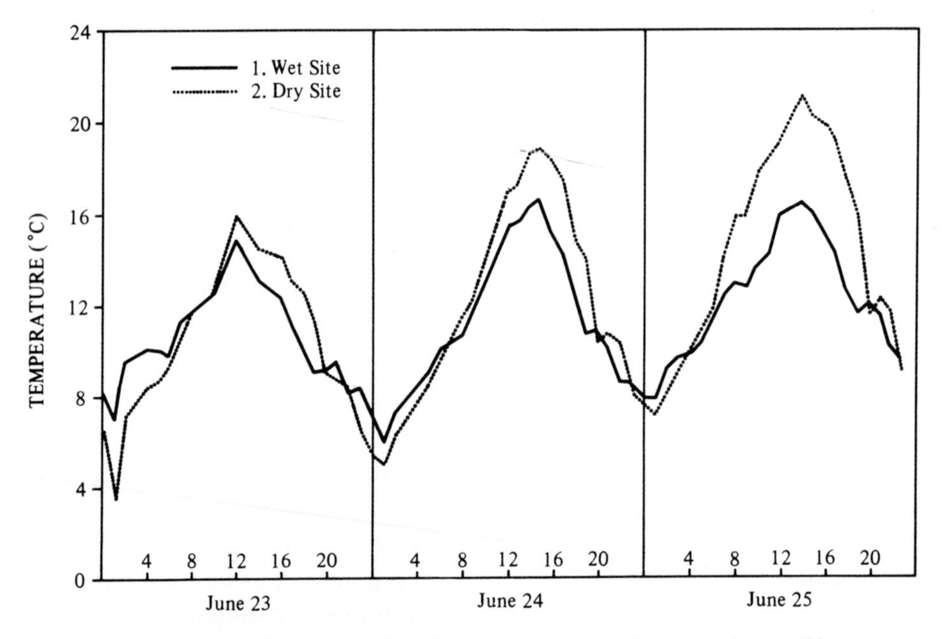

Figure 3.6 Diurnal patterns of surface temperature at the same sites as Figure 3.5 (from Smith, 1977).

amounts of incoming solar radiation. Interestingly, while $K\uparrow$ is similar at both sites (i.e. the albedo is similar), the net radiation at 1 is higher – this is due to the lower surface temperatures there which reduce the outgoing longwave radiation (as shown in equation (3.3)).

3.5.2 A buffer layer model

In terms of their significance to the ground thermal regime, Luthin & Guymon (1974) visualised these boundary layer interactions in terms of a buffer layer model, comprising the vegetation canopy, ground cover and snow cover, interposed between the atmosphere and the ground (Figure 3.7). Atmospheric mass and energy flows together with the geothermal heat flux constitute the boundary conditions, with the vegetation canopy, snow cover (when present) and surface organic layer acting as buffers between the atmosphere (climate) and the ground. In areas of little vegetation or snow-cover – such as can be found in the Canadian Arctic Islands, for example – the linkage between air temperatures and ground temperatures is more direct.

In Figure 3.4, the influence of the buffer layer is described in terms of

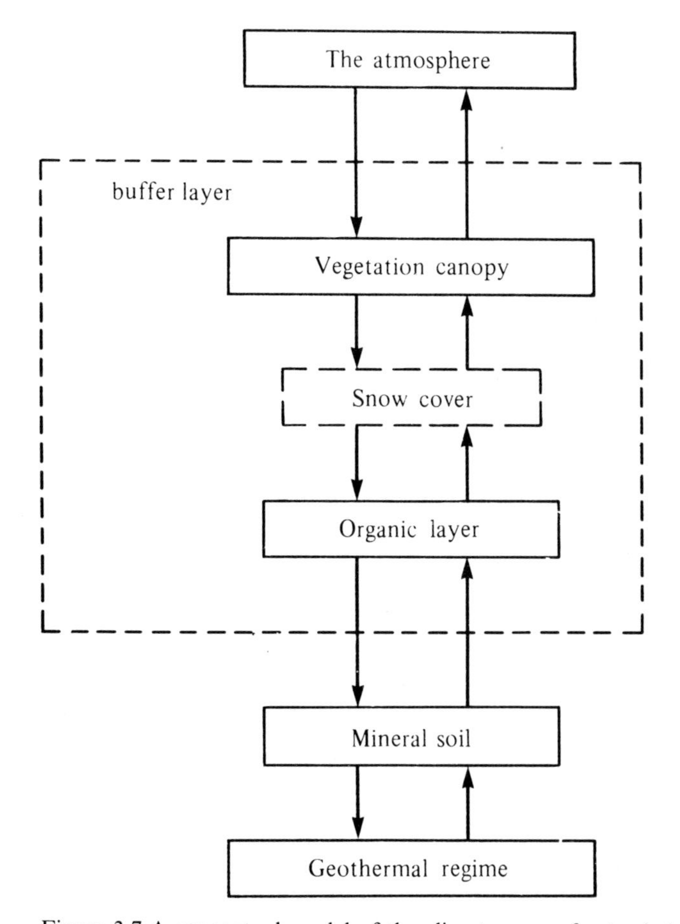

Figure 3.7 A conceptual model of the climate–permafrost relationship (based on Luthin & Guymon, 1974).

various site-specific factors, which depend on vegetation, topography and snow cover. The exact partitioning of the radiative surplus or deficit (Q^*) between Q_H, Q_{LE} and Q_G is governed by the nature of the surface (albedo, roughness, soil moisture, etc) and the relative abilities of the atmosphere (turbulence) and the ground (thermal conductivity) to transport heat.

The primary influences of the vegetation canopy are the reduction of solar radiation reaching the ground surface and the variable effects on the depth and persistence of snow cover (e.g. Luthin & Guymon, 1974; Rouse, 1982). In addition, interception of precipitation and transpiration by the canopy influence the ground thermal regime through the water balance (Riseborough, 1985). The role of the vegetation canopy has been investigated by comparative observational studies. For example, Rouse (1984) found that summer soil temperatures beneath an open spruce forest were lower than adjacent tundra, presumably as a result of radiation interception by the canopy, higher evaporation from the wetter surface and the large aerodynamic roughness of the forest producing greater turbulent exchange with the atmosphere. Annersten (1964) concluded, however, that the direct effect of vegetation *per se* was far less important than its role as a snow accumulator – a view supported by the results of Smith (1975) and those of Rouse (1984). Figure 3.8 shows that, although cooler in summer, the near-surface forest soils are considerably warmer than in the nearby tundra in winter, due to the forest acting as a snow fence and trapping a deep blanket of snow. As a result, the forest soil temperatures are more than 3 °C warmer on an annual basis.

Snow is a very important factor to the ground thermal regime, since it presents a barrier to heat loss from the ground to the air in winter (see section 4.4.3). In the Schefferville area Nicholson & Granberg (1973) found the variation in mean annual ground temperatures to be determined primarily by the snow depth, with variations in summertime conditions being less important. Smith (1975), in a study of ground temperature variations in the Mackenzie Delta, found that the warmest sites were so because of higher minimum (winter) temperatures, as influenced by snow cover. In some locations the mean annual ground temperature was raised above 0 °C because of snow accumulation (Figure 3.9). Calculations showed that the outflow of heat from the ground at these sites during the winter was only one-fifth to one-tenth of that at a site with only 25 cm of snow, and that permafrost was actively degrading as a new, higher surface temperature became established.

Near the southern fringes of permafrost distribution, snow cover alone

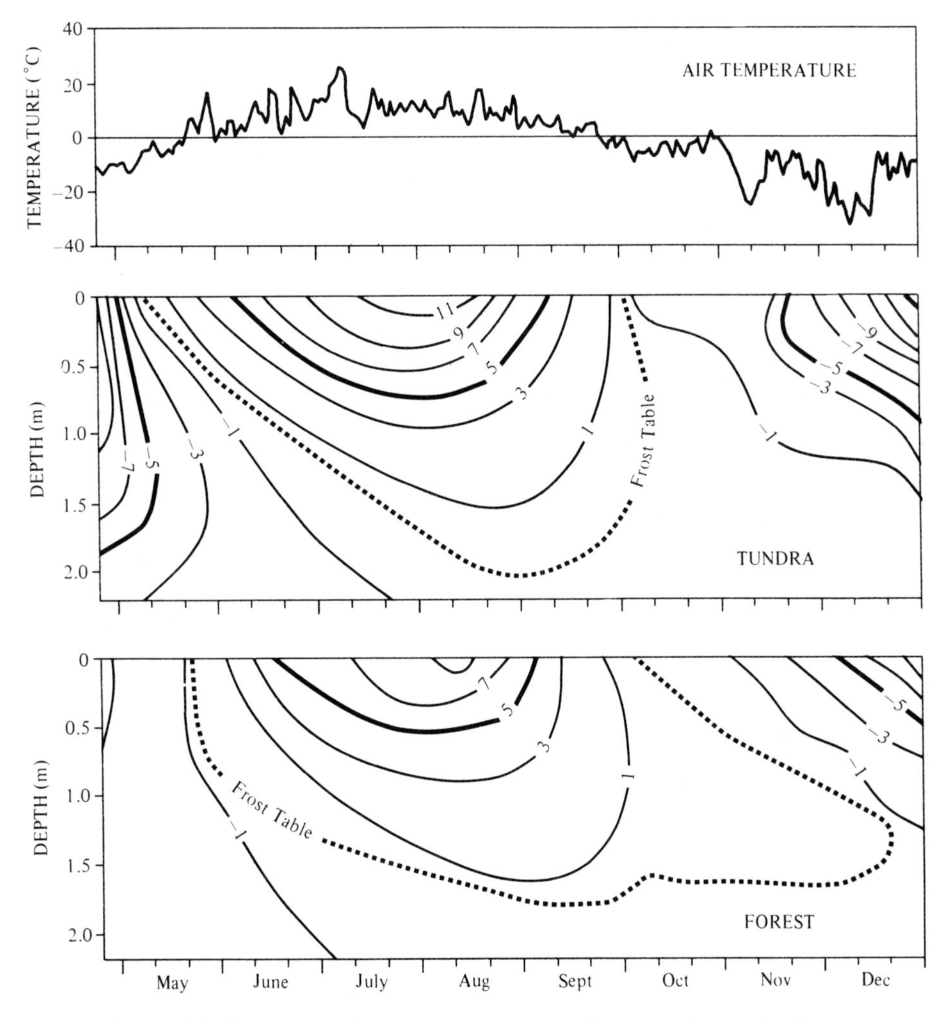

Figure 3.8 The ground thermal regime at a tundra and a forest site (from Rouse, 1984).

may be *the* critical local factor determining whether permafrost is present or not. In the colder regions of more widespread permafrost, it influences the depth of the active layer. Also, in regions of heavy snowfall, lake and river ice will not be so thick, so that even shallow water bodies may not freeze through. Such is the case in the Mackenzie Delta, where this has an important effect on the local distribution of permafrost (Smith, 1976).

Finally, the influence of organic material on the ground thermal regime in permafrost areas is well documented in the literature (see section 4.4.3). Generally, its presence leads to lower mean annual temperatures, which is attributed mainly to:

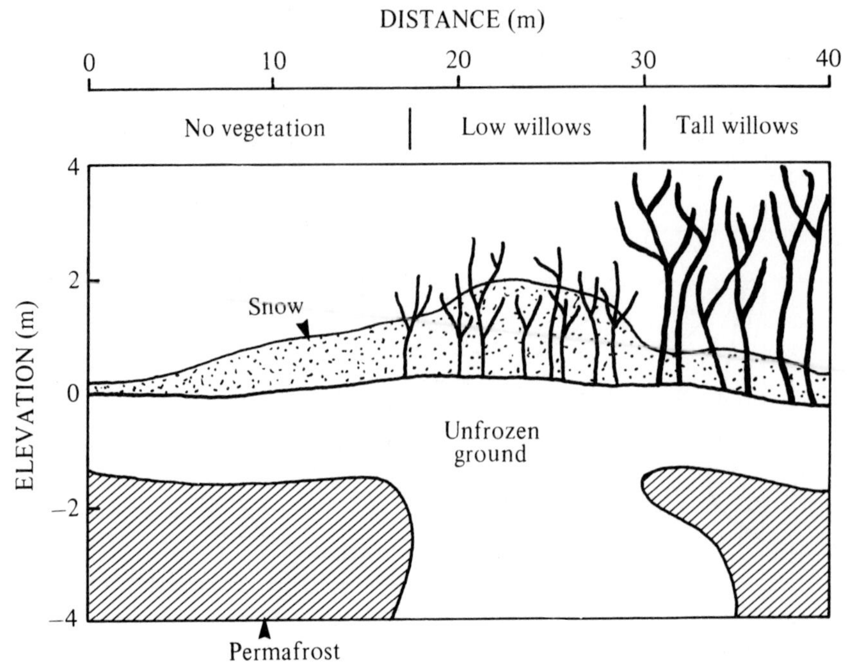

Figure 3.9 Details of permafrost configuration beneath a perennial snowbank (from Smith, 1975).

1. the seasonal variation in conductivity (low in summer, high in winter);
2. the evaporative regime, which promotes cool summer temperatures.

In addition, the results of Nelson *et al.* (1985) indicate that non-conductive modes of heat transfer may be important in the thermal buffering effect of organic material, which keeps ground temperatures low in summertime.

R. J. E. Brown (1963) reported ground temperature data from the Norman Wells area that reveal a decrease in temperature with increasing thickness of moss and peat. In a later paper (R. J. E. Brown, 1965), he concluded that variations in the vegetation canopy *per se* were a relatively minor influence on the ground thermal regime compared to the surface organic layer (see also Riseborough, 1985).

3.5.3 Effects of surface changes

It should be apparent that changes in the thermal regime of the ground, such as associated with degradation and formation of permafrost, can result from changes in the surface conditions as well as from fluctuations in climate. Most documented instances of disturbance to the natural environment are associated with human activities, but natural events such as fire are also important. Any removal, damage or compaction of surface materials (vegetation, peat, soil) will alter the balance of surface energy transfers; in general this will lead to an increase in the mean summer surface temperature (Figure 3.10). This increase will eventually be accompanied by melting in the upper layer of permafrost. For example, Mackay (1970) describes the example of an experimental farm site at Inuvik that was cleared of spruce and birch in 1956. The depth of thaw prior to clearing was about 36 cm, but by 1962 the active layer had deepened to 183 cm.

In winter, changes in snow cover accumulation, as might result from barriers, structures and depressions, can lead to significant warming (or cooling) of the ground. Figure 3.11 shows how the erection of snow fences had an immediate effect on ground temperatures and how the warming was maintained even against natural cooling trends.

Figure 3.12 shows the progressive degradation of permafrost as a result of (controlled) surface disturbance, over a 26-year period. Throughout the period, mean climatic conditions remained more-or-less constant, as revealed by the stability of permafrost conditions in the undisturbed area. The surface treatments in the cleared and stripped areas altered both the summer and winter surface temperature regimes. The bare surface would be the warmest in summer, but because of lower snow accumulation it would cool off more in winter (cf. Figure 3.1). The net result is that while perma-

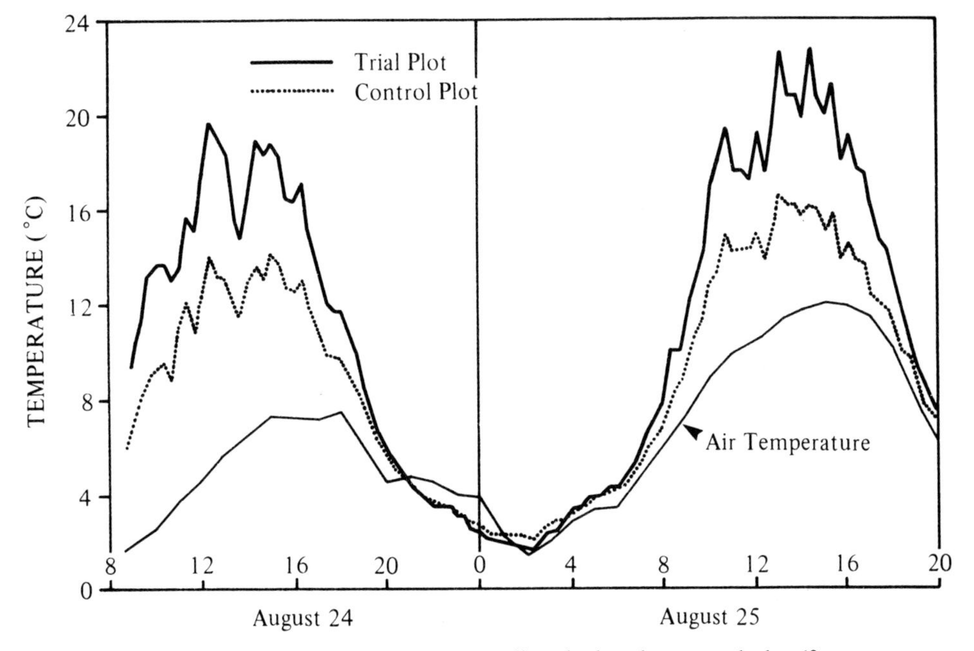

Figure 3.10 Surface temperatures at a disturbed and a control plot (from Nicholson, 1978). The vegetation was removed from the disturbed plot.

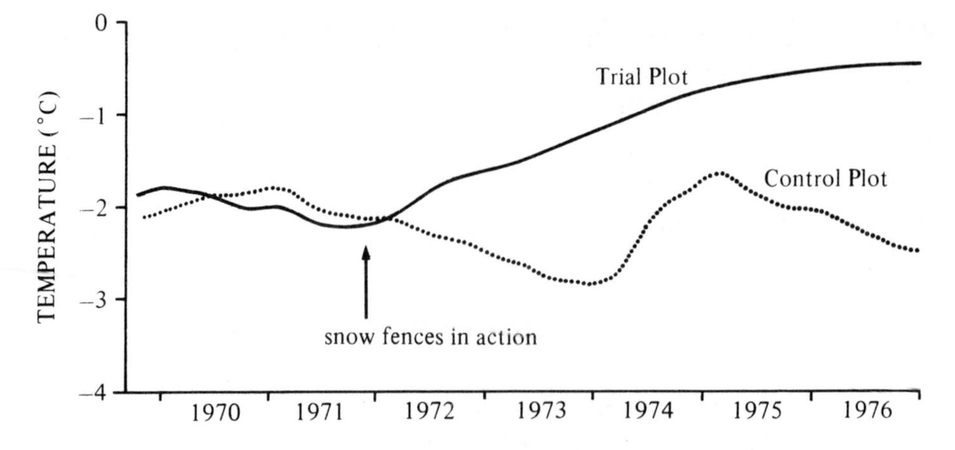

Figure 3.11 12-month running mean ground (10 m) temperatures at two adjacent plots with different snow covers (from Nicholson, 1978).

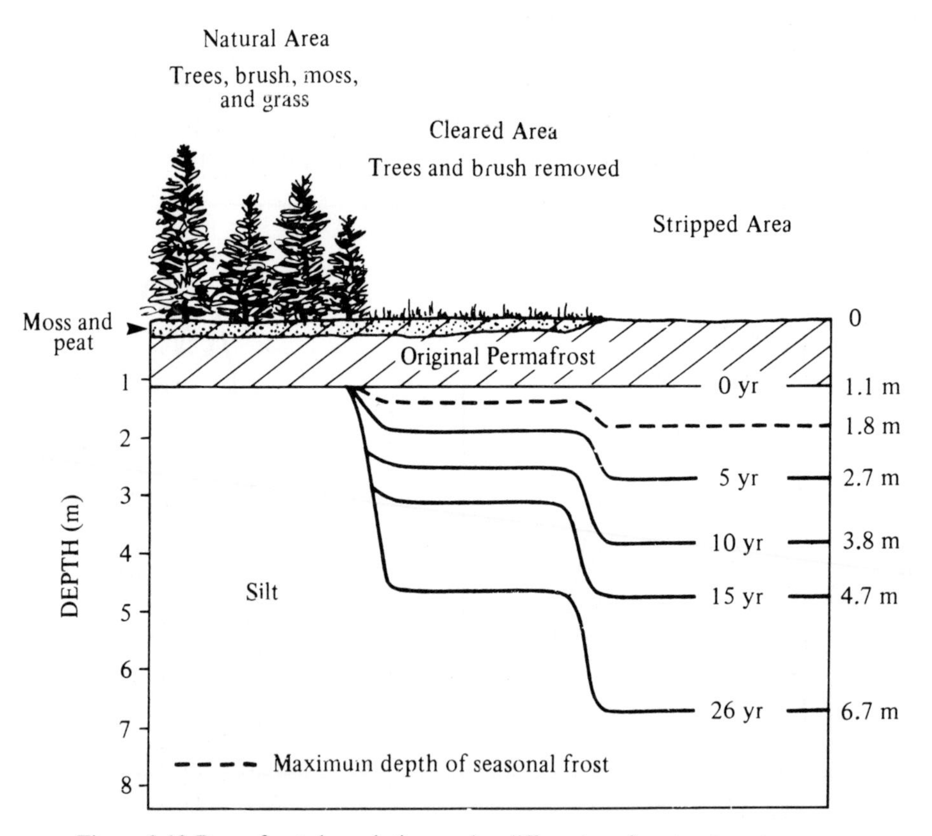

Figure 3.12 Permafrost degradation under different surface treatments over a 26-year period (from Linell 1973).

frost degradation was greatest in the stripped area, the depth of seasonal frost was also somewhat greater.

If the degrading permafrost is supersaturated (ice-rich), and if the excess water is able to drain away, a permanent subsidence of the ground surface results. This is illustrated in Figure 3.13. Since the soil layer near the top of permafrost is frequently ice-rich, ideal conditions for subsidence may be widespread. *Thermokarst* (see also Chapter 5), which refers to the melting out of ground ice leading to ground subsidence and thaw lakes, is a vivid expression of the effect of changes in the surface energy balance and consequence changes in ground thermal conditions. The resulting thermokarst depressions may act as snow traps in winter, promoting further warming of the ground. Where the depressions collect water, additional effects take over (see section 4.5.1). When thawing occurs on slopes, other changes can happen in the form of failures and mass movements of various kinds (see section 5.5).

Mackay (1970) reviewed the effects of various activities, related to oil and gas exploration, on permafrost degradation and thaw subsidence (see also Brown & Grave, 1978). These authors also point out that marked effects result from natural forest or tundra fires. Mackay (1977a) followed changes in the active layer at Inuvik after a forest–tundra fire; the depth of thaw increased rapidly for the first 4 to 5 years, and was still increasing after 8 years. Secondary effects of fire on permafrost terrain result from thaw

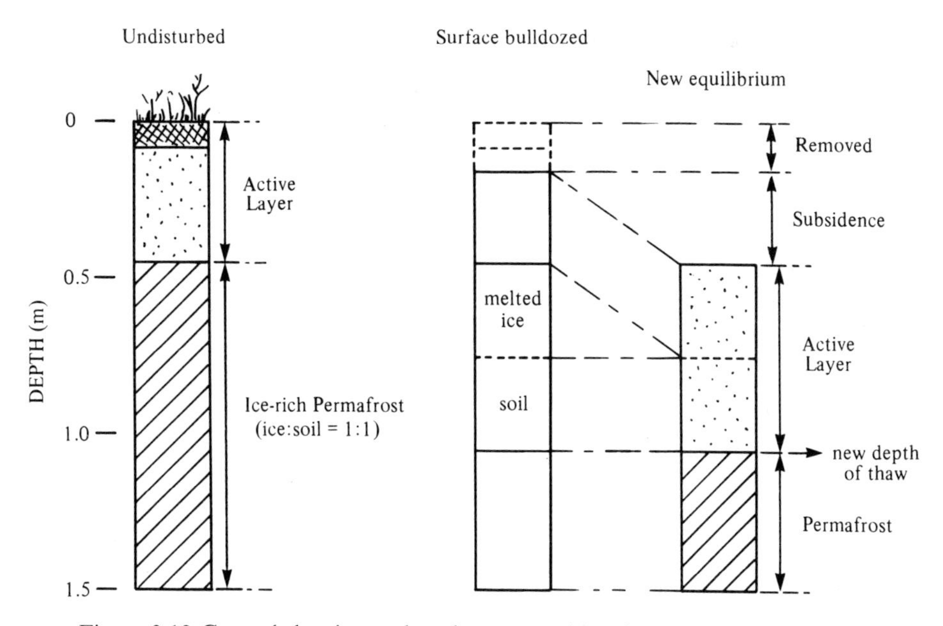

Figure 3.13 Ground thawing and settlement resulting from surface disturbance (after Mackay, 1970). The permafrost is assumed to have 50% excess ice.

subsidence and erosion, particularly where fire lines have been cleared with bulldozers (Heginbottom 1973).

Rouse (1976) compiled a time sequence of the microclimatic changes resulting from forest fire. The most obvious effect of fire is on the tree canopy, but a more significant effect may be the partial or complete destruction of the organic layer at the soil surface. Both the canopy and the organic layer serve to insulate the mineral soil from the heat of summer, the net result being a lower mean annual soil temperature. Destruction of these by fire causes an increase in soil temperature.

Figure 3.14 shows a decrease in net radiation following a fire, even though the albedo is reduced. This is a result of the increase in the surface temperature (as shown in equation (3.3)), which itself is due to the substantial reduction in evapotranspiration (cf. Figures 3.5 and 3.6). All this leads ultimately to an increase in soil temperature – *the important consequence for*

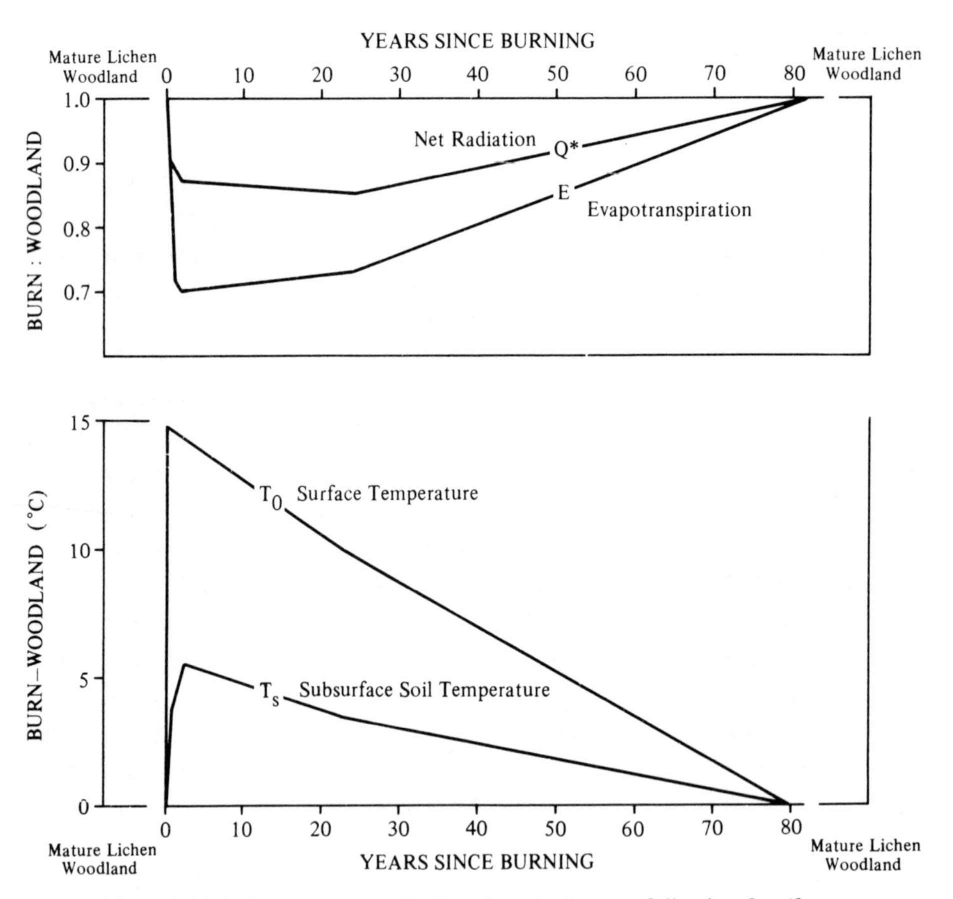

Figure 3.14 A time sequence of microclimatic changes following fire (from Rouse, 1976).

ground thermal conditions thus being the surface temperature at which the new balance of energy transfers is achieved. The heat of the fire, as such, is largely inconsequential to ground warming. The example in Figure 3.14 shows that all the terms gradually return to their original values as the vegetation is restored with time. However, if thermokarst processes are active, some of the changes will be irreversible.

While the foregoing examples all deal with ground warming and permafrost degradation, there are occasions where surface changes bring about ground cooling. For example, permafrost is absent, or lies at great depth, beneath many of the lakes, ponds and rivers in the north. Lake drainage or river shifting will expose these unfrozen sediments to freezing air temperatures, leading to ground freezing and the growth of permafrost (e.g. Smith, 1976; Mackay, 1979). In practical terms, where highways and airfields are built, the clearance of snow in winter may serve to stabilise the thermal condition of the permafrost, perhaps resulting even in some aggradation. Unfortunately, this may also lead to icing problems on highways (see section 2.2.4).

3.6 Possible effects of climatic change

Since permafrost is a thermal ('climatic') condition, it is potentially sensitive to climatic change. Currently, there is growing interest in the impact of human activity on the levels of CO_2 and other radiatively active trace gases in the atmosphere, and their likely effect on climate. Climatologists have shown that air temperatures would increase because of the so-called 'greenhouse effect', although it is not clear by how much. Studies demonstrate, however, almost universal agreement on a warming concentrated in the polar regions, along with increased precipitation there; for a doubling of the atmospheric CO_2 content, some models predict that annual air temperatures throughout the Arctic could increase by 3° to 6 °C, and perhaps even more in some areas (Harvey, 1982; Hansen *et al.*, 1984). However, this would undoubtedly take many decades, at least.

While such a trend has not yet been clearly identified in the meteorological record (e.g. see Weller, 1984), the magnitude of such changes in climate would produce serious and far-reaching environmental and engineering problems in permafrost regions, and for the arctic environment as a whole (McBeath, 1984; French, 1986; Goodwin *et al.*, 1984). However, it is difficult to deal with this question in any precise way yet, since most of the interrelationships are poorly known.

Permafrost is unique in earth material terms, since it exists close to its melting point. Most discontinuous permafrost is either relict, or in such delicate balance that climatic or other environmental changes can have drastic disequilibrium effects. Tens of thousands of square kilometres of permafrost are warmer than −3 °C, and we can expect that most of it would eventually disappear under the climatic warming predicted, although complete degradation would certainly take many centuries. Thie (1974) reported a decrease in the areal extent of permafrost from 60 to 15% over the past several centuries in the discontinuous permafrost region of Manitoba. Suslov (1961) wrote that the permafrost at Mezen (NE of Archangel) has retreated northward at an average rate of 400 m/yr since 1837, and according to Bird (1967), Sumgin, in 1934, wrote that in the USSR generally, the southern limit of permafrost was probably receding, as a result of climatic amelioration.

In the zone of continuous permafrost, ground temperatures, which may be many degrees below 0 °C, would rise with a climatic warming, but perhaps no major changes in regional distribution of permafrost would occur. Meanwhile, however, we could expect progressive deepening of the active layer with the melting of shallow ground ice and ensuing thaw subsidence. In some areas, this would undoubtedly create severe maintenance and repair problems for roads, airports, buildings, pipelines, etc. Greater depths of gravel padding would be needed to preserve permafrost under roads and structures.

In addition to permafrost degradation *per se*, *any* change in the temperature could cause a major change in the strength and deformation properties of frozen ground, *even without thaw*. The effect of temperature is important not only because of its influence on the deformational mechanisms in the ice but also as it determines the amount of unfrozen water in frozen soil (Chapter 9). There could be problems with the bearing capacity of piles, which are widely used in northern construction, as permafrost warms and adhesion forces decline. As well, the creep rates of (ice-rich) permafrost slopes would increase and slope stability would be decreased.

It is important to realise that climatic changes would also affect various earth surface processes which are characteristic of the permafrost environment, such as ice wedge cracking, frost heave, mass movements and creep. This may not be easy to determine, however, since according to Washburn (1980) the climatic relationships of permafrost processes are poorly known. This aspect has been reviewed briefly by Smith (1986). It has been hypothesised that thaw lakes in the western Arctic expanded during periods of warmer climatic conditions (Carson, 1968; Rampton, 1973). However, the sensitivity of thaw lakes to climate (change) is not well understood, although careful observations of lakes over a 10- to 20-year period might yield valuable information on growth patterns as a function of climatic variation.

Consideration of climatic change should not be confined to temperature alone, nor simply to a change in mean annual conditions. According to Hansen *et al.* (1984), while the mean annual temperature in northwestern North America could be 7 °C higher in a 2 × CO_2 world, increases in winter temperatures (up to 11 °C) would be 3- or 4-times greater than for summer (2 to 3 °C). In addition, there will be changes in precipitation (10 to 50% higher in summer, 60% higher in winter). Various earth surface processes in the permafrost environment could be affected by *increased rainfall*, but more importantly, perhaps, changes in *snow cover* would complicate the effect of climatic warming on ground thermal conditions. While increased snow depths would partly offset any adjustment in ground thermal conditions to higher winter air temperatures, calculations made by Goodrich (1982b) showed that a doubling of snow cover *per se* from 25 to 50 cm increased the minimum ground surface temperature by about 7 °C and the mean annual surface temperature by 3.5 °C (see section 4.4.3). In terms of ground temperatures, variations in snow cover are most critical at shallow depths, and the precipitation increases of as much as 60% in fall and early winter, predicted in the Hansen model, could be a significant factor in permafrost degradation, particularly in marginal areas.

Caution must be exercised in simply extrapolating a warming trend in the atmosphere to the ground, however. As discussed in section 3.5, permafrost conditions are affected by the nature of vegetation, soil and snow conditions, as well as climate. Riseborough (1985) concluded that the ground thermal regime in the boreal forest, where a surface organic layer is present, could be considerably shielded from the effects of climatic change. The interception effects of the forest canopy, but, more importantly, the thermal resistance of the moss and peat, serve to isolate the ground from the atmosphere. Ground thermal conditions, therefore, may be highly buffered from atmospheric (climatic) changes, so that any response in the permafrost conditions may be very slow to develop. In addition, ground thermal conditions will certainly be further affected as vegetation itself changes in response to climatic change. Finally, the thermal stability of permafrost is aided by the widespread occurrence of ice-rich ground near the present permafrost table. These considerations indicate the difficulty in assessing the sensitivity of local permafrost conditions to climatic change on a regional scale (see Smith & Riseborough, 1983). Nevertheless, Lachenbruch & Marshall (1986) have reported that throughout much of the 100 000 km^2 region of northernmost Alaska the temperature in the upper two metres of permafrost has generally increased about 2 °C or more during the last several decades to a century.

Finally, we must recognise that since more than simply climatic conditions determine the surface temperature regime, ground thermal conditions can change for a variety of other reasons.

3.7 Summary

Climate sets the stage for ground thermal conditions over the earth, but other factors complicate any otherwise simple relationship. The physical conditions at the earth's surface intervene to buffer the effects of the atmosphere on the ground, and serve to influence the surface temperature regime. Such conditions can change dramatically over quite short distances, and therefore it is the *microclimate* which is of ultimate importance to particular, local ground thermal conditions. Additionally, the thermal properties of the ground materials determine the pattern of variation in ground temperatures, together with the influence of the local geothermal heat flux. Apart from introducing considerable spatial variation in ground thermal conditions, changes in surface conditions arising from natural or human factors can lead to the degradation or formation of frozen ground. Finally, since permafrost is ultimately a 'climatic' condition, it is potentially sensitive to climatic change, surface buffering effects notwithstanding. There is evidence of this occurring in the past and we may anticipate further changes in the future, especially if the predictions of CO_2-induced warming are fulfilled.

4

The ground thermal regime

4.1 Introduction

The previous chapter discussed how the surface temperature is governed by climatic and microclimatic conditions. In turn, the surface regime is the major factor affecting the ground thermal regime. Once the surface temperature regime is known, the thermal regime of the ground may generally be analysed without further reference to climate. However, ground thermal properties and, at greater depths, the heat flowing from the earth's interior, serve to modify the effects of surface temperature (Figure 3.4).

The behaviour of soils in cold regions is strongly influenced by temperature and therefore the analysis of ground thermal regimes is of importance in many problems of scientific interest. In addition, the thermal interaction of engineering structures with frozen ground must be understood to allow for their proper design. While cold is generally seen as the singular feature of high latitudes, it is often the problems resulting from thaw that are of vital concern to engineering design. The problem of thaw settlement is not purely thermal, however. When soils thaw, meltwater is produced at a rate controlled by thermal processes, whereas the dissipation of this water depends on the discharge capacity of the soil (e.g. Nixon & Ladanyi, 1978).

Ground temperatures are, for the most part, determined by *conductive heat transfer*, although localised circulation of groundwater can occur particularly in areas of seasonal frost and discontinuous permafrost. In general, therefore, ground temperatures can be analysed in terms of *heat conduction theory*. The ground is not a simple solid, however, but comprises different layers whose thermal properties vary with mineral composition, organic content, density, moisture content and temperature (see Farouki, 1981). Moisture may occur as vapour, ice and unfrozen water.

Under freezing and thawing conditions the analysis of temperature changes in the ground is further complicated by the phase change relations of water. Changes in temperature alter the phase composition of soil moisture and can cause dramatic changes in thermal properties. In particular, the heat capacity is dominated by the *latent heat of fusion*. When dealing with short-term transient problems these phase change effects must be included, or the rate of ground thermal changes may be greatly over estimated.

It is not the aim of this chapter to review the range of techniques used for analysing specific ground thermal problems; for this, the reader is referred to treatments by Carslaw & Jaeger (1959), Jumikis (1977), Goodrich & Gold (1981), Lunardini (1981) and Goodrich (1982*a*), among others. Rather, this chapter applies heat conduction theory to understanding the significant features of the ground thermal regime in cold regions (cf. Gold & Lachenbruch, 1973).

Even though much insight and practical value can be gained from the application of heat conduction theory to ground thermal conditions, this overlooks the true complexity of soil freezing, which arises from the fact that frozen soils can contain appreciable amounts of unfrozen water at temperatures down to several degrees below 0 °C (see Chapter 7). As a result, a temperature gradient in frozen soil establishes a gradient of water potential which will induce water movements. A comprehensive analysis of soil freezing must ultimately deal with this coupling of heat and moisture flows, and these aspects are discussed in Chapter 8.

4.2 Heat flow in the ground

The basic traits of the ground thermal regime can be understood by considering the ground as a homogeneous medium in which heat flows by conduction in the vertical direction only. The amount of heat that flows by conduction is given by:

$$Q_G = -K(\mathrm{d}T/\mathrm{d}z) \tag{4.1}$$

where the 'constant' of proportionality, $K(\mathrm{W\,m^{-1}\,K^{-1}})$ is the *thermal conductivity* (see section 4.3). (In reality, in the presence of a temperature gradient in moist soil, whether frozen or not, some heat transfer takes place by mass flow and hence K is really an effective thermal conductivity.) The negative sign in equation (4.1) indicates that the heat flow is in the direction of decreasing temperature. Equation (4.1) also indicates that for a given temperature gradient, more heat will flow through a material of higher conductivity (or, for a given value of heat flux, the temperature gradient must increase as the thermal conductivity of the material decreases).

Under steady-state conditions, $\mathrm{d}T/\mathrm{d}z$ is constant if K is constant – i.e. the temperature profile is linear with depth – and equation (4.1) can be integrated to give an expression for the temperature at any depth, z, as follows:

$$T_z = T_s + (Q_G/K)z \tag{4.2}$$

where T_s is the surface temperature.

In reality, the process of heat conduction in the ground is more complex than this, partly because steady states are rarely achieved, since the surface temperature is continually changing, and partly because the natural variations in soil conditions leads to variations in thermal properties. In addition, the thermal properties of frozen soils change significantly with temperature. The main limitation of heat conduction models is the degree to which such natural conditions can be represented.

4.2.1 The heat conduction equation

While the *rate* of heat transfer depends on the thermal conductivity, the temperature change experienced by the soil as a result depends upon the *volumetric heat capacity*, C ($\mathrm{J\,m^{-3}\,K^{-1}}$). When a unit volume of a substance changes temperature by $\mathrm{d}T$, the change in heat content is simply the product ($C\mathrm{d}T$). We assume, for now, that C itself does not vary with temperature (but see section 4.3.2).

With expressions for heat flow (equation (4.1)) and heat content in hand, we may write the equation for the heat balance of a soil layer, of unit cross-sectional area, and thickness $\mathrm{d}z$. The heat flowing into the layer must be balanced by the heat flowing out and the change in heat content of the layer:

Rate of heat flow in = Rate of heat flow out + Rate of change of heat storage

or:

$$K(\mathrm{d}T/\mathrm{d}z)_i = K(\mathrm{d}T/\mathrm{d}z)_o + C(\mathrm{d}T/\mathrm{d}t)\mathrm{d}z \tag{4.3}$$

where t is time. Rearranging equation (4.3) leads to:

$$\mathrm{d}T/\mathrm{d}t = (K/C)(1/\mathrm{d}z)((\mathrm{d}T/\mathrm{d}z)_o - (\mathrm{d}T/\mathrm{d}z)_i) \tag{4.4}$$

which, in differential form, is the heat conduction equation:

$$\frac{\partial T}{\partial t} = \kappa \frac{\partial^2 T}{\partial z^2} \tag{4.5}$$

where κ ($=K/C$) is the *thermal diffusivity* ($\mathrm{m^2\,s^{-1}}$) – i.e. the coefficient of heat diffusion. A high value for thermal diffusivity implies rapid and large changes in temperature. Equation (4.5) expresses the fact that the rate of change of temperature at any depth is proportional to the divergence of heat flow (as expressed by the curvature of the temperature profile at that depth). In other words, where there is a curved profile in the ground, the temperature will change most rapidly (largest $\partial T/\partial t$) at the depth where the curvature ($\partial^2 T/\partial z^2$) is greatest. In the case of steady-state conditions ($\partial T/\partial t = 0$), the equation reduces to the form in equation (4.1).

Equation (4.5) furnishes a general solution to determining the temperature variations, $T(z, t)$, in a solid body and the solution of any particular problem in heat conduction must satisfy this equation, or some form of it. In addition, T must satisfy not only equation (4.5) but also the particular initial and boundary conditions of the problem.

Analytical (i.e. mathematically exact) solutions to equation (4.5) are available for a wide variety of initial and boundary conditions and are described in standard texts (e.g. Ingersoll, Zobel & Ingersoll, 1954, Carslaw & Jaeger, 1959, Lunardini, 1981). However, their applicability is restricted to problems of limited complexity, and where transient effects may be neglected or are unimportant.

In its simplest application, the heat conduction equation assumes the ground is entirely uniform, and thus a single value only for the thermal diffusivity is required. This condition is rarely met, if ever, in nature, since soil thermal properties normally vary with depth, moisture conditions and temperature. In particular, the assumption is not acceptable for freezing and thawing problems over short (engineering) time scales, where the transient effects of phase change are important. For problems such as these, numerical solution methods must generally be used. In this case, the differential terms are replaced by small finite differences, Δz and ΔT, and a series of simultaneous solutions formed and solved by computer. Complex initial and boundary conditions, and complex problem geometries can also be accommodated in numerical models. Lunardini (1981) and Goodrich (1982*a*) provide an introduction to these methods, and can refer the reader to many specialised texts that are available.

Since many natural processes occur very slowly, problems of a long-term (geomorphological) nature can often be analysed adequately without phase change considerations, using simple analytical solutions to the heat conduction equation (e.g. Lachenbruch, 1957*a*, *b*, 1959; Mackay, 1963; W. G. Brown *et al.*, 1964; Smith, 1976). Furthermore, even simpler steady-state solutions may be adequate in some situations, when $\mathrm{d}T/\mathrm{d}t$ is very small.

4.3 Thermal properties

A knowledge of thermal properties is required for the proper interpretation and analysis of thermal conditions in the ground, and for carrying out thermal calculations. For steady-state conduction problems (i.e. no change of temperature with time), only the thermal conductivity, K ($W\,m^{-1}\,K^{-1}$), need be considered. This is a measure of the quantity of heat that will flow through a unit area of the substance per unit time, under a unit temperature gradient – i.e. $J\,s^{-1}\,m^{-2}\,(K\,m^{-1})^{-1}$, or $W\,m^{-1}\,K^{-1}$. Thus it determines the rate of heat transfer. In transient problems, the thermal diffusivity, κ, arises, but in situations where phase change occurs, the latent heat of fusion, L_f ($MJ\,kg^{-1}$) dominates.

The presence of ice and water close to their transition temperature has a dominant effect on the thermal properties of frozen soils. When water changes to ice, its conductivity increases four-fold, its mass heat capacity decreases by half, and it releases heat equivalent to that required to raise the temperature of an equal volume of rock by about 150° (Gold & Lachenbruch, 1973). The situation is made more complex by the unfrozen water content relationship of frozen soils. Since the volume fractions of ice and unfrozen water in a soil are temperature dependent, the thermal properties of soils can change significantly with small changes in temperature below 0 °C. Because of these effects, soil moisture content plays a decisive role in the thermal conditions of frozen earth materials.

In addition to the special circumstances associated with freezing, the generally variable nature of soil conditions precludes the unique specification of the thermal properties for an entire profile or soil formation, and the dependence on soil type, density, and water content, as well as temperature must be considered. These relationships are generally determined by laboratory experimentation, although the complex nature of frozen materials presents some unique problems. The methods in use were initially developed for unfrozen soils or less complex materials such as metals, and modifications and limitations become necessary.

Unfortunately, there is little published information available on the temperature-dependency of soil thermal properties in the range between 0 to -3 °C, although for many practical situations this is the range of greatest concern. In many cases, therefore, estimates and approximations must be used. Various methods for calculating the thermal conductivity of frozen materials are available, and these have been reviewed by Farouki (1981, 1982). In addition, Smith & Riseborough (1985) analysed the errors that can occur in thermal calculations as a result of simplifying assumptions about soil thermal properties.

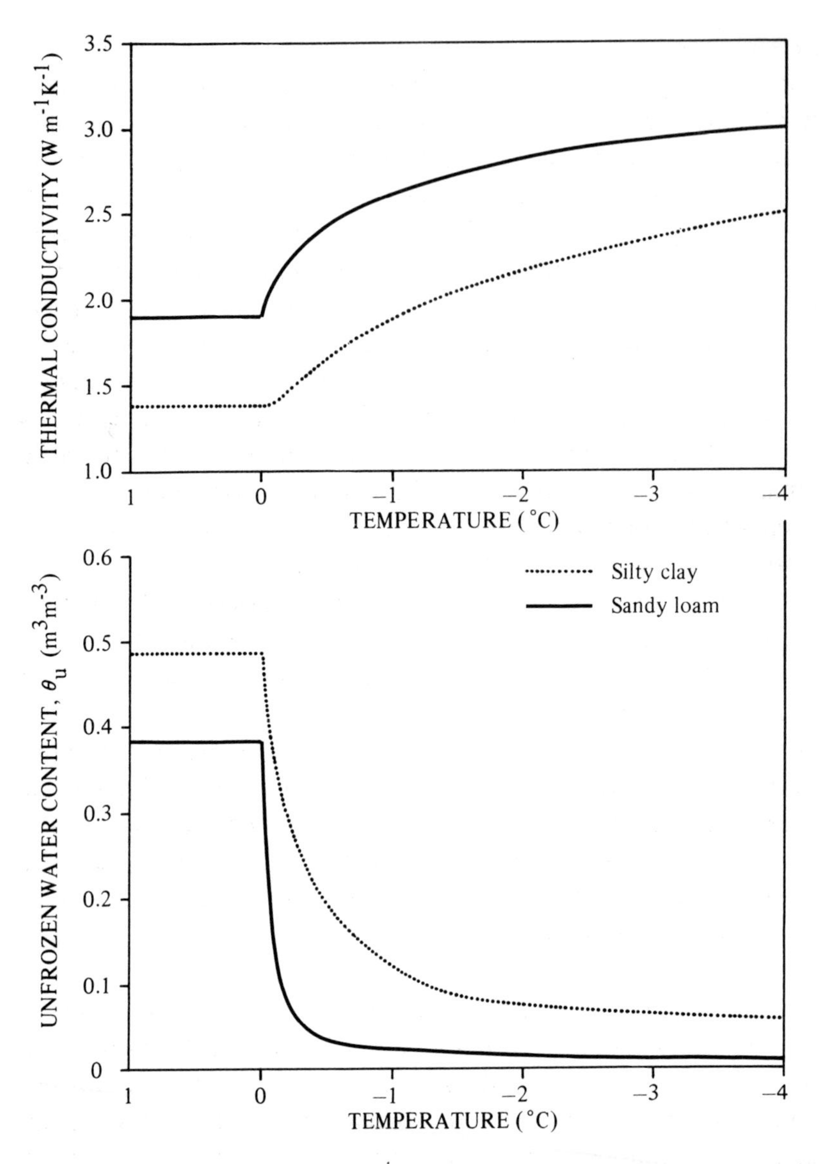

Figure 4.1 (**a**) Variation of thermal conductivity with temperature for a sandy loam and a silty clay. (*b*) Unfrozen water content characteristics for the same soils.

4.3.1 Thermal conductivity

Since the conductivity of ice is higher than for water, frozen soils have higher thermal conductivities than unfrozen soils. However, since the volume fractions of ice and water are temperature-dependent, there can be no single value for the thermal conductivity of a frozen soil (provided it contains water). Figure 4.1 shows the variation of thermal conductivity with negative temperature for a silt and a clay soil, together with the unfrozen water content curve for each. In the case of the silt, most of the water freezes by about −1 °C, and the conductivity increases quite abruptly over this range. Below about −2 °C, the conductivity is relatively constant, since as Figure 4.1(*b*) shows, the unfrozen water content changes very little below this temperature. The change in unfrozen water content is more gradual for the clay, and this is reflected in the conductivity curve. These examples illustrate the importance of the unfrozen water content characteristics of the soil to the thermal properties.

Figure 4.1 also illustrates the importance of soil mineralogy. The thermal conductivity of quartz is much higher than for other soil minerals (Table 4.1), and the higher quartz content in the case of the silt leads to higher conductivity values overall.

The first comprehensive testing of the thermal conductivity of soils in the frozen and unfrozen states was that of Kersten (1949). This work predated much research into the behaviour of water in soil materials, so that some of the results are now seen to be questionable. Specifically, the values were obtained using a steady-state method, which we now know is not suitable for thermal conductivity tests on fine-grained materials at freezing temperatures.

There are three basic conductivity test procedures for use on earth materials: they are known as the Guarded Hot Plate (GHP), Divided Bar (DB), and Thermal Probe (TP) methods (see Farouki, 1981). Each method involves heating the test material in some way, which raises unique problems in the case of frozen materials containing water – i.e. the latent heat release associated with temperature change, and variations in thermal properties associated with temperature-dependent phase composition. Thus steady state methods requiring prolonged and/or substantial heating of the specimen (GHP and DB) will cause a change in the properties being measured. In addition, the potentials which develop in unfrozen water films (or gradients in vapour pressure at low water contents) can result in the redistribution of moisture when a temperature gradient is present, leading to complications in heat transfer and inhomogeneity in the sample. When the effects of liquid and vapour transport are present in a test, one can then only determine an apparent conductivity (e.g. see Kay *et al.*, 1981).

Table 4.1. Thermal properties of soils and their constituents

		Density ($kg\,m^{-3}$)	Mass Heat Capacity ($J\,kg^{-1}\,K^{-1}$)	Thermal Conductivity ($W\,m^{-1}\,K^{-1}$)	Thermal Diffusivity ($\times 10^{-6}\,m^2\,s^{-1}$)
(a) Soil constituents					
Quartz		2660	800	8.80	4.14
Clay minerals		2650	900	2.92	1.22
Organic matter		1300	1920	0.25	0.10
Water (0 °C)		1000	4180	0.56	0.13
Ice (0 °C)		917	2100	2.24	1.16
Air		1.2	1010	0.025	20.63
(b) Unfrozen Soils	Water Content ($m^3\,m^{-3}$)				
Sandy soil	0.0	1600	800	0.30	0.24
(40% porosity)	0.2	1800	1180	1.80	0.85
	0.4	2000	1480	2.20	0.74
Clay soil	0.0	1600	890	0.25	0.18
(40% porosity)	0.2	1800	1250	1.18	0.53
	0.4	2000	1550	1.58	0.51
Peat soil	0.0	300	1920	0.06	0.10
(80% porosity)	0.4	700	3300	0.29	0.13
	0.8	1100	3650	0.50	0.12

Compiled from Monteith (1973) and other sources.

In thermal conductivity determinations, then, the duration of the test will affect the results, and as a general rule the test should be kept as short as possible, although the limits depend on the moisture content of the sample. For example, steady state methods are likely to induce substantial moisture migration in wet fine-grained soils near 0 °C, although they are suitable for dry materials, those with low porosity, or for fine-grained soils at low temperatures where most of the moisture is immobilised as ice. The effect of moisture redistribution is less in transient state tests, since these are short enough to limit the amount of redistribution which can occur. According to Lachenbruch (1957c) the time of the test should be kept short when liquid water is present in the specimen. Riseborough *et al.* (1983) have developed a method for determining the thermal properties of frozen soils at temperatures close to 0 °C, using a thermal probe, while E. Penner (1970) presents a variety of results for lower temperatures.

Where certain physical data are available, it is possible to substitute a calculation method for the direct determination of thermal conductivity. Various methods exist, but that developed by Johansen (1972, 1973, 1977) is the most useful. It is applicable to frozen, mineral soils containing ice and water, and requires information on the physical properties, phase composition and quartz content of the material. The latter is important because, as stated, quartz has a higher thermal conductivity than other soil-forming minerals (Table 4.1).

4.3.2 *Heat capacity*

In order to describe the heat content of a substance, we need to know its heat capacity, expressed on either a mass or volume basis. The mass heat capacity, c ($\mathrm{J\,kg^{-1}\,K^{-1}}$), is the amount of heat required to change the temperature of 1 kg of the substance by 1 K. For a given amount of heat supplied, changes in temperature will be greater in a material with a low heat capacity. If we multiply the mass heat capacity by the density of the substance, ρ, we obtain the volumetric heat capacity, C, ($\mathrm{J\,m^{-3}\,K^{-1}}$). In changing its temperature by an amount $\mathrm{d}T$, a unit volume of substance will experience a change in heat content of ($C\mathrm{d}T$).

In the case of composite materials such as soils, a weighted average value for heat capacity must be used:

$$C_{\mathrm{s}} = X_{\mathrm{m}} C_{\mathrm{m}} + X_{\mathrm{o}} C_{\mathrm{o}} + X_{\mathrm{w}} C_{\mathrm{w}} \tag{4.6}$$

where X is the volume fraction of soil minerals, organic material and water respectively (the influence of air content being negligible). The thermal properties of soil constituents are given in Table 4.1. Quartz, feldspar and clay minerals, the main solid constituents of most soils, have similar heat capacity values. Organic matter has a higher specific heat, but being much less dense has a volumetric heat capacity that is similar to the mineral components. As a result, most dry soils have a heat capacity in the range 1.0 to 1.5 $\mathrm{MJ\,m^{-3}\,K^{-1}}$ near 0 °C. Since the heat capacity of water is 4.2 $\mathrm{MJ\,m^{-3}\,K^{-1}}$ (at 0 °C), the heat capacity of a soil increases substantially when it is wet, the increase being linear with the increase in water content.

In frozen soils, the temperature dependence of the unfrozen water content means that changes in heat storage are dominated by latent heat effects, especially within a few degrees below 0 °C. The complete change in heat storage is then given by:

$$\Delta H = \int_{\mathrm{T_i}}^{\mathrm{T_f}} C\mathrm{d}T + L_{\mathrm{f}}' \tag{4.7}$$

(Lunardini, 1981), where the first term refers to the (sensible) heat capacity of the soil and the second term is the latent heat capacity of the soil. As a consequence of the unfrozen water content relationship (such as shown in Figure 4.1(*b*)) latent heat of fusion is released over a range of negative temperature, such that equation (4.7) becomes:

$$\Delta H = \int_{T_i}^{T_f} C\mathrm{d}T + \int_{T_i}^{T_f} \rho_w L_f \mathrm{d}\theta_u \qquad (4.8)$$

where $\mathrm{d}\theta_u$ is the change in the unfrozen water content between T_i and T_f. Thus changes in heat storage in the soil will include a large latent heat component over the entire range of temperature where there is a significant slope ($\mathrm{d}\theta_u/\mathrm{d}T$) to the unfrozen water content curve. In clays, for example, this range can extend to many degrees below 0 °C. Once again, this illustrates the importance of the unfrozen water content characteristics of the soil.

The latent heat effect is often described in terms of an '*apparent heat capacity*' for the soil. An example will explain this. Let us consider 1 m^3 of moist clay soil at an initial temperature of 1 °C and with a heat capacity of 3 MJ m^{-3} K^{-1} in the unfrozen state. To change its temperature to 0 °C, we must remove 3 MJ of heat. If it is further cooled to some temperature below 0 °C, an amount of latent heat is released corresponding to the volume of soil water that freezes over the particular range of negative temperature. In the case of the clay soil shown in Figure 4.1(*b*), this would amount to some 63.5 MJ m^{-3} over the temperature range from 0° to −1 °C. This heat tends to warm the soil, and it must therefore be removed if the soil is to cool further. In this way the heat capacity *appears* to be much larger than would be the case in the absence of any phase change. On the other hand, the apparent diffusivity is very much smaller. If phase change effects are ignored in thermal analyses, the sequence of ground thermal changes will be distorted.

Since the rate at which soil water turns into ice is temperature dependent, the apparent volumetric heat capacity of the soil varies with temperature, as follows:

$$C_a(T) = C_s(T) + \rho_w L_f (\mathrm{d}\theta_u/\mathrm{d}T)_T \qquad (4.9)$$

As indicated, the volumetric heat capacity of frozen soil, $C_s(T)$, also varies with temperature and the change in the volume fractions of ice and water. However, its influence is quite minor compared to the second term, which can be evaluated by taking the slope of the unfrozen water content curve at temperature T (Figure 4.2(*a*); compare Figure 4.1(*b*)).

Heat capacity values for any frozen material can be determined using the equations described in this section. Phase composition data are required to calculate heat capacity values for materials containing ice and water. Smith

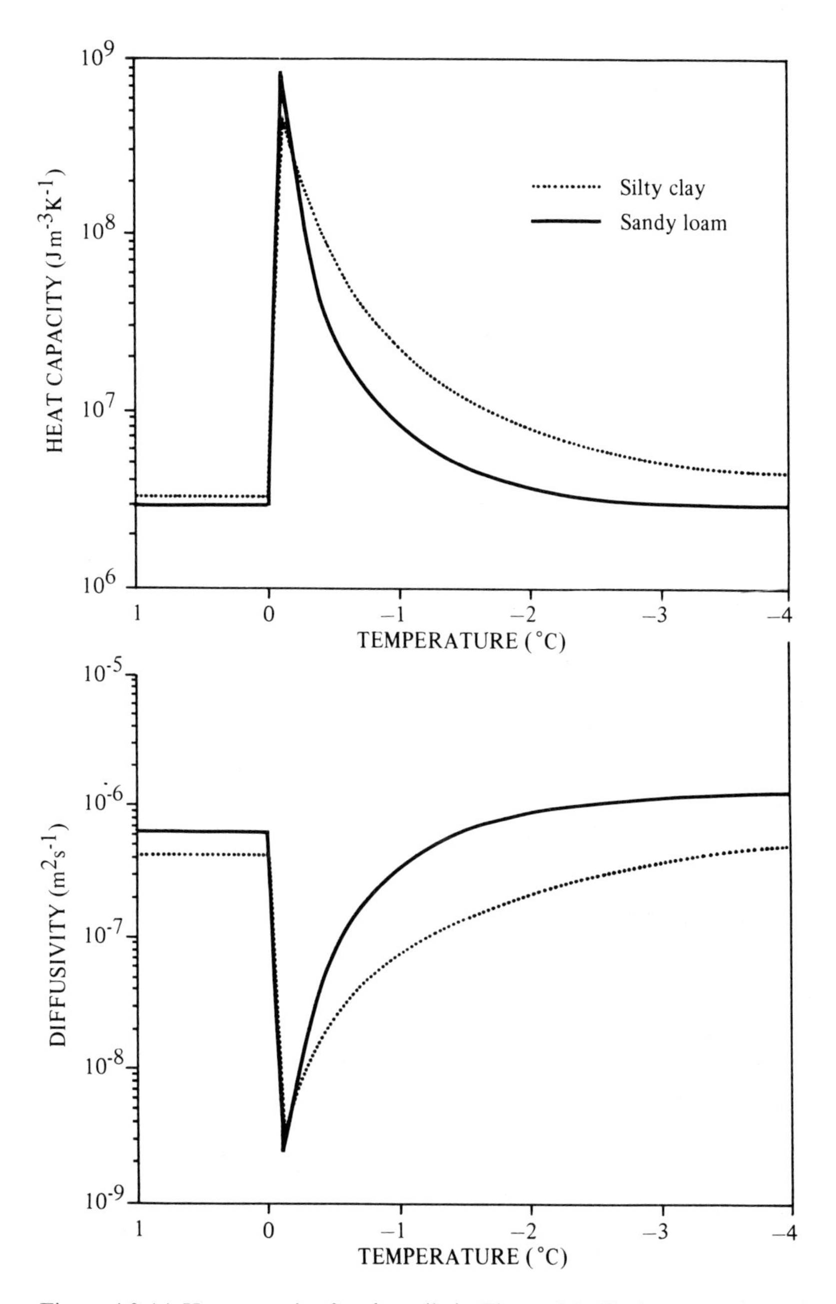

Figure 4.2 (*a*) Heat capacity for the soils in Figure 4.1. (*b*) Apparent thermal diffusivity for the same two soils.

& Riseborough (1985) have illustrated the importance of accurate heat capacity values to thermal calculations.

4.3.3 Thermal diffusivity

Since the thermal diffusivity is equal to the ratio (K/C), its variation with water content and temperature depends on the interplay of these two defining properties. When water is added to dry, unfrozen soil, the conductivity increases more rapidly than the heat capacity, so that the diffusivity also increases with water content. At high water contents, however, the increase in conductivity levels off while the heat capacity continues to increase at a constant rate. As a result, the diffusivity may start to decrease (Table 4.1.) Thus the diffusion of heat may be impeded at low water contents by the low thermal conductivity of the soil, and at high water contents by the large heat capacity.

The thermal diffusivity of frozen soils is highly temperature dependent and is dominated by the heat capacity term, especially within the range 0° to −3 °C. For example, Figure 4.1(*a*) shows that the conductivity of the clay soil increases by a factor of almost two between −0.05 and −3 °C. Over the same temperature range, the apparent heat capacity decreases by a factor of 75, with the result that the apparent thermal diffusivity, κ_a, increases by two orders of magnitude (Figure 4.2(*b*)). With decreasing temperature, the effect of increasing conductivity and decreasing apparent heat capacity results in an increasing diffusivity. A noteworthy feature of Figure 4.2(*b*) is the change of κ_a over a wide temperature range for the clay, indicating continuing phase change.

4.4 Ground thermal conditions

4.4.1 Geothermal heat flow and permafrost thickness

A simple application of equation (4.1) can be made to the case of the steady-state one-dimensional flow of heat from the earth's interior to the surface. Under uniform ground conditions, this can be written as:

$$Q_G = K.Gg \qquad (4.10)$$

where Gg is called the *geothermal gradient* ($K\,km^{-1}$). The temperature at any depth in the ground, z, is simply given by:

$$T(z) = T_s + Gg.z \qquad (4.11)$$

(cf. equation (4.2)). Where the long-term mean surface temperature, T_s, is below 0 °C, permafrost will be present, with an equilibrium depth (z_p) at

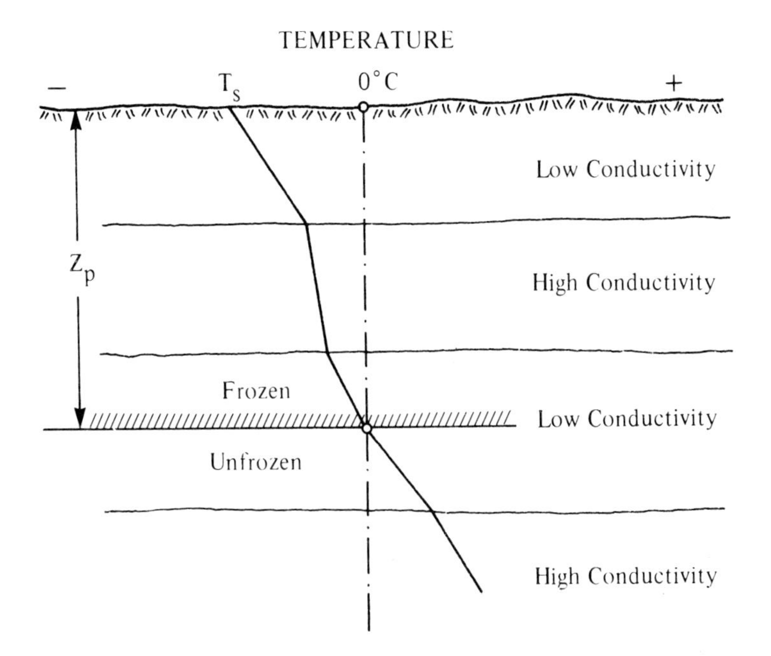

Figure 4.3 Influence of thermal conductivity on the geothermal gradient (from Terzaghi, 1952).

which the temperature increase due to the geothermal gradient just offsets the amount by which 0 °C exceeds T_s:

$$z_p = (0° - T_s)/Gg \tag{4.12}$$

In this case it should be noted that, because of the high pore pressures existing at substantial depths in the ground, there will be a temperature difference between the base of *ice-bearing* permafrost and the 0 °C isotherm. If we assume that the pore pressure is hydrostatic, then at a depth of 600 m the freezing point depression is about −0.44 °C (Osterkamp & Payne, 1981). A freezing point depression can also result from the presence of solutes in the pore water and soil particle effects (see section 7.1.2).

In nature, the thermal conductivity normally varies with depth at any location, as soil and rock materials change. Thus the geothermal gradient also varies, and the change in the slope of the temperature profile affects the permafrost thickness (Figure 4.3). From equation (4.10) we have:

$$Gg = Q_G/K \tag{4.13}$$

and equation (4.12) then becomes:

$$z_p = T_sK/Q_G \tag{4.14}$$

taking the absolute value of T_s. Equation (4.14) expresses the important

Table 4.2. Variation in permafrost thickness with T_s and K

T_s(°C)	Q_G ($W m^{-2}$)	K ($W m^{-1} K^{-1}$)	z_p (m)
−2	0.04	1.5	75
−1.5	0.04	1.5	56.3
−1	0.04	1.5	37.5
−0.5	0.04	1.5	18.8
0	0.04	1.5	0
−2	0.04	2.5	125
−1.5	0.04	2.5	93.8
−1	0.04	2.5	62.5
−0.5	0.04	2.5	31.3
0	0.04	2.5	0
−2	0.04	4.0	200
−1.5	0.04	4.0	150
−1	0.04	4.0	100
−0.5	0.04	4.0	50
0	0.04	4.0	0

fact that the thickness of permafrost is equally sensitive to the thermal conductivity and geothermal heat flux, as it is to the surface temperature. If K is large, then Gg will be small, and permafrost will be thick; in materials with a low conductivity, permafrost will be thinner.

In this simplified picture it is apparent that the thickest permafrost will occur under a combination of low surface temperature, low heat flow and relatively high thermal conductivity. For a mean surface temperature, T_s, of −5 °C, Judge (1973) calculated a range of 100 to 550 m in the thickness of permafrost, for variations in Gg. Where T_s is −10 °C, the equivalent values are doubled (200 to 1100 m).

Variations in surface temperature and the geothermal gradient may be particularly important in the marginal areas of permafrost. For example, for a mean surface temperature of −1 °C and a local variation of ±1 °C, a heat flux of 0.04 $W m^{-2}$ and thermal conductivity values of 1.5, 2.5 and 4 $W m^{-1} K^{-1}$ the thickness of permafrost could vary from 0 to 200 metres (Table 4.2). For northern Canada, Judge (1973) calculated that permafrost is thickest beneath the pre-Cambrian rocks of northern Baffin Island, Boothia Peninsula and Victoria Island, where it reaches 1000 metres or more; in general, however, values of around 500 metres apply. It should be appreciated that the development of permafrost to such great thicknesses involves geological time. According to Kudriavtsev (1965), the occurrence of permafrost up to 700 m thick in western Siberia indicates that it has probably existed during the entire Quaternary period. In contrast, in the formerly glaciated areas of the western Canadian Arctic, for example, permafrost is much thinner and is only a few thousand years old. Washburn (1979, pp.37–40) presents a table of permafrost thickness data for many locations throughout the Arctic.

Lachenbruch *et al.* (1982) have explained the variations in permafrost thickness along the Arctic coast of Alaska, where the heat flux is fairly uniform, by differences in T_s and the thermal conductivity of the ground materials. At Prudhoe Bay (T_s about -11 °C) permafrost is about 650 metres thick, whereas at Barrow ($T_s = -12$ °C) it is only about 400 metres thick, because of a lower thermal conductivity. At Cape Thompson, where T_s is only -7 °C, permafrost is still about 360 metres thick, because of the relatively high conductivity of the ground materials there.

Where the ground is not homogeneous with depth, the problem outlined above can still be solved quite easily for steady-state conditions (see Lunardini, 1981; Lachenbruch *et al.*, 1982).

4.4.2 Effects of climatic change

Under steady-state conditions, the mean annual ground temperature profile is linear with depth (assuming that thermal conductivity is constant). A conspicuous feature of temperature profiles observed in many northern boreholes, however, is a distinct *inversion* in the upper 100 metres or so, with near-surface temperatures being significantly warmer than those obtained by simple upward extrapolation (Figure 4.4). Widespread 'deviations' in near-surface temperatures such as this can be explained by climatic change, although similar effects can be produced locally by changes in surface conditions, such as a forest fire, or submergence, for example.

Using a heat conduction model, Lachenbruch & Marshall (1969) analysed borehole temperature data from Alaska, and concluded that the mean annual surface temperature had increased in the range of 2 to 4 °C during the last few decades to a century. If maintained, such a change would eventually lead to a decrease in permafrost thickness (Figure 4.4), and in the marginal areas some permafrost could well disappear. As the envelope of seasonal temperature fluctuation (see section 4.4.3) shifts to warmer values, the active layer will deepen. At the same time, permafrost thins from below due to the melting by geothermal heat.

Suppose that a change in the climate results in a shift of the mean annual surface temperature from some initial, stable value, T_s, to a new value ($T_s + \Delta T_s$). After thermal equilibrium is re-established in the permafrost, the new temperature profile will be (after Gold & Lachenbruch, 1973):

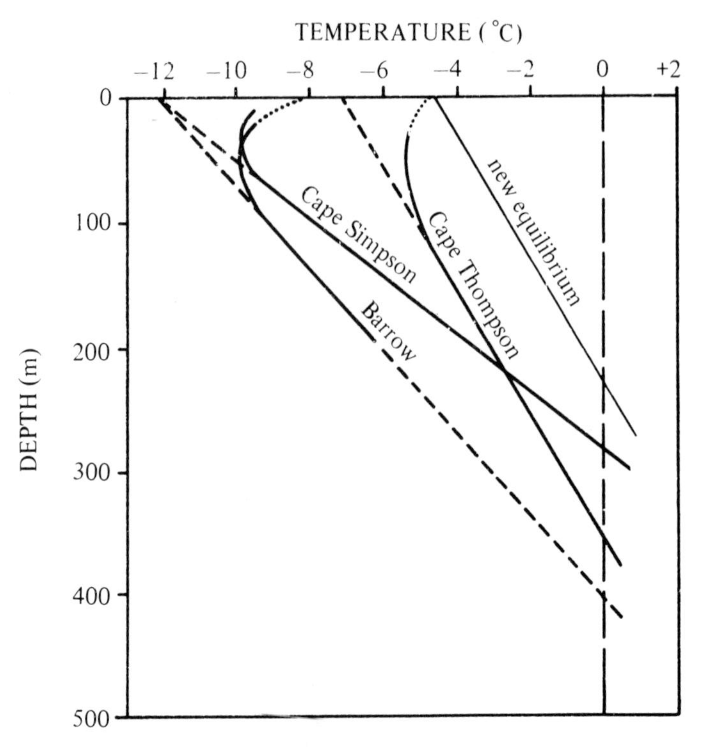

Figure 4.4 Temperature borehole profiles from the arctic coast of Alaska (from Lachenbruch *et al.*, 1982).

$$T(z) = T_s + Gg.z + \Delta T_s \tag{4.15}$$

and the permafrost will have thinned from the bottom by $(\Delta T_s/Gg)$. For $Gg = 20\,\mathrm{K\,km^{-1}}$, and $\Delta T_s = 4\,°\mathrm{C}$, this would amount to 200 m. However, as Lunardini (1981, pp.138–142) demonstrates, the time required for this new equilibrium to be reached is very long – maybe as much as 10 000 years or more, depending on the thermal diffusivity and depth. He calculates that surface temperature fluctuations of short period (a century or less) will not affect the thickness of permafrost significantly. Indeed, Lachenbruch *et al.* (1982) calculated that the climatic warming of the last 100 years or so has been sufficient to melt a total of only 0.8 m at the bottom of permafrost during the period. The main destabilising effects of short-term climatic change will be those resulting from the melting of shallow ground ice as the surface temperature change(s), will propagate into the ground. The transient temperature profile is given by:

During the transition towards a new steady-state condition, the effects of surface temperature change(s), will propagate into the ground. The transient temperature profile is given by:

$$T(z, t) = T_s + Gg.z + \Delta T(z, t) \tag{4.16}$$

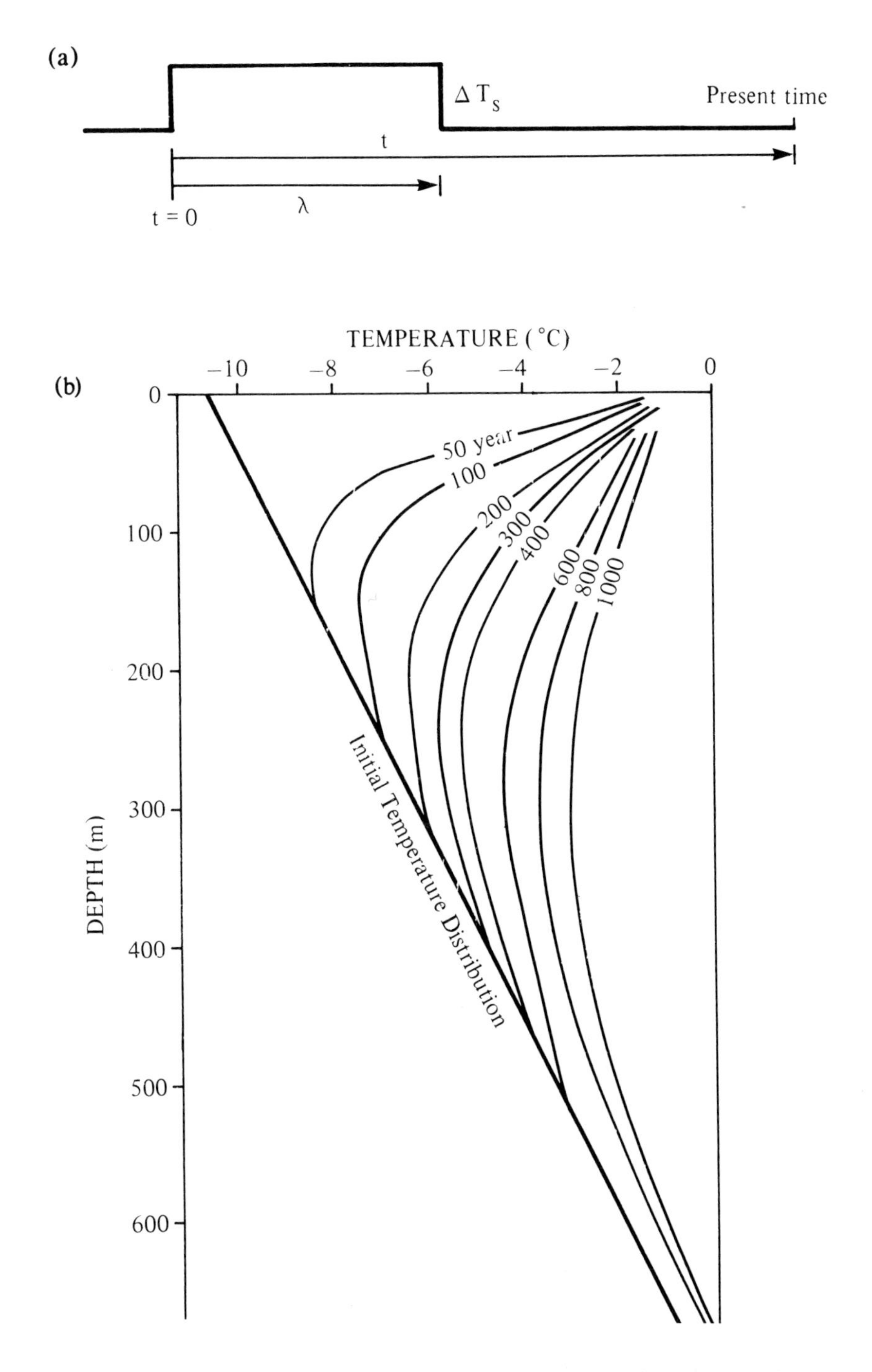

Figure 4.5 (*a*) Illustration of a step change model for climatic change. (*b*) Sequence of geothermal response to a surface temperature step change (from Molochuskin, 1973).

Where temperatures remain quite cold (say below $-2\,°C$ or so), the transient departure of the temperature profile from equilibrium can be analysed quite adequately without reference to phase change complications. Birch (1948) developed a general formulation for the problem of climatic change in terms of simple step functions applied as the boundary condition to equation (4.5). In the case of a single episode as illustrated in Figure 4.5(*a*), the term $\Delta T(z,t)$ in equation (4.16) is given by:

$$\Delta T(z,t) = \Delta T_s \left[\operatorname{erf}\left(\frac{z}{(4\kappa(t-\lambda))^{1/2}} \right) - \operatorname{erf}\left(\frac{z}{(4\kappa t)^{1/2}} \right) \right] \qquad (4.17)$$

where erf( ) is the error function and λ is defined in Figure 4.5(*a*). Equation (4.17) expresses the fact that ground temperatures respond to surface changes according to the thermal diffusivity, the depth in the ground and the elapsed time since the surface change (Figure 4.5(*b*)). If we refer back to Figure 4.4, we see that the effects of the recent climatic warming have penetrated most deeply at Cape Thompson (highest diffusivity) and least at Cape Simpson (lowest diffusivity).

By superimposing the results for different episodes the effects of arbitrary changes in surface temperature on ground thermal conditions can be calculated. Thus a variety of past climatic changes can be represented in differential fashion throughout the ground temperature profile.

In principle, a detailed analysis of the present temperature record with depth can be used to examine hypotheses concerning past climatic patterns, although local environmental effects can distort or obscure the climatic 'memory' of the ground. For example, Cermak (1971) describes a method which analyses the departure of the temperature gradient from the equilibrium condition. Very precise values for temperature (to within a few millidegrees) and thermal diffusivity are required. He applied his method to estimate the magnitude and duration of the post-Glacial climatic optimum in Central Canada, and illustrated how different models of past climatic change can be examined using temperature borehole data (see also Gold & Lachenbruch, 1973; Lachenbruch & Marshall, 1986).

4.4.3 Seasonal temperature variations

Whereas the position of the bottom of permafrost is determined by T_s, K and Q_G, under processes that act over long periods of time, the position of the top of permafrost (the depth of the active layer) is controlled by the seasonal fluctuation of temperature about the mean annual value, specifically by the warmest temperatures of the year. The seasonal fluctuation about the annual mean also determines the depth of winter freezing in the more temperate zones.

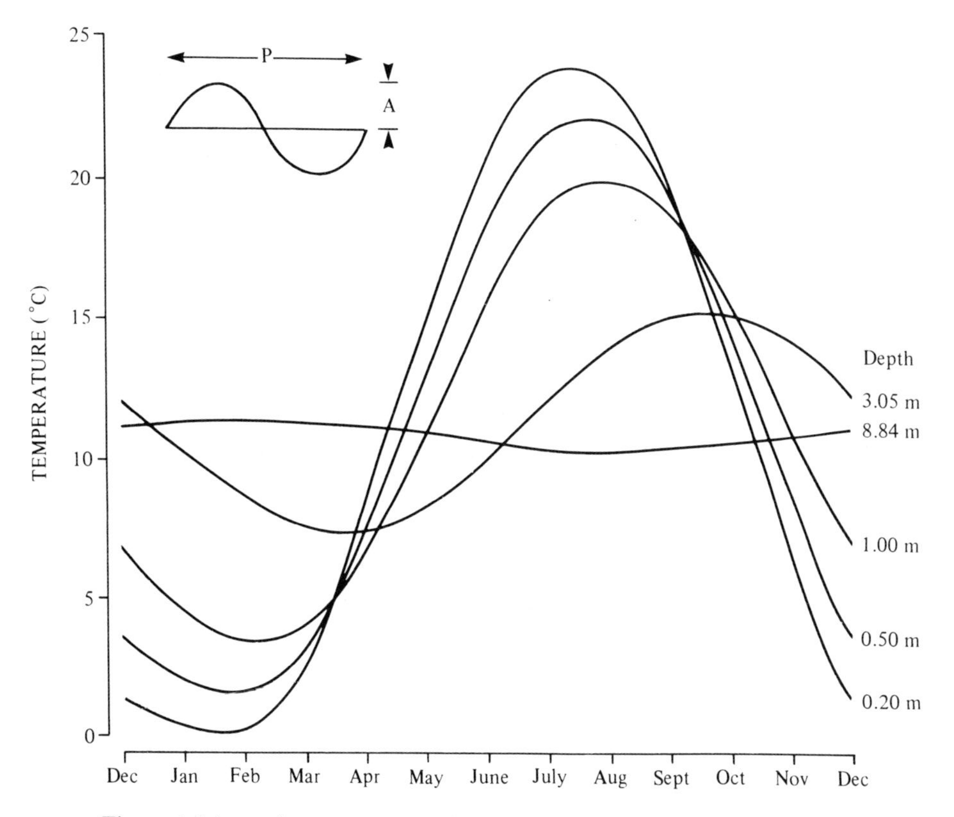

Figure 4.6 Annual temperature variations in unfrozen soil (from Carson & Moses, 1963).

The systematic seasonal fluctuation of solar radiation and air temperature imparts a temperature wave to the ground surface that propagates downward into the earth. The temperature at depth oscillates with the same (annual) frequency as the surface temperature, but with an amplitude that diminishes and a phase lag that increases with depth (Figure 4.6). Eventually, a point is reached where the temperature remains essentially constant, except for any long-term drift related to climatic change, for example. This depth of *'zero' annual amplitude* may be 10 to 15 metres in soils, but can be much deeper in rock.

The top of permafrost coincides with the depth where the *maximum* annual temperature is 0 °C. In addition, the seasonal variations of temperature are responsible for periodic thermal stresses in the surface ground layers that cause them to crack, with important geomorphological consequences such as ice-wedge development (Chapter 6). Periodic temperature changes may also be important to creep activity in surface layers of the ground (Chapter 9).

The general features of the thermal regime in the layer of annual variations – such as the exponential *attenuation* of the seasonal wave *with depth*, and the *lag in phase* – can be analysed using the heat conduction equation (4.5) with a sinusoidal surface temperature variation. In this case, the temperature at any depth, z, is given by (Ingersoll, Zobel & Ingersoll 1954, pp.45–57):

$$T(z, t) = \overline{T}_z + A_s e^{-z(\omega/2\kappa)^{1/2}} \sin\left[\omega t - \left(\frac{\omega}{2\kappa}\right)^{1/2} z\right] \tag{4.18}$$

where $\omega = 2\pi/P$, P is the period of the wave (one year), and A_s is the amplitude of the surface temperature wave (Figure 4.6). Time t is counted from the date in spring when the surface temperature wave passes through its mean annual value. The expression:

$$A_z = A_s e^{-z(\omega/2\kappa)^{1/2}} = A_s e^{-z(\pi/\kappa P)^{1/2}} \tag{4.19}$$

represents the amplitude of the temperature wave at depth z, and the term:

$$t = z\left(\frac{1}{2\kappa\omega}\right)^{1/2} = \frac{z}{2}\left(\frac{P}{\pi\kappa}\right)^{1/2} \tag{4.20}$$

is the lag of the wave with depth. From equation (4.19), we see that the depth of zero temperature change (i.e. where $A_z = 0$) technically occurs at $z = \infty$. Normally, we take an operational definition for the depth of 'zero' annual amplitude as that where the change is less than 0.1°, say. Alternatively, we can take the depth where the annual wave is delayed by exactly one year from that at the surface. This is given by:

$$z = (\pi\kappa P)^{1/2} \tag{4.21}$$

The effect of the thermal diffusivity on seasonal fluctuations in ground temperatures can be illustrated by applying equation (4.18) with different

Table 4.3. Effect of thermal diffusivity on the amplitude reduction and lag of the annual wave at depths of 1 and 3 m

κ ($m^2 day^{-1}$)	A_1/A_s	A_3/A_s	t_1 (days)	t_3 (days)
0.010	0.40	0.06	54	162 (snow, peat)
0.025	0.56	0.17	34	102
0.045	0.65	0.27	25	76
0.065	0.69	0.34	21	63
0.085	0.73	0.38	18	55 (frozen soil)
0.110	0.76	0.43	16	49 (rock)

values for κ. This is equivalent to imposing the same surface temperature variation on different lithologies. The result (shown in Figures 4.7(a) and (b)) reveals greater damping of variations for the low diffusivity case; at a depth of 8 metres, the annual amplitude is virtually zero. (In these examples, a mean annual surface temperature of $-5\,°C$ was assumed, and the geothermal gradient was ignored). Figure 4.7(c) shows the envelope of annual temperature variation for the same two cases. The seasonal range is greatly reduced in case 1, and a much greater active layer depth results in case 2, under the identical surface temperature conditions.

The dependence of the amplitude reduction and phase lag on the thermal diffusivity is illustrated further in Table 4.3. Temperature waves are attenuated less and suffer a shorter lag in materials with high diffusivity, such as rocks.

Conversely, if seasonal ground temperature data are available, they can be used to determine the thermal diffusivity. If we take the ratio of the amplitudes at two depths, z_1 and z_2, we have:

$$\frac{A_2}{A_1} = e^{-(z_2 - z_1)(\pi/\kappa P)^{\frac{1}{2}}} \tag{4.22}$$

from which κ can be calculated as follows:

$$\kappa = \frac{\pi}{P}\left(\frac{z_2 - z_1}{\ln A_1/A_2}\right)^2 \tag{4.23}$$

Alternatively, we can compare the lag at two depths:

$$t_2 - t_1 = (z_2 - z_1)\left(\frac{P}{4\pi\kappa}\right)^{1/2} \tag{4.24}$$

from which κ can again be calculated:

$$\kappa = \frac{(z_2 - z_1)^2}{(t_2 - t_1)^2} \cdot \frac{P}{4\pi} \tag{4.25}$$

These equations have been widely used, and a review of these and other methods can be found in Horton, Wierenga & Nielsen (1983).

The results of the periodic (annual) temperature model have been discussed widely in the literature. They apply equally to other variations, such as diurnal for example, although, as the equations describe, the period of temperature variation affects the depth of propagation and the variation at any depth. Ultimately, the practical value of the relationships presented depends on the extent to which the surface temperature regime can be

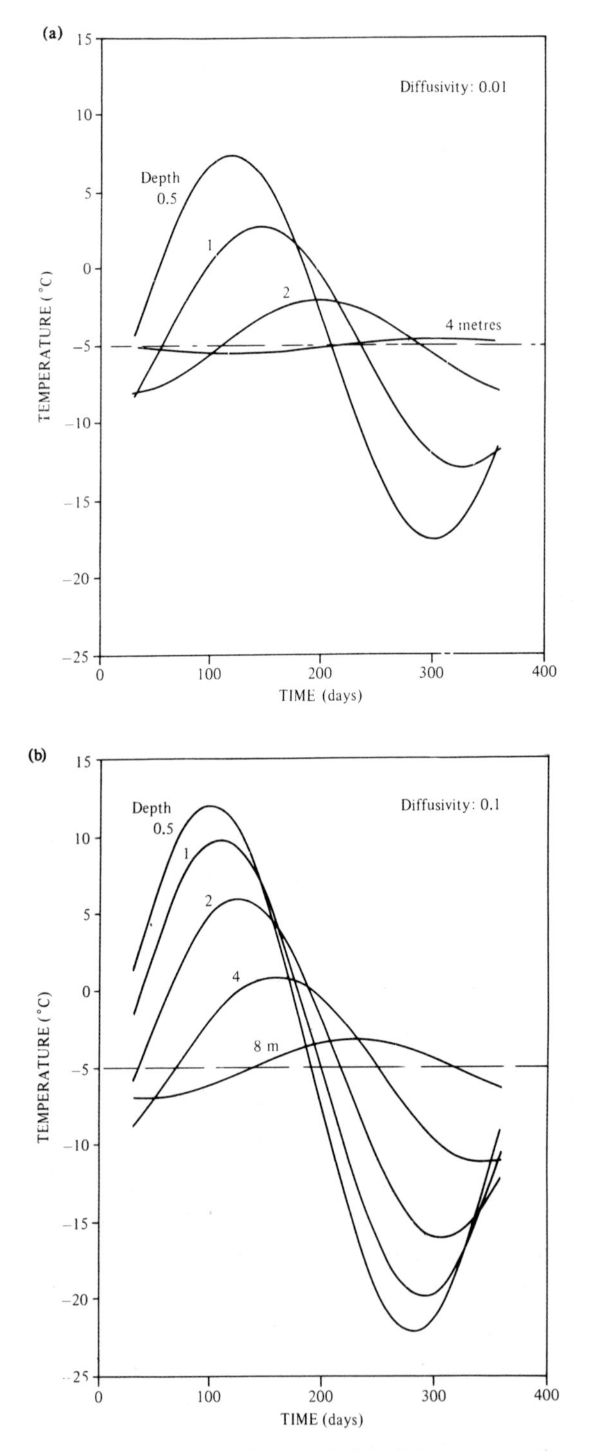

Figure 4.7 Influence of thermal diffusivity on annual ground temperature variations.

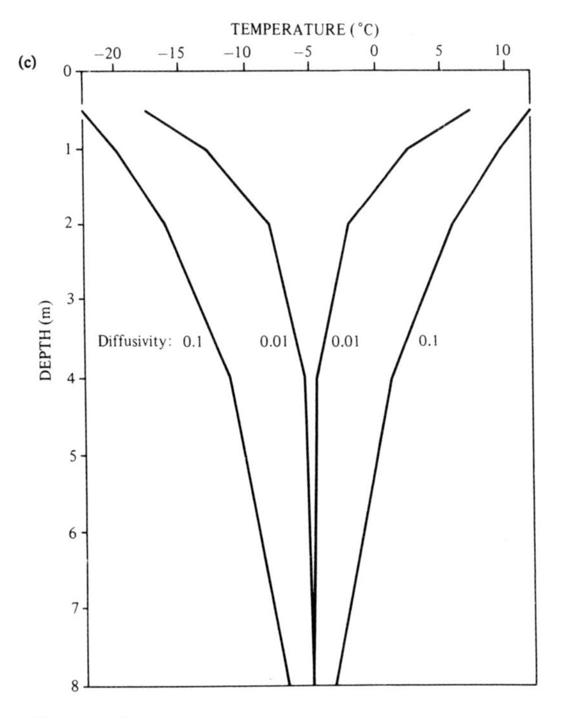

Figure 4.7 (*cont.*)

represented by a simple sine wave. For diurnal temperatures especially, this is often far from the case. However, the relationships can provide valuable insights into the nature and significance of periodic variations in ground thermal conditions.

An important departure from the simple temperature model described above arises from the effects of periodic freezing and thawing in the surface layer of the ground. In the autumn, the surface temperature drops sharply below 0 °C, and temperatures throughout the active layer fall to near 0 °C shortly after this. Subsequently, however, ground temperatures may fall slowly for a prolonged period, because large amounts of latent heat originating from the freezing of water within the active layer must be removed from the ground. The result is that the ground temperature remains near 0 °C for some period. This is terminated when most of the water is frozen, and rapid cooling then ensues. This effect, termed the '*zero curtain*', is more prolonged in soils with high water contents. Sometimes one can note a prolongation of the zero curtain to lower temperatures, indicating that appreciable amounts of soil water have freezing points below 0 °C (see section 7.1.2).

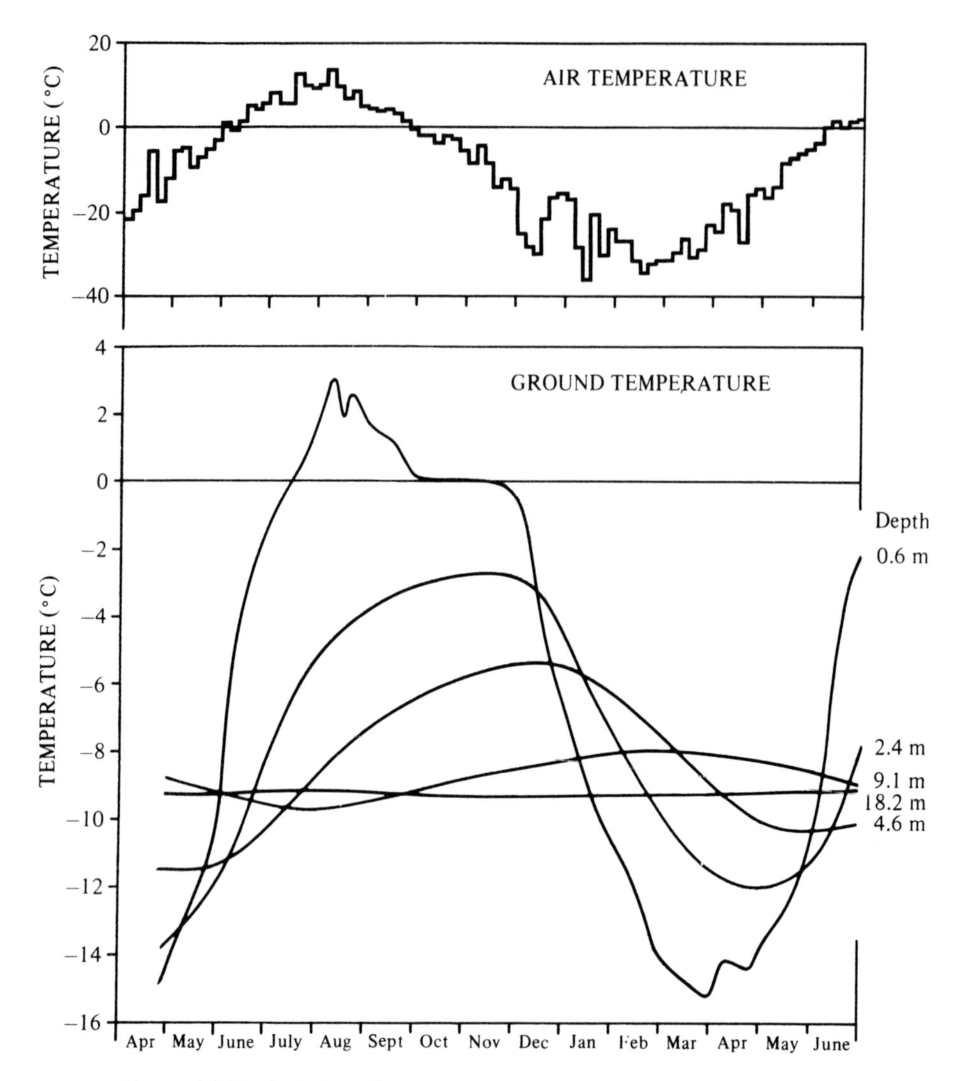

Figure 4.8 Typical air and ground temperatures at Barrow, Alaska (after Lachenbruch *et al.*, 1962).

The zero curtain effect produces some asymmetry in ground temperature curves, although this vanishes, along with high frequency fluctuations due to weather (noticeable in the air temperature record), at increasing depths (Figure 4.8). In Figure 4.8, the annual temperature range also decreases from 18° at 0.6 m to 0.1° at 18.2 m. The figure also shows that the minimum temperature at 4.6 m occurs in the spring, and at 9.1 m in mid-summer. This phase lag can sometimes result in the spring and summer freeze-up of wells and buried conduits in regions of discontinuous permafrost (Lachenbruch *et al.*, 1962).

Where substantial phase change is involved, use of equations (4.23) and (4.25) may not be useful for determining the thermal diffusivity, κ from ground temperature data, but another approach is possible using the heat conduction equation (4.5). By evaluating $(\partial T/\partial t)$ and $(\partial^2 T/\partial z^2)$ at a particular time and depth, and by taking the ratio of these terms, the apparent thermal diffusivity, κ_a, is determined (see Takagi, 1971, McGaw *et al.*, 1978). This approach incorporates the effects of latent heat, and can provide estimates of κ_a as a function of temperature. It can also be used to reveal whether non-conductive forms of heat transfer are operative in the ground (e.g. Nelson *et al.*, 1985).

THE ROLE OF PEAT AND SNOWCOVER

The importance of snow cover and peat to ground thermal conditions is widely recognised. Their influence can be explained with reference to the pattern of seasonal temperature variation. Snow is an *insulator* compared to other natural materials, and is a leading factor in protecting the ground from heat loss in winter. Its net effect is to raise mean annual ground temperatures; Gold (1963) concluded that snow cover was the principal reason why annual average ground temperatures can be many degrees warmer than the mean air temperature in cold regions. Snow amelioration measures were widely used in the north-eastern USSR to mitigate ground freezing in agriculture, as well as in open pit mining (Klyukin, 1963). Where temperatures are close to 0 °C, snow cover can be responsible for the absence of permafrost in certain locations (e.g. see Smith, 1975). Conversely, peat, which also acts as an insulator, is commonly associated with the *existence* of permafrost at certain locations in marginal areas. Unlike snow cover, of course, a peat layer is present the year round, but its conductivity varies seasonally with moisture conditions. When it is dry, as it frequently is in the summer, the conductivity is low and thus the soil beneath is shielded from the heat of summer. Consequently, the mean annual ground temperatures are lower than otherwise. Even if the peat remains wet during the summer, the resulting predominance of evaporation in the energy regime will lead to lower surface, and hence subsurface, temperatures (see section 3.5.1).

The snow cover interposes a layer of low thermal diffusivity between the air and the ground, serving to isolate the ground from the extreme temperature changes of the air (compare the two sites in Figure 4.9). Since heat exchange takes place at the snow surface rather than the ground surface, the range of annual ground surface variation is reduced and ground temperatures are higher, not only in winter, but also on an annual basis. In Figure 4.10, an increasing divergence between the air and ground surface tempera-

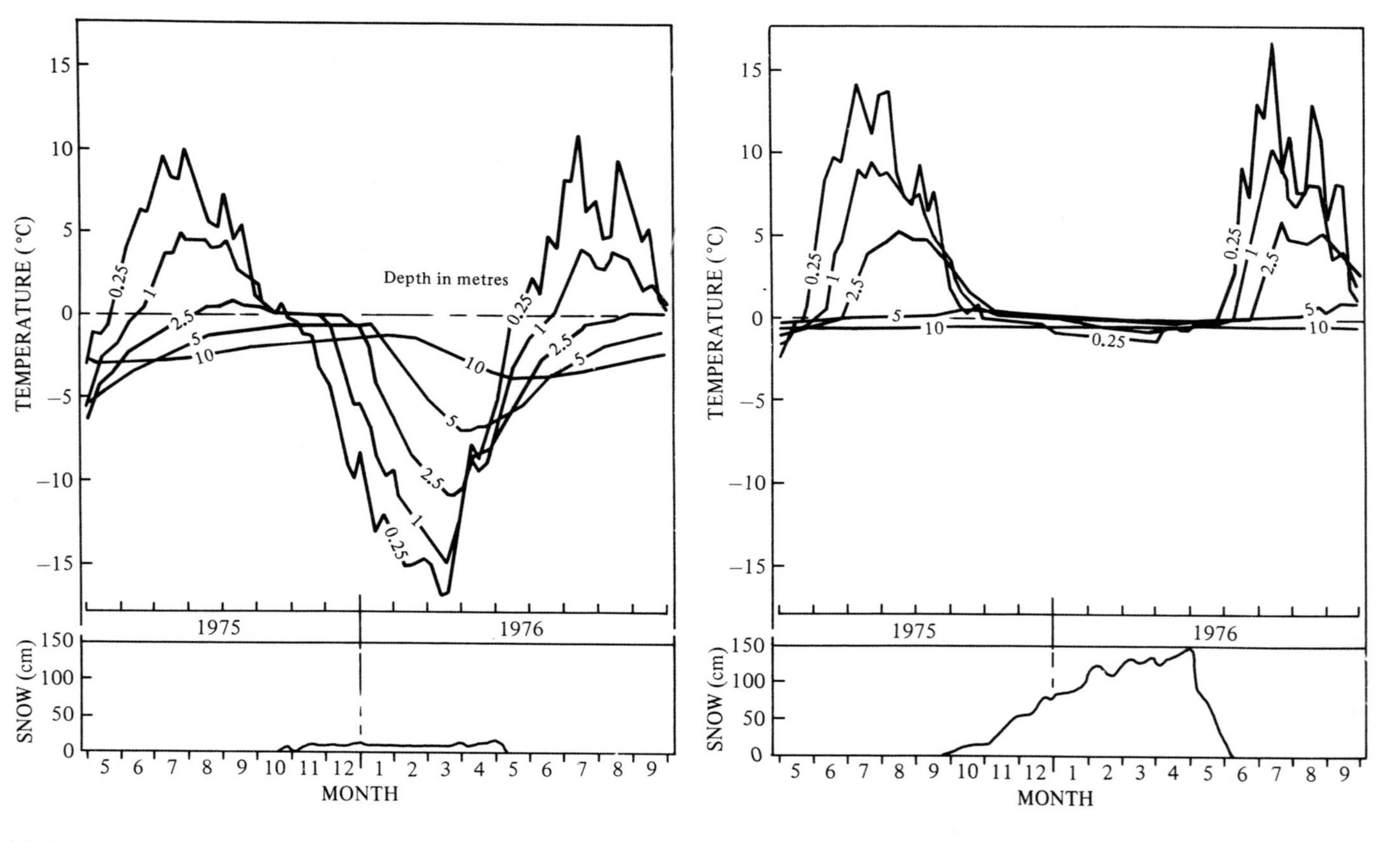

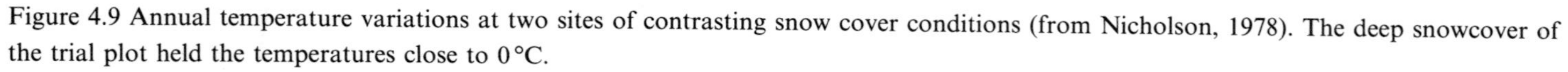

Figure 4.9 Annual temperature variations at two sites of contrasting snow cover conditions (from Nicholson, 1978). The deep snowcover of the trial plot held the temperatures close to 0°C.

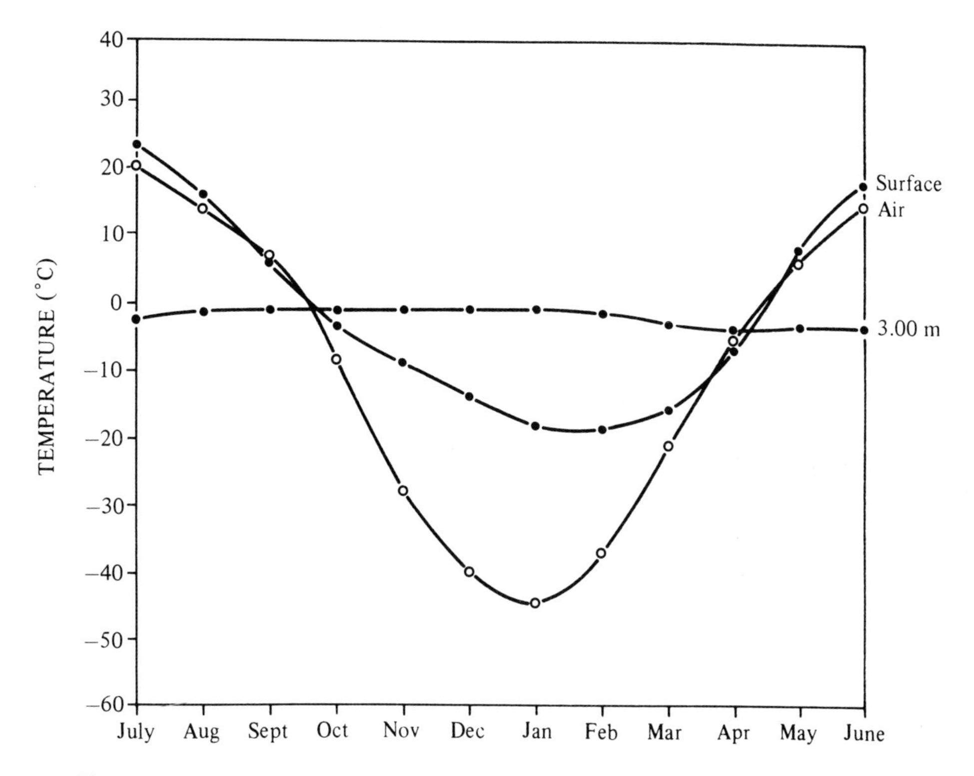

Figure 4.10 Air, surface and ground temperature regimes for Yakutsk (data from Pavlov, 1973).

tures occurs through the winter with the seasonal build up in snow depth. The large difference between the air and ground surface temperatures in winter (20° or more) contrasts with their closeness in summer, when heat exchange takes place at the ground surface. As a result, there is virtually no cooling evident at a depth of 3 m in the ground. In the Mackenzie Delta, where the mean daily air temperature is below −20 °C for almost 6 months in winter, the 1-metre ground temperature beneath 120 cm of snow did not fall below −0.2 °C (Smith, 1975).

This effect of snow results from its very low thermal conductivity, which in turn depends on the density. For the density range 100 to 400 kg m^{-3}, the relationship shown in Figure 4.11 can be used. At a density of 200 kg m^{-3}, the snow has a conductivity of 0.11 W m^{-1} K^{-1}, which compares to values of 2 to 3 for frozen soil. The heat capacity of the snow (applying equation 4.6) would be 0.42 MJ m^{-3} K^{-1}, and the diffusivity 0.01 m^2 day^{-1}. This compares to values up to 0.1 m^2 day^{-1} for frozen soil. Referring to Table 4.3, we see that the annual surface temperature wave would be reduced to 40% beneath 1 m of snow. For snow of higher density, the damping effect would be less.

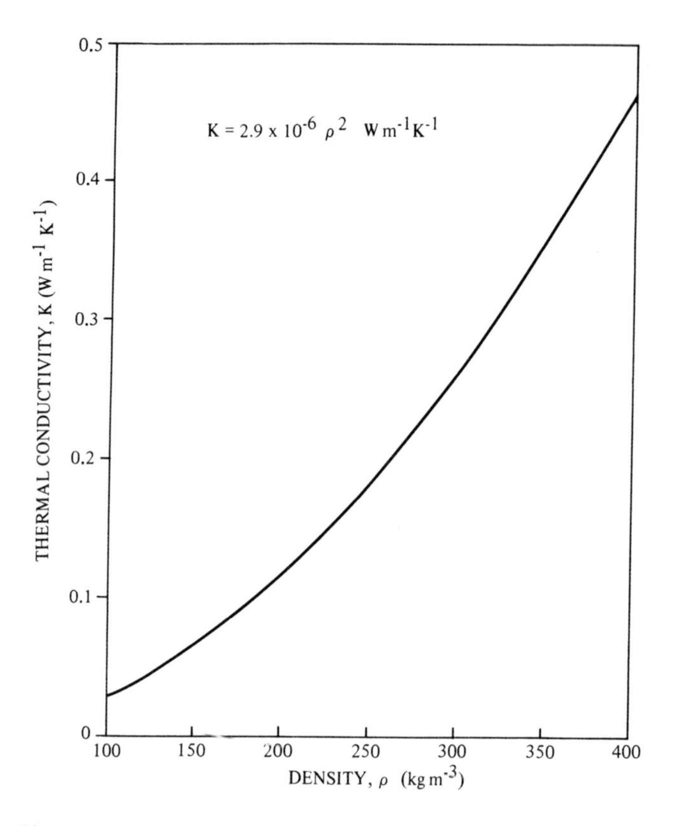

Figure 4.11 Relationship between thermal conductivity and density of snow (based on a relationship in Goodrich, 1982*b*).

Smith (1975) measured ground surface temperatures in mid-winter in the Mackenzie Delta, and found a relationship with snow depth (Figure 4.12). The range in temperatures shown – about 20 °C – is twice as great as the maximum variation observed in summer (which result from differences in vegetation). Figure 4.12 indicates that increases in snow cover become less effective above 50 or 60 cm. Figure 4.13 shows snow and ground temperature profiles on a typical day in early winter at two sites near Schefferville, Quebec. There is a characteristically steep temperature gradient immediately below the snow surface, with a marked inflection at a depth of 50 to 60 cm, below which temperatures are fairly uniform. Together, Figures 4.12 and 4.13 imply that variations in snow depth are more critical for ground temperatures where the snow cover is thin.

Goodrich (1982b) analysed the general features of snow cover/ground thermal interactions, using a heat conduction model. He concluded that mean annual ground temperatures are most strongly influenced by an accumulation of snow in autumn and early winter, and by the maximum

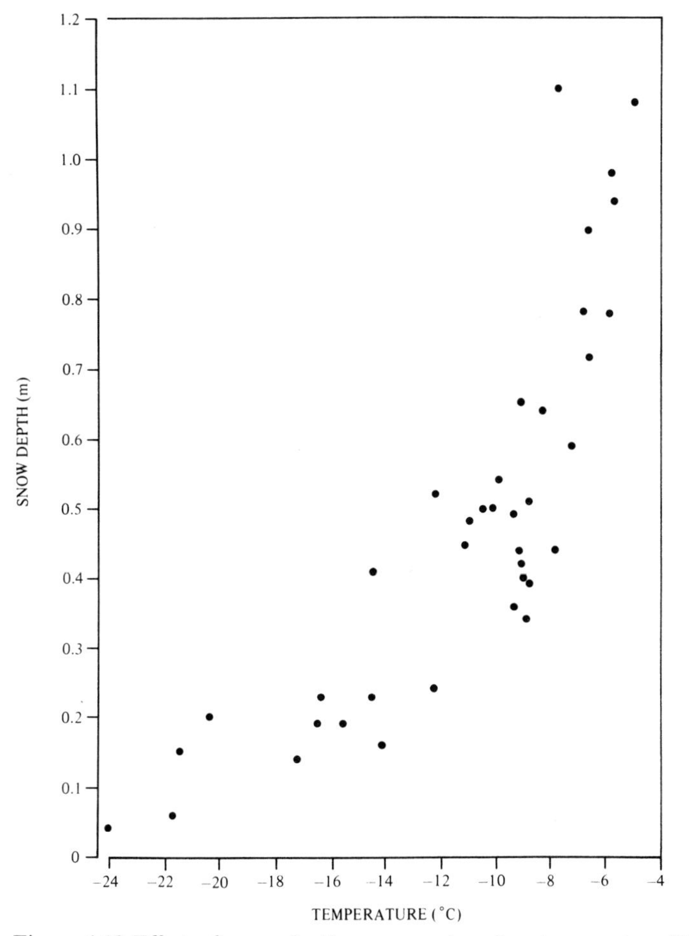

Figure 4.12 Effect of snow depth on ground surface temperature (from Smith, 1975).

depth attained over the winter. In his calculations, a doubling of the snow cover from 25 to 50 cm increased the minimum ground surface temperature by about 7° (from −19.7° to −12.3 °C), and the mean annual surface temperature by 3.5° (from −7° to −3.4 °C). If the 50 cm of snow built up within 30 days in autumn, the minimum temperature would be only −2.8 °C, the mean annual temperature would be +1.1 °C, and permafrost would degrade.

Finally, Mackay (1984b) has shown that snow depth plays an important role in ice-wedge cracking. A rapid drop in the ground temperature is necessary to cause cracking, and this is impeded by the insulating properties of a deep snow cover. He found that a depth of 60 cm was sufficient to prevent ice-wedge cracking in an area of active ice wedges on Garry Island, NWT.

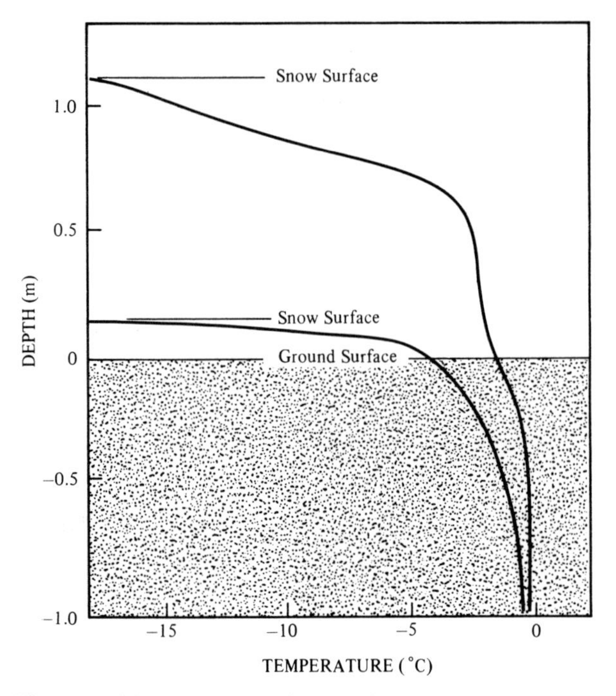

Figure 4.13 Mean snow and ground temperatures at two sites for a typical day in winter (from Nicholson & Granberg, 1973).

The influence of the organic layer on ground thermal conditions has also been well documented in the literature (e.g. Nakano & Brown, 1972, Luthin & Guymon, 1974, Zoltai & Tarnocai, 1975, Fitzgibbon, 1981), and the presence of permafrost in marginal areas is frequently associated with peat. For example, R. J. E. Brown (1973) reports that in the Yellowknife area, the greatest local extent of permafrost is in peatlands, and that the mean annual temperature at a depth of 15 m ranges from about 2 °C in granite to − 1.0 °C in spruce peatland (see Figure 3.2). This effect arises because of the *marked seasonal variation in the thermal properties* of peat, which imparts an asymmetry to the pattern of seasonal temperature variations. In summer, when the surface layer is usually dry (because of evaporation), the thermal conductivity is very low and warming of the ground is inhibited. However, in the fall the peat becomes quite moist, because of the much reduced evaporation rate. Further, when it freezes the conductivity becomes even higher and the ground can cool rapidly. The net effect, in contrast to snow, is that mean annual ground temperatures under peat are (much) lower than under adjacent areas without peat.

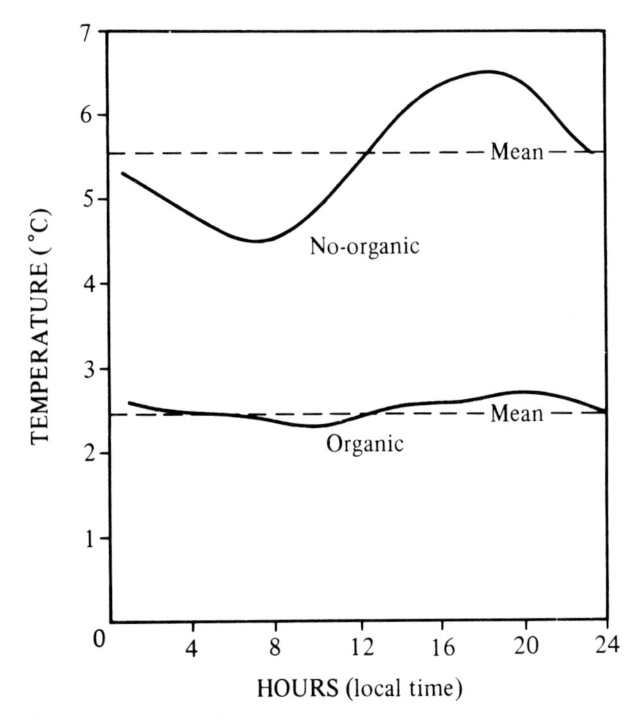

Figure 4.14 Mean diurnal 10-cm ground temperature regimes at adjacent sites in the Mackenzie Delta (from Smith, 1975).

Figure 4.14 shows the temperatures measured at a depth of 10 cm over the summer at two sites only 1 m apart, in a spruce forest in the Mackenzie Delta. At one site there was a 10-cm layer of organic material at the surface, whilst at the other there was simply bare mineral soil. The mean daily temperature is 3 °C warmer at the bare site and the diurnal range 5 times greater. R. J. E. Brown (1963) writes that after removal of the moss and peat from the ground surface, in the arctic region of the Yenisey River valley, the depth of thaw increased by 1.5 to 2.5 times.

Goodrich (1978) used a heat conduction model to demonstrate the significance of temperature-dependent thermal properties – with peat providing an extreme case – to ground thermal conditions. He showed that an offset occurs in the mean annual temperature profile as a result of the seasonal change in the thermal conductivity. When the frozen conductivity is higher, the mean annual temperature profile shifts to colder values down to the depth to which the property change takes place (Figure 4.15(*a*)). The effect is seen to increase as the ratio of frozen to thawed conductivities increases. R. J. E. Brown & Péwé (1973) point out that the conductivity of peat can also change throughout the year as a function of the water

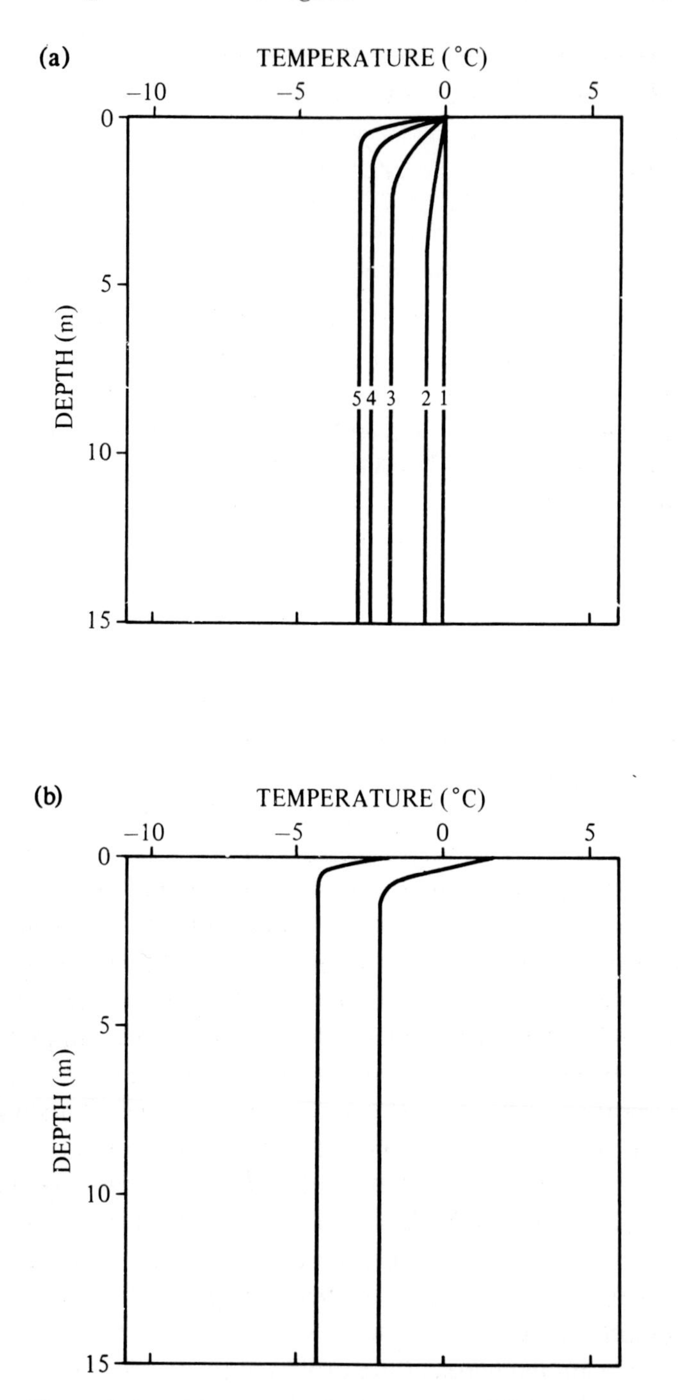

Figure 4.15 (*a*) Mean annual temperature profiles in relation to the ratio of frozen to unfrozen thermal conductivity. (*b*) Significance of the thermal offset near 0 °C. (From Goodrich, 1978). Profiles are shown without the influence of the geothermal gradient.

balance; the ratio of wet to dry thermal conductivity is about 30 for peat (Fitzgibbon, 1981). Since this seasonal trend is generally the same as for the frozen/thawed variation, the effect mentioned above is reinforced. Recently, Nelson *et al.* (1985) have demonstrated that non-purely conductive processes of heat transfer (i.e. vapour transport) may be responsible for the thermal buffering effect of peat layers in summertime.

Overall, these effects can lead to the existence of permafrost in locations where the mean annual *surface* temperature is actually above 0 °C (Figure 4.15(*b*)); see also Lindsay & Odynsky, 1965; Zoltai, 1971, for example).

In concluding this section, the reader is reminded that it is rare that any single factor alone can explain local ground thermal conditions. As discussed in Chapter 3, the ground thermal regime results from the interaction of climatic, surface and subsurface factors, and this is ultimately responsible for the (considerable) variations in ground temperatures which occur locally.

4.5 Lateral variations in ground temperatures

Until this point, we have considered only how ground temperature varies one-dimensionally with depth beneath the surface. This view would be quite adequate if it happened that the earth's surface temperature was uniform over large areas. In Chapter 3, however, it was emphasised that natural variations in microclimatic and terrain conditions introduce significant variations to the surface temperature regime within even a small area.

An important problem, therefore, both from a scientific and practical (engineering) viewpoint, is to determine the 'disturbance' of ground temperatures that result when the temperature within some finite boundary differs from that of the surrounding area. Such conditions could pertain to the presence of natural features such as shorelines, lakes and rivers, or to modifications of the surface as a result of vegetation removal, buildings, or highways. At geomorphological time-scales, when conditions approach the steady state, the purely transient effects of latent heat become negligible and such problems can be analysed by means of simple heat conduction models. However, as stated previously, detailed engineering analyses of transient problems, for time-scales measured in decades, generally require more complex numerical approaches.

The theoretical aspects of three-dimensional heat conduction in a semi-infinite medium disturbed by surface effects, have been treated by Lachenbruch (1957*a*, *b*) and Carslaw & Jaeger (1959), while W. G. Brown (1963) and Lunardini (1981) present a number of worked examples.

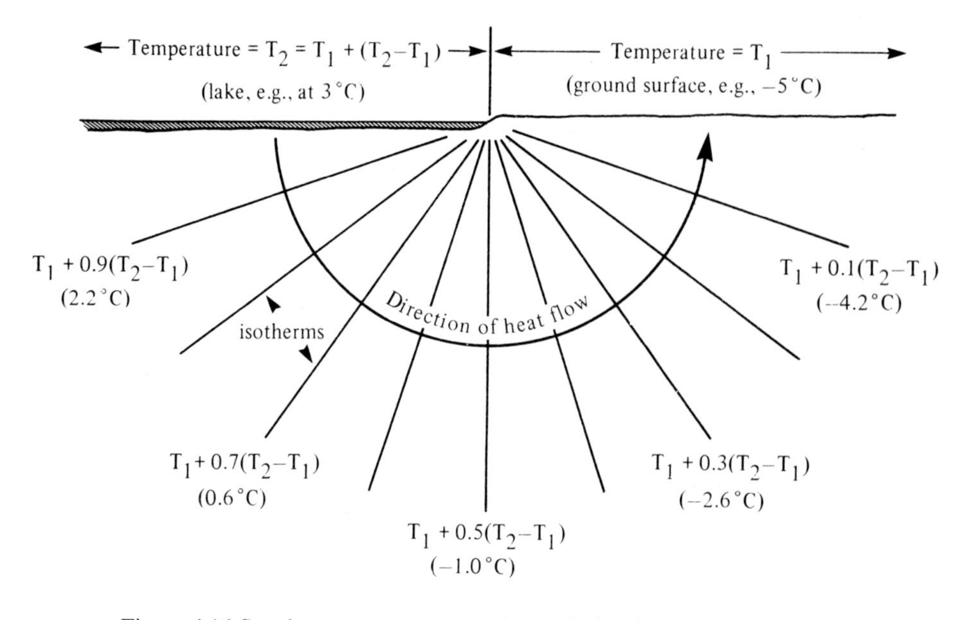

Figure 4.16 Steady-state temperatures beneath the shoreline of a large lake (from W. G. Brown, 1963).

4.5.1 Influence of water bodies

In high latitudes, bodies of water that do not freeze to the bottom in winter have a marked effect upon ground temperatures and the local configuration of permafrost. This arises from the fact that the mean annual bottom temperature must be greater than 0 °C, whereas the temperature of the neighbouring land surface may be – 5 °C or lower. The presence of a water body thus constitutes a heat source, giving rise to anomalous heat flow and temperature conditions in the ground (Figure 4.16). According to the results of Brewer (1958a, b), Johnston & Brown (1964) and Smith (1976), where the mean annual air temperature is as low as – 10° to – 15 °C, even relatively shallow lakes do not freeze to the bottom. This is especially so where the snow cover on top of the ice is sufficient to limit downward freezing. Wherever water bodies remain unfrozen at depth, permafrost is affected (Figure 4.17). The size and temperature of the water body are important factors (see below). Hopkins *et al.* (1955) reported that permafrost is absent or lies at great depths beneath lakes and ponds throughout Alaska. The thermal effects of water bodies constitute the greatest local departures of ground temperatures from any systematic geographical patterns determined by climate.

Under steady-state conditions, the normal geothermal temperature

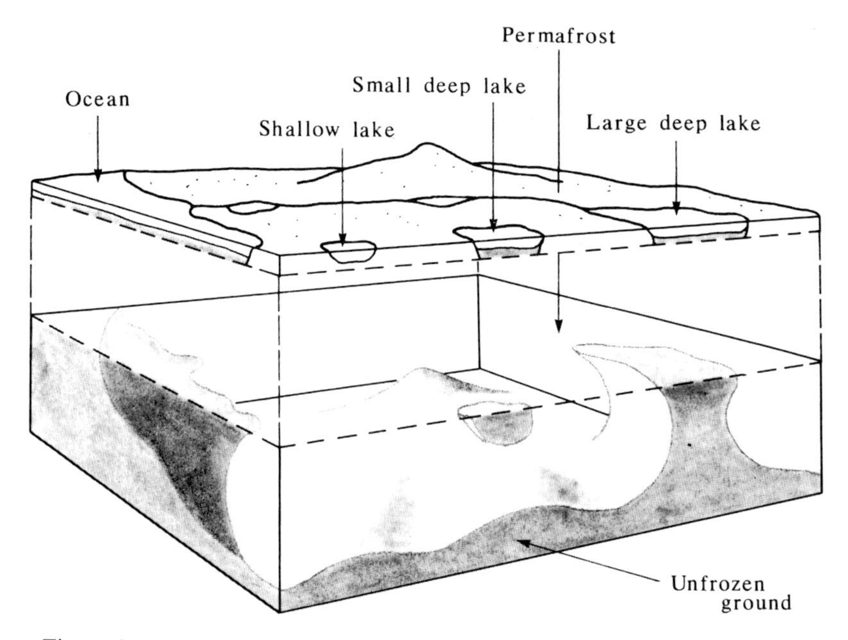

Figure 4.17 Schematic representation of permafrost configuration beneath water bodies (from Gold & Lachenbruch, 1973).

profile (equation (4.11)) is modified in the vicinity of a water body as follows:

$$T(z) \;=\; (T_s + Gg.z) + (T_w - T_s).\ \Phi(x, y, z) \qquad (4.26)$$

where the second term on the right-hand side is the thermal disturbance at the point (x, y, z) in the ground due to the water body, or water bodies nearby (as illustrated in Figure 4.16). This term can be calculated for any arbitrarily-shaped area, by dividing it into sectors of a circle, θ (Figure 4.18(*a*)), and summing the effects. In this case:

$$\Phi(x, y, z) \;=\; \sum \frac{\theta}{360}\left[1 - \frac{z}{(z^2 + R^2)^{1/2}}\right] \qquad (4.27)$$

(Lachenbruch 1957a). By repeating the procedure for various (x, y, z)s, a complete picture of the temperature field created by the disturbance can be compiled. For locations lying outside the boundary of the feature, the disturbance is obtained by subtracting the effects due to sectors of radii R_2 from those with radii R_1 (Figure 4.18(*b*)). The method is easily extended to more complex areas (Figure 4.18(*c*)). Lachenbruch (1957b) has described special solutions for estimating the thermal effects of rivers and ocean shorelines.

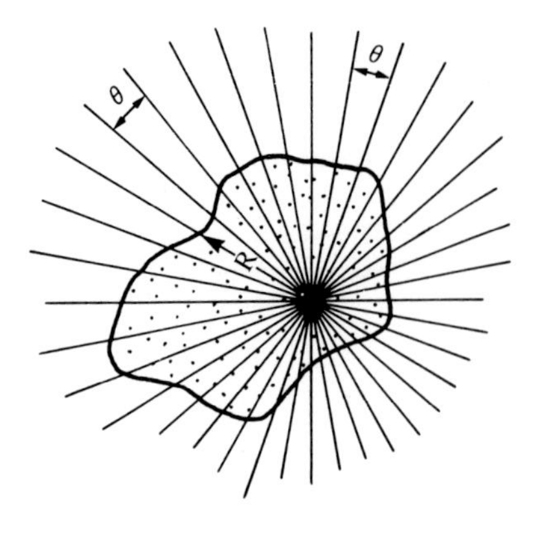

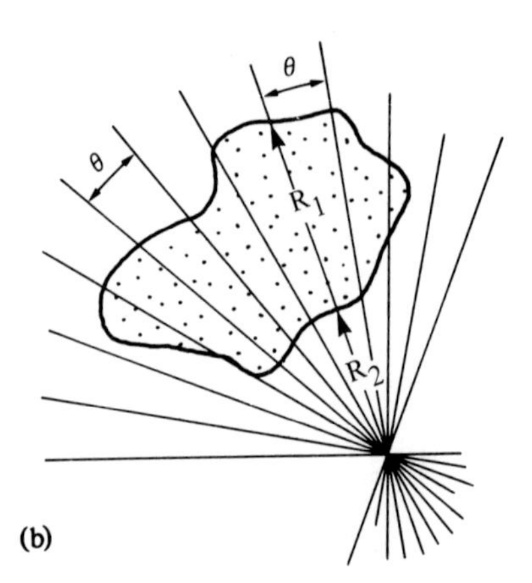

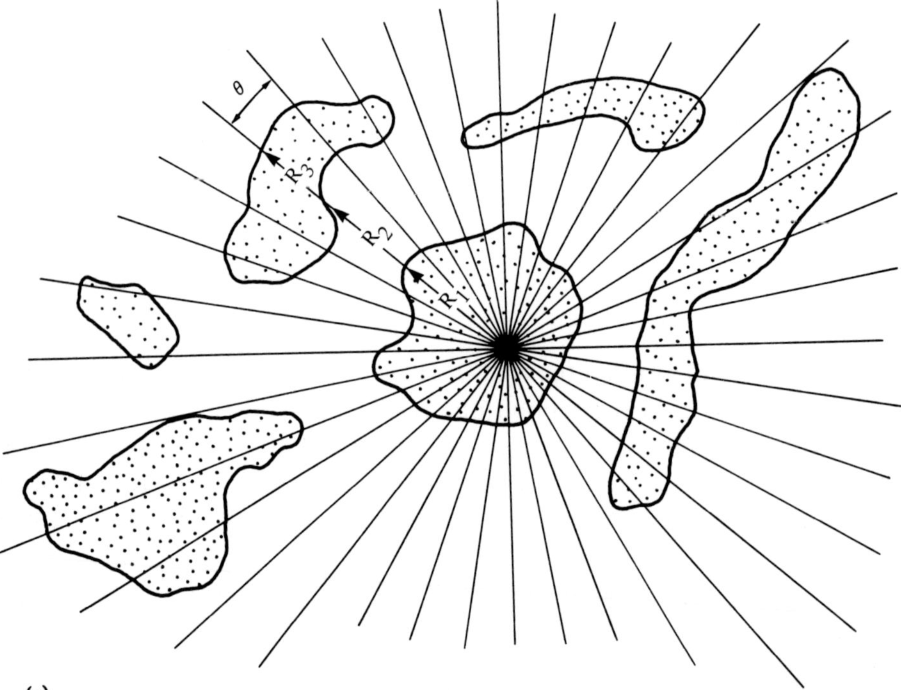

Figure 4.18 Method of dividing a given surface area into sectors of circles (from W. G. Brown, 1963).

Using the approach outlined above, Smith (1976) calculated ground temperatures and permafrost configuration within an area of the Mackenzie Delta (Figure 4.19). In this example, permafrost shows a steeply plunging surface at the edge of the river and the large lake. Beneath the

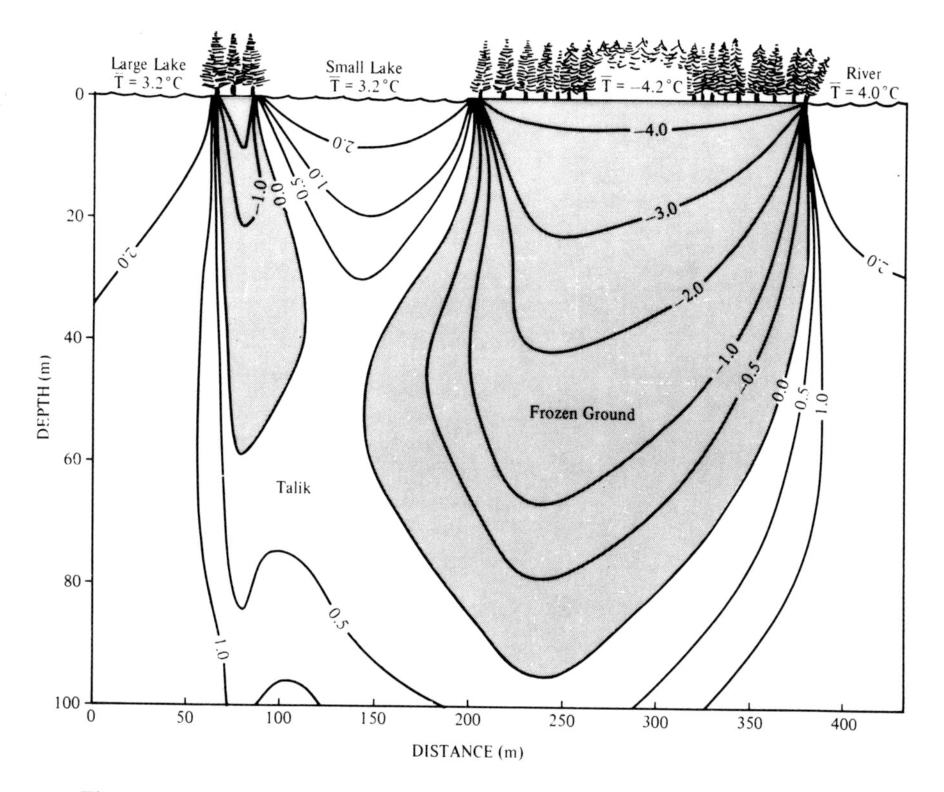

Figure 4.19 Calculated temperatures and permafrost configuration (from Smith, 1976).

small lake, the thermal effect is sufficient only to form an hour-glass-shaped *talik*. The maximum permafrost thickness is 95 m, but beneath the narrow isthmus it is only 60 m. Equations (4.26) and (4.27) were used by W. G. Brown *et al.* (1964) to calculate the ground temperatures beneath a small lake in permafrost, and Mackay (1963) used the simplified form for a circular lake, in his explanation of pingo formation.

In the case of a circular lake, of radius R, the steady-state thermal disturbance beneath the centre can be calculated from:

$$\Phi(z) = \left[1 - \frac{z}{(z^2 + R^2)^{1/2}}\right] \tag{4.28}$$

This expression is plotted in Figure 4.20. Using this, we may calculate the ground temperature disturbance at a depth of 20 m below the centre of a circular lake which has a radius of 100 m. R/z is equal to 5 and from Figure 4.20, Φ is equal to 0.8. If we assume that the mean lake bottom temperature, T_w, is 3 °C, and that for the surrounding land area, T_s is -7 °C, then:

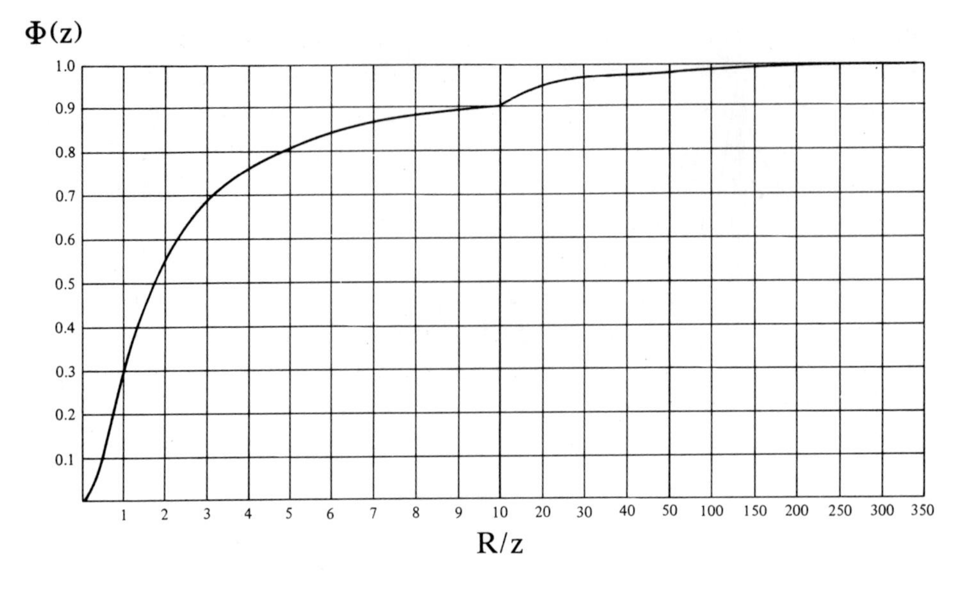

Figure 4.20 Function for the steady-state temperature disturbance of a circular lake (from W. G. Brown, 1963).

$$\Phi(T_w - T_s) = (0.8).(10\,°C) = 8\,°C$$

That is, the ground at this location is 8° warmer than it would be in the absence of the lake. The actual temperature can be calculated from equation (4.26). Assuming a geothermal gradient of 20 K km^{-1}, the temperature would be:

$$T_{20} = (-7 + 20 \times 0.02) + 8 = 1.4\,°C$$

and thus the ground at this point would be unfrozen.

4.6 Summary

In summary, the thermal conditions in the ground can be viewed as the combined outcome of a number of effects. First, the flow of heat from the earth's interior, together with the thermal conductivity of earth materials, establishes a steady state profile (increase) of temperature with depth. Superimposed upon this are the seasonal (periodic) variations of temperature in the upper 10 to 20 metres of the ground induced by the annual regime of climate and modified by snow cover and ground thermal properties. Where the thermal properties are strongly temperature dependent, the seasonal fluctuation may impart a characteristic offset to the mean temperature profile. Superimposed upon this long-term steady state picture are the transient effects associated with shifts in the surface temperature regime as a result of climatic change, vegetation change, submergence or geomorphic activity for example. A number of such effects, operating at different time scales, may be present simultaneously. Finally, lateral variations in surface temperature, as a result of terrain and/or hydrological conditions, or microclimatic differences, produce distinct spatial variations in ground thermal conditions.

Each of these effects can be considered in terms of heat conduction theory. Together, the expressions presented in this chapter provide insights into understanding the relationship between the temperature of the ground and the dynamic climatological and geomorphic processes that modify the surface terrain and energy regime.

5

The forms of the ground surface 1: slopes and subsidences

5.1 Freezing and thawing and displacements of the ground

The thermodynamic and mechanical properties of freezing soils and rocks find expression in the appearance of the natural terrain. For over a century textbooks have described characteristic features of the ground surface associated with cold climates. Yet to this day there are features whose precise origin remains in doubt. Shaping of the ground surface by running water, the erosive effects of wind, or the instability of sloping surfaces induced by changing moisture or other conditions are common to temperate and cold regions. But those terrain forms exclusively associated with cold climates are of unusual interest and may be highly significant indicators of the special properties of freezing soils.

The development of slopes is, in the simplest view, a consequence of gravitational forces. The same applies to the subsidences of the ground which follow the thawing of excess ice. Such movements are the topics of this chapter. In the following chapter a variety of phenomena associated with essentially level ground in cold regions are considered.

5.2 Instability of soil on slopes: overview

The term *solifluction* has been used to describe many kinds of downslope soil movement in cold climates. It has also been applied to forms, superficially similar, in warm regions. There are, however, various flow features having a lobate or terrace-like form (Figures 1.10, 5.1), with 'fronts' tens of centimetres or even a metre or more high and revealing a downslope movement of only a few centimetres or less per year (Washburn, 1979, C. Harris, 1981), which are unique to the cold regions. It is useful to restrict the term solifluction to such features, in which freezing of the soil is essential. 'Congelifluction', 'gelifluction' and 'frost creep' are terms which have been used for the same, and for other phenomena. These terms have often been defined genetically, even though the origins of some of the

Figure 5.1 Solifluction terrace formed by slow movement of soil down a slight slope (near Reinheim, Dovrefjell, Norway).

features are still in doubt. For example 'solifluction' has been defined as the flow of saturated materials – yet there is often uncertainty as to when the movement occurs and whether saturation (with ice or water) is necessary or sufficient. The use and definition of solifluction advocated here is more exclusive.

Rapid and often more conspicuous soil movements are also common in the arctic and sub-arctic and while some may call these 'solifluction' they are better described as landslides, mudflows (or '*thaw slides*', '*thaw mudflows*') and similar terms, because of the obvious correspondence to features in temperate regions. The processes of ice accumulation and its thawing are an important element in the mechanics of most landslides and mudflows in cold regions, but the effects can be largely understood from the principles of soil mechanics applying also in temperate (non-freezing) conditions.

A simple approach to the mechanics of slopes* (without regard to

* For a more detailed discussion of slope mechanics, see Williams (1982 – for a summary account) and Carson & Kirkby (1972); Craig (1974) or other texts for engineers, demonstrate engineering procedures.

freezing) is to assume the slope to be 'ideal': that is, uniform in materials and geometry. The weight, γz of a column of soil (Figure 5.2(*a*)) gives a vertical stress on the sloping plane, of $\gamma z \cos\beta$

where

γ = unit weight of soil $\mathrm{N\,m^{-3}}$

(note that $\gamma = \rho g$ where ρ is density, g is gravitational acceleration)

z = depth, m

β = angle of slope

The *shear stress* acting parallel to the straight slope and thus tending to cause movement is given by (Figure 5.2(*b*)):

$$\tau = \gamma z \cos\beta \sin\beta \qquad (5.1)$$

while the stress perpendicular to the same plane, the (total) *normal stress*, σ, is given by:

$$\sigma = \gamma z \cos^2\beta \qquad (5.2)$$

The shear stress is resisted by the shear strength of the soil, S:

$$S = C + \acute{\sigma} \tan\phi \qquad (5.3)$$

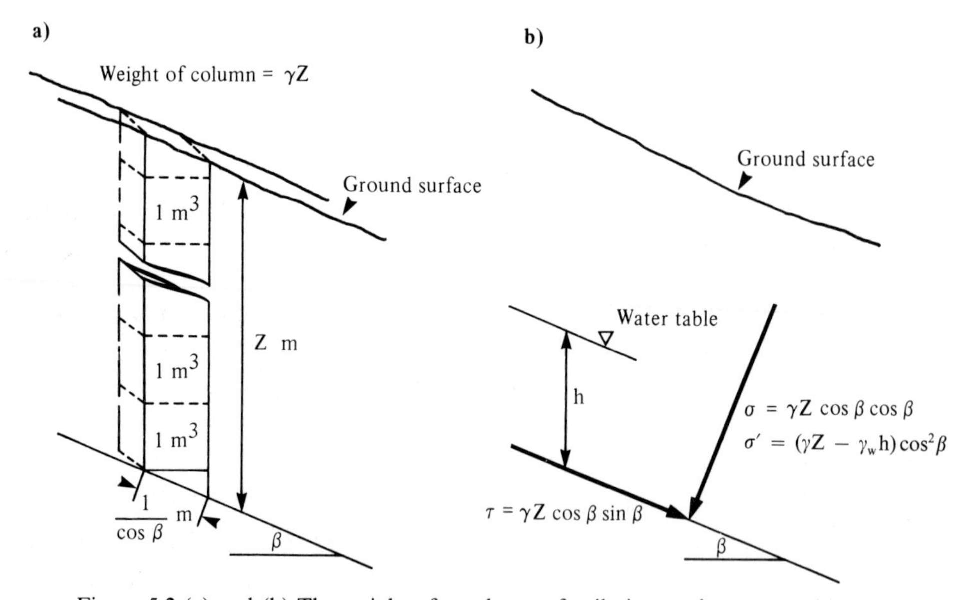

Figure 5.2 (a) and (b) The weight of a column of soil gives a shear stress (τ) on a plane in the slope, according to equation (5.1). The weight also gives a stress (σ) normal to the plane (equation (5.2)), which is modified ($\acute{\sigma}$) by the pressure of water (equation 5.4)).

where

C = cohesion (which may be absent)

$\acute{\sigma}$ = effective normal stress

ϕ = friction angle

movement occurs when:

$$\tau > S$$

Equations (5.1, 5.2) illustrate certain principles of the mechanics of slopes. If movements extend deeply in a short slope, or if the slope is irregular in form or changing in slope angle, shear stresses may have to be determined in more detail. However, slope movements in cold regions are commonly fairly shallow and these principles are applicable, albeit with modifications, in many cases.

The strength (or resistance) S, of a soil is dependent on the value of tan ϕ – the so-called '*angle of internal friction*' which is a property of the soil and whose value can be determined by testing. It is rarely more than 40° or less than 15°. The quantity $\acute{\sigma}$, is the effective normal stress which depends on the weight of the material and the pressure of water, u, in the pores of the soil (which may be a suction and, thus, a negative value). In fact: $\acute{\sigma} = \sigma - u$ and so in Figure 5.2,

$$\acute{\sigma} = (\gamma z - \gamma_w h) \cos^2 \beta \quad (5.4)$$

where γ_w = unit weight of water.

In soils the effect of (positive) pore water pressure is, therefore, to reduce the normal stress and thus to reduce the strength. Pore water pressure depends, of course, on various factors which are considered with regard to thawing of soils in section 5.5.2. 'Excess pore water pressure' is probably the largest single cause of landslides, mudflows and other abrupt movements in general. This is also true for the cold regions. Nevertheless, such a simple analysis is not satisfactory for most solifluction occurrences.

5.3 The mystery of solifluction

Movements by solifluction are relatively shallow – perhaps only a few centimetres, but more often extending 0.5 to perhaps 2.0 metres. In some cases the movements are reported for slopes lacking any particular micro-relief and with only the unevenness of surface which is always expected where the soil is disturbed by frost heave. Solifluction is not restricted to active layers underlain by permafrost and is common to tundra regions outside the permafrost regions.

Commonly, the displacement as a function of depth (the 'vertical velocity profile') is concave downslope (Figure 5.3) but it is occasionally convex, especially when permafrost is present. Both kinds of profile may occur in proximity. Solifluction frequently occurs on slopes of low angle – sometimes of only a few degrees. Washburn (1979) gives a good, illustrated review.

The very slow protracted movements, undisturbed vegetation and apparently stable surface forms, and the often very low slope angles characterize solifluction. In spite of the apparent stability, D. J. Smith (1987) has measured annual displacements of several cm in features similar to those in Figure 5.1. Mass movements due to excess water pressure, by contrast, tend to be sudden and disruptive, and at least fairly rapid. If slope angles are low, shear stresses are low and for movement to occur, according to equations (5.1) and (5.3), the pore water pressures have to be particularly high and, as a result, the soil quite weak and soft. Observations on a solifluction slope of 8° (Williams 1966b), where movement was recorded year by year, showed that pore water pressure was not high. The friction angle of the material was estimated at more than 30° and the ground was quite firm although wet during the thaw period. For 5 or 10° slopes, pore pressures would have to be artesian (water flowing upwards and through the surface) if they were to cause a loss of strength sufficient to cause a flow of the soil. In fact, the process of thaw is unique in liberating much water which occupies less volume than it did before (when it was ice).

The exact timing of the movements in solifluction (as defined above) has never, it seems, been established. The numerous experimental observations have always involved observations of displacements after they have occurred. Further uncertainty concerns the manner in which the soil advances, wavelike, over itself or, more precisely, over the vegetation cover. This effect gives a buried organic layer which can provide an estimate of the rate and extent of the movement (Figure 5.4). Such layers may extend many metres back and may represent vegetation overrun during thousands of years. The fronts (cf. Figure 5.1) of such 'waves', terraces or lobes often contain an accumulation of boulders and stones the reason for which is not known.

It is apparent that a number of effects related to freezing or freezing and thawing, should be examined as possible causes of solifluction.

Figure 5.3 Deformation of a plastic tube due to solifluction, after three years in an 8° slope. The horizontal marks in the excavation are at one foot (30.4 cm) intervals (from Williams, 1966b).

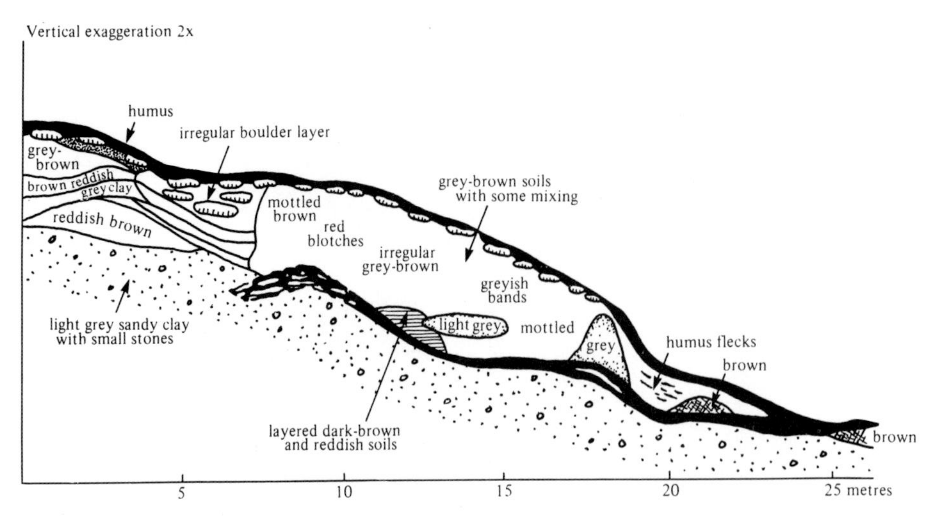

Figure 5.4 Profile through a large solifluction lobe (or 'terrace') showing a layer of buried vegetation. (From Williams, 1957).

5.3.1 Origin of small movements

A century ago Davison (1889) proposed that frost heave and thaw produce a downslope movement because lifting by heave occurs, generally, perpendicular to the slope while a dropping-back at thaw occurred vertically. The downslope component, L is given by:

$$L = H \tan \beta \qquad (5.5)$$

where H = frost heave. If the angle of slope, β, is small, the downslope movement would be small even if the heave is quite large.

Soil particles, and even small stones, lying on the ground surface can be lifted by needle ice crystals (see section 2.5). When thaw commences the particle may fall the millimetres or centimetres from the crystal to the soil surface in accordance with Davison's idea. More often it seems, the crystals bend prior to thaw, under the weight of the attached particle and the downslope displacement of the particle is enhanced – although still small (Higashi & Corte, 1971).

Davison's concept is described, quite ambiguously, as frost creep. It has been widely assumed to be relevant, not only to discrete particles on the surface, but to the near surface layers as a whole. A moment's thought, however, leads one to doubt that a soil, a continuum of adhering particles, could behave in a manner similar to single, separated particles. The settling of the soil at thaw (its consolidation) involves the particles coming together with a tendency to return to their *original position on a slope*, a fact which was demonstrated by Washburn (1979) who measured the 'retrograde'

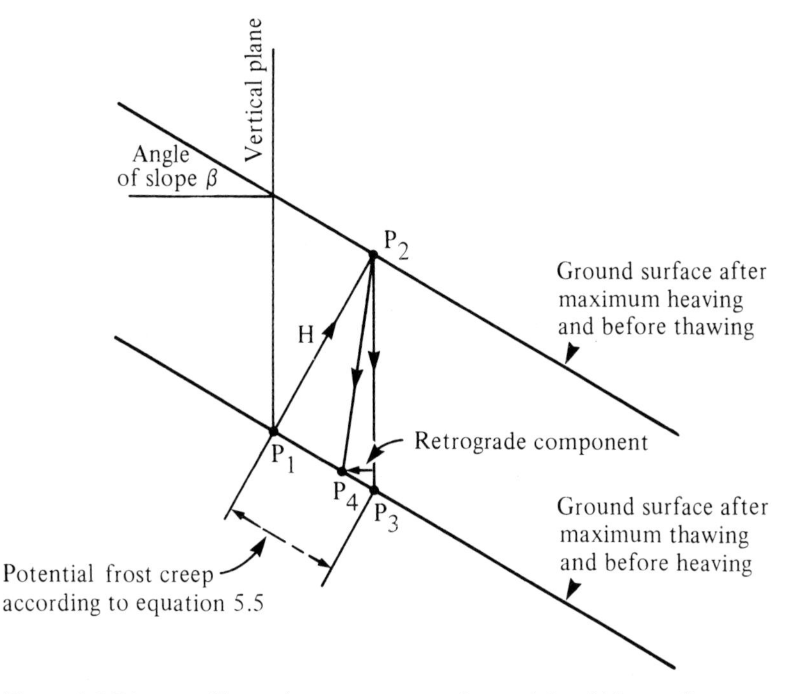

Figure 5.5 Diagram illustrating movements of a particle within a soil mass, as a result of heaving and subsequent consolidation, on a slope. Frost creep according to Davison's (1889) theory is shown as well as the retrograde movement actually occurring during thaw and consolidation. Figure modified from Washburn (1979).

movement (Figure 5.5). Davison himself commented on this possibility. Were his concept to apply exactly, the adherence of the particles to each other (the strength of the soil) would have to be effectively nil – or, alternatively, discrete masses of soil might move downslope as a whole but only if they became separated by openings from adjacent soil. The former is clearly not the case and the latter has not been reported.

It is unfortunate that Davison's concept is referred to as creep. Since his time, creep has become widely used in science and technology as a term referring to slow deformation of a material, at stresses below those necessary to provoke immediate flow or other disruption of the integrity of the material. Creep so defined, has nothing to do with lifting things and dropping them. Furthermore, it is also used in reference to soils (Carson & Kirkby, 1972), and creep of frozen soil is considered in detail in section 5.3.2 and Chapter 9. The confusion is made worse because the very situations where equation (5.5) is often invoked – solifluction movements especially – are those where creep in the proper or more general scientific sense, indeed occurs.

Observed movements on solifluction slopes are slow. Even so, the displacements over one year, or over one passage from winter (frozen) to summer (thawed), are often greater than L in equation (5.5). Washburn observed such movements which occurred just prior to (and perhaps simultaneous with) the retrograde movement, and referred to them as gelifluction. Any downslope movement occurs only because downslope stresses are greater than the resistance (in this case, a resistance against *slow* movement, which must be less than that against rapid movement).

The frost heave of the slope does little to the shear stresses in the downslope direction (as equation (5.1) shows) but the whole process of ice segregation and heave and subsequent thaw and settlement is, of course, associated with drastic changes in the soil strength. The transitory movements which result vary from millimetres to metres or more. There seems no reason why the magnitude of the displacements should have anything to do with equation (5.5).

Mackay (1981) draws attention to the concentration of ice lenses formed just above the permafrost during the late summer (see section 8.6). The melting of these lenses results in a distinct shear zone (Rein & Burrous 1980) and is apparently responsible for so-called plug flow. The greater part of the thickness of the active layer is included and the velocity profile is convex downslope. The movement of such 'plugs' (the localised shear) is probably due to transient high pore water pressures following melting of the excess ice. If the plug itself remains relatively intact this cannot be true of the neighbouring soil which probably undergoes deformation to accommodate the movement of the 'plug'. Whether or not such movements are considered as solifluction or simply a particular form of thaw slide (see section 5.5, below) is a matter of definition. Quite frequently, saturated weak masses flow some tens of centimetres per year. These are probably best regarded as mudflows due to high pore water pressures rather than solifluction in the narrower sense defined in this chapter.

In some cases, vertical or near vertical ice layers and the associated discontinuities at thaw, may cause movement of soil plugs and Mackay (1980a, 1981) suggests that annual contraction and expansion of the active layer in ice-wedge polygons (section 6.4.1) may also give a component of downslope movement.

In addition to such effects there are a variety of processes which occur on the microscopic or near-microscopic scale. These are revealed by characteristic structures observed in thin sections viewed under the microscope. So-called 'stress cutans', accumulations of fine particles above gravel-sized or other larger particles, represent a sorting process resulting from repeated freezing and thawing. Vesicles – essentially cavities formed by air coming out of solution during freezing – are quite persistent, and isolated soil particles may also be pushed ahead of ice segregations as they form. Numerous small displacements, apparently where small lenses of ice have existed and where shearing has occurred, have been observed in recently thawed soils as well in material thawed from permafrost thousands of years ago (Van Vliet-Lanoe 1985, Coutard & Mucher 1985 – see also section 2.2.8). All these features seem to be widespread and thus are a quite general effect of freezing and thawing. In soils, freezing and thawing occurs over a range of temperature, of course, and the importance of the processes within the *frozen* soil is examined further below and in Chapter 9. The features become more and more developed as freezing and thawing cycles are repeated. The processes whereby they are formed are not known in detail but collectively they constitute multitudinous small differential movements, displacements, of material.

Each such displacement, whether of a single particle relative to its surrounding particles, or of groups of particles across a small shear plane, represents an overcoming of shear resistance by shear forces. Whether the resistance is locally decreased, or whether the disturbing forces have been locally increased (as in the growth of an ice lens for example) is immaterial. What is important in the present context is that each such event gives the possibility of an, albeit very small, component of movement downslope. Resistance is overcome, so movement occurs, and the stresses responsible (whose direction will depend on their origin) *must* be modified by the ever-present component of shear stress in the downslope direction. This downslope stress (arising from gravity) has not caused the displacement, in most cases, but it must have an influence on the direction of the movement once it is initiated.

Thus one is lead to the conclusion that the microstructures produced by freezing and thawing in soils are the evidence of the nature of creep processes in such soils. Indeed, because the structures are likely to be destroyed if the soil is subjected to more widespread and rapid flow, they may be the indicator of precisely those soils where movement is, dominantly, the result of creep in the correct sense of the word. It seems that the creep will occur during and following the freezing of the soil and during and following thaw – depending on the precise nature of the microscopic displacement. Finally, as discussed below, creep occurs in *frozen* ground apparently as a result of several processes. Thus there arises the possibility of creep in solifluction features occurring, one way or another, throughout much of the year.

5.3.2 Creep in the frozen state

The term 'creep', as noted above, refers to continuing slow deformations occurring as a result of stresses less (often much less) than those necessary to overcome strength, as normally understood, that is, in the short term. Frozen ground shows 'classical' creep behaviour which resembles that of a number of common materials, and notably, glacier ice. The ice is a hard strong material yet flows, slowly, even though the slope is slight. The creep behaviour of frozen soils is more marked and extensive than that of unfrozen soils.

The creep of ice has been studied over many years, and ice-rich soil often behaves in a rather similar fashion (Morgenstern 1985). The creep properties of frozen soil are discussed in more detail in Chapter 9. McRoberts (1975) using data for ice, has calculated the movements that might be observed at the surface of slopes, for various slope angles and thicknesses of frozen soils (Figure 5.6), as a result of creep (see also section 9.8). The creep is assumed to occur at all depths so that the surface is carried along by the cumulative effect of all movements below. The thickness of material

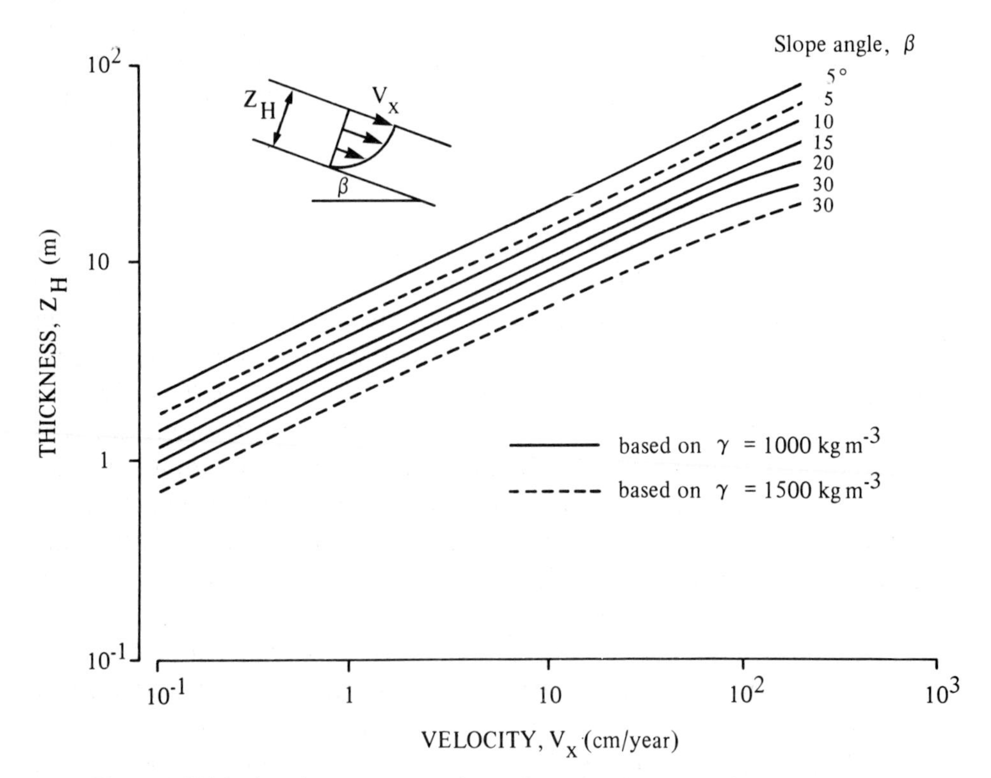

Figure 5.6 Velocity of movement at the surface of a slope, as a function of slope angle and thickness of frozen soil, calculated on the assumption that the creep of frozen soil is identical to that of ice (after McRoberts, 1975).

affected is also important because it determines the shear stresses, which increase with depth (see equation (5.1)).

The values in Figure 5.6 indicate that where there is at least several metres depth of ice-rich soil, movements at the surface similar to those observed in solifluction occur even on slopes of only a few degrees. However, solifluction usually involves only to 1 to 2 metres thickness (that is, equivalent to the height of the visible lobe or terrace front), and the rate of surface movement in Figure 5.6, for such a thickness, is an order of magnitude too small. In fact, the data on which the figure is based may well be inaccurate to that extent. McRoberts points out that rates of creep increase near 0 °C, that is, near the melting point. Solifluction commonly involves fairly warm frozen ground. Even more important, Figure 5.6 is based on observations for ice. Frozen ground is essentially at its freezing 'point' for several degrees below 0 °C – in that ice and unfrozen water coexist. Not only would accelerated creep occur to lower temperatures than in ice – the particular nature of frozen soil suggests moisture movement and regelation, and the displacement of particles augment the creep in the manner discussed in the previous section, and in Chapter 9. Thus it is reasonable to conclude that creep of frozen ground is the cause of at least some solifluction. That creep is due to the creep properties of the contained ice, as well as to processes following from partial freezing and thawing.

5.4 Rock glaciers

There is a conspicuous group of features in which creep of frozen soil seems to be the main, if not exclusive, cause of movement. This is the *rock glaciers* which, characteristically, appear as an assemblage of boulders and boulder-rich soil, but in form resemble a valley glacier or a cirque glacier. The downslope fronts may be very steep and tens of metres high. The soil material in the fronts must be strong enough to resist the high shear stresses that would otherwise cause collapse to a lower angle. Yet the surface of a rock glacier when 'active' (i.e. moving) is advancing at tens of centimetres or a metre or so per year and the movements extend to many metres depth (Washburn, 1979, reviews many observations). An extraordinary hundred metres per year is reported by Corte (1976) for a rock glacier in South America. Although shallow movements, falls of blocks, small slumps and slides may mark the rock glacier surface and particularly its front, there seems no doubt that the creep of the frozen core is the main source of the movement. The core may contain large more or less clean ice masses, or alternatively the ice may occur in the soil pores only. Active rock glaciers have mean temperatures between 0 °C and −1 °C or −2 °C, and rock glaciers are rather common in middle latitudes. This temperature in the frozen core favours, of course, movement by creep. Inactive rock glaciers (which may closely resemble active ones) have presumably ceased to have a frozen, or perhaps a sufficiently ice-rich, interior. The origin of the material in characteristic rock glaciers is often in doubt. Some rock glaciers may indeed be the rocky remnants of a heavily loaded true valley glacier; perhaps more often the rock glacier has been supplied with boulders and soil material originating on adjacent valley sides with the ice content coming from water flowing in sub-surface drainage paths.

Features resembling rock glaciers although of different origin are ice-cored terminal moraines (Østrem 1963a), which are recognised by the waviness or wrinkling of their surface when viewed from the air. The ice core originates by the burial of snow or occasionally, glacier ice, under rock debris (moraine material) at the glacier margin. It is reasonable to assume that creep occurs in these features as well. They might be regarded as a special form of rock glacier. Apart from obvious fossil rock glaciers, many accumulations of boulders are found in mountainous regions which, in their alignment and situation, seem likely to have moved downslope at some time in the past. Sometimes water flows through such accumulations and, conceivably, even annual freezing could give sufficient accumulation of ice that creep and thus displacement of the boulders occurs. Some rock glaciers show recent rapid movements, restricted to a part of the feature, which are apparently due to local thawing and weakening (Johnson 1983). It may be that flowing water together with impeded drainage due to ice masses is responsible, and leads to local high pore water pressures. The stresses on the frozen mass may be increased as a result of the transfer of stresses from such weakened material. When parts of a slope are weakened, the total downslope forces must be carried by the parts which remain intact. Accordingly, the creep of the frozen material may be accelerated as well.

5.5 Effects of thawing: landslides and slumps

Characteristic thaw landslides are shown in Figure 5.7. The exposed surface of shear on the upslope edge, the saturated and semi-fluid nature of the moving masses and the disrupted vegetation means that the features are immediately recognisable to anyone even remotely aware of the nature of landslides. There is a variety of other forms referred to as mud flows, slides and slumps in which similarly the effects of excess water and high pore water pressures are evident. Such forms are quite distinct in appearance from the slower more stable solifluction type of movement (which quite often goes unrecognised by those unfamiliar with terrain in cold regions).

Figure 5.7 Landslides due to thawing of permafrost (foreground). The presence of numerous landslides of this type, together with many water-filled depressions, characterises thermokarst (from Mayo, Yukon, Canada – Photo: C. R. Burn).

Usually the flow is limited in depth by the presence of frozen ground; it can involve metres of displacement over only a few days. Only rarely are the slides deep-seated such as to involve shearing of the frozen ground itself (Isaacs & Code 1972, p.155; Mackay, J.R. & Matthews 1973).

There is, however, a range of size and form, and especially in association with patches of late-lying snow, terrace-like features occur which are saturated with water during the thaw but move only centimetres or tens of centimetres downslope per year. There is uncertainty as to whether the latter features are fully explained by the effects of elevated pore pressure following from rapid thawing of the soil and ice coupled with loss of cohesive strength, or whether they involve other mechanisms associated more properly with solifluction. Movements of several centimetres during spring, on a hummocky sparsely wooded slope with a complete vegetation cover were observed by Wu (1984). The pore pressures during thaw varied from point to point, and, because the slope was quite steep (30°), were sufficient to explain the movements. Pore pressures depend of course on the level of water in the ground (the relative position of the water table (Figure

5.2)) and are increased if the water is entrapped so that the weight of overburden bears on it.

In general, tongues of soil and boulders, if elongated downslope, and if the slope is steep, are the result of more rapid movements. These movements are also due to weakening caused by high pore water pressure associated with excess moisture from melting ice inclusions and snow. *Avalanche boulder tongues* are somewhat similar in form but consist of material entrained by snow avalanches.

As in other climates, landslides and mud flows occur on valley sides, often following erosion by water at the foot of the slope (Chapter 8). The removal of such basal support and steepening of the slope increases the shear stresses. In the case of thaw landslides overlying permafrost, initial movements, destroying or removing the vegetation cover, modify the energy exchange of the ground surface and result in accelerated thawing and an extension of the affected part of the slope. The presence of permafrost or, in the absence of permafrost, of the still-frozen deeper parts of the seasonally frozen layer, in the spring or summer, retards drainage of melt water – whether that originating from ice in the soil or from snow or ice on the surface.

Such retarded drainage and the supposition that the high water content material would necessarily be weak or fluid constituted for many years the generally accepted explanation for most mass movements in periglacial regions. During the late stages of seasonal thaw, in the absence of permafrost, frozen ground is discontinuous and itself permeable and thus does not represent a complete barrier (Kane & Stein 1983b), but it is clear that the drainage is commonly reduced and modified. Flow of water parallel to the surface is promoted by the particularly high permeability of thawed soils in that direction, following from the orientation of the melted ice lenses. It is not the state of saturation, but the pore water pressures and consequent effective stresses which determine whether rapid movement occurs. These are also dependent on the rate of thaw (considered in section 5.5.2).

5.5.1 Loss of cohesive strength

Apart from the high pore water pressures which may be produced, freezing and thawing has other radical effects on the strength of the newly-thawed soil. The term C in equation (5.3) refers to cohesion, which arises because of forces of attraction between adjacent particle surfaces or between particle surfaces and cations and adsorbed water molecules, with or without the added effects of cementing materials. Cohesion has the effect of 'stickiness'. It depends on the closeness of the packing of the particles as well as the mineralogical, chemical and other conditions, including the degree of consolidation of the soil. When frost heaved, soils have a greater void ratio (bonds have broken and surfaces have separated) and this tends to persist through thaw. Of course, so long as a continuous framework of ice persists, this gives a significant strength which is largely cohesive (see section 9.2). The unique looseness and relative weakness of soil following complete thaw is well-known and is an advantage in agricultural practice. This loss of strength may, especially if the slope angle is steep, be great enough that downslope movement occurs. High pore water pressures are then not the primary cause for the movement, although they contribute towards it.

The loss of cohesive strength in clay soils occurs in spite of a significant consolidation of the individual clay layers or aggregates that lie *between* the layers of ice (see below, section 5.5.2). The sites of ice layers become discontinuities along which there is no cohesive strength. If these discontinuities are abundant and continuous the soil in bulk behaves as through lacking cohesive strength.

Clay-rich and therefore strongly cohesive soils are uncommon in present-day cold regions but soil deposits believed to have moved during earlier cold periods are frequently clay-rich (C. Harris, 1981). These are examples of *cryoturbates* (section 2.2.8), which are characterised by 'turbulent' flow patterns in profiles. Cryoturbation may also occur on level ground in certain forms of patterned ground (section 6.3.2). Clays, in general, are more likely to become unstable the greater their thickness on the slope. Hutchinson (1974) sought to explain depths of over two metres of what he called 'solifluction' material, on low-angled slopes, in England, in this way. His analysis assumes such depths of active layer occurred, in the past, and that the newly thawed material would have been similar to mechanically remoulded, or disturbed, 'undrained' clay. The condition implied, that of water effectively trapped in the saturated material, is one of high pore water pressure. Such a condition is known to be the cause of slow mudflows, in non-freezing situations, in which a reduced, residual cohesion is also significant.

5.5.2 Soil consolidation and strength during thaw

When uniformly coarse-grained soils thaw the water can drain relatively freely. In any case there will be little or no segregation ice in such soils. Unless other forms of excess ice are present little volume change occurs and at least after a small amount of drainage the pore water pressures approach equilibrium with the groundwater and hydrological environment. There is little loss of strength and little movement of soil downslope. Indeed, areas with such coarse soil material usually lack any indications of movement.

The behaviour of fine-grained, that is, frost-susceptible soils on thawing is much more complicated. These are the soils which show frost heave and ice segregation, and in which there is also considerable unfrozen water to several degrees below 0 °C. The volume reduction and settlement and eventual loss of strength which occurs on thawing involves the process of *consolidation.*

This term refers to the loss of volume of a soil by loss of water and an attendant decrease in the size of the soil pores. The soil minerals themselves are not compressed. The term was applied originally only to unfrozen soils. The weight of overburden or of structures on the ground can lead to consolidation as does drying of the ground. Unless the process is fully understood and controlled, damage occurs to buildings, highways and other works. The volume decrease of a compressible ('consolidatable') soil occurs until any component of pressure in the pore water due to the weight of overlying soil material or surface structure is dissipated, that is, until the pore water has come to equilibrium and water ceases to move out of the consolidated material. More specifically, the amount of consolidation (measured by the voids ratio: (vol. of voids/vol. of solids)) depends on the effective stress existing when the consolidation is completed. Consolidation is discussed in basic soils engineering texts, and some texts on soil physics, for example, Williams (1982). The consolidation associated with the thawing of soils is, however, unique in several ways.

The freezing of a soil produces an effective stress and consequently a consolidation of the soil matrix. This effective stress follows from the suction developed in the pore water (see sections 7.3, 7.4) and the elevated pressure of the ice. Frozen, compressible soils consist, therefore, of markedly consolidated aggregates, that may be microscopic, or perhaps finger-sized nodules (Chamberlain & Gow 1979) of various shapes, separated by the lenses or other bodies of ice. These structures are discussed further in sections 2.2.8 and 2.2.9. If frost heave has occurred, there will be excess ice.

On thawing, when excess ice is present, there will be water liberated which, unless it drains as fast as the ice melts, will come to bear part of the weight of the soil above. The water pressure rises as a consequence, the effective stress decreases and there is a loss of strength of the soil mass (see equation (5.3)). This can occur even without excess ice, because the consolidated aggregates do not reabsorb all the water removed from them.

The loss of strength (if any) will depend on the rate of water liberation (from the thawing ice) and the rate at which this water drains away. The latter is determined by the rate of consolidation of the saturated soil and the permeability of the soil. In an attempt to quantify these effects, Morgenstern & Nixon (1971) proposed the *thaw–consolidation ratio* R:

$$R = \frac{a}{\sqrt{C_v}}$$

a is a constant characterising the rate of thaw. The thickness thawed χ equals $a\sqrt{t}$ (the depth of the thaw increases proportional to the square root of time t – see section 4.4.3). C_v is the consolidation coefficient and

$$C_v = \frac{k}{\rho_w m_v},$$

where k is the permeability and m_v the coefficient of compressibility. The latter is the compression of unit thickness of soil due to a unit increase of pressure. The concept of the consolidation coefficient is quite basic in soil mechanics (see e.g. Terzaghi and Peck, 1967).

Because the pore water pressures developed during thaw are a function of R, so is the maximum angle a slope may have. In Figure 4.8 (from McRoberts & Morgenstern 1974), the maximum angle is shown as a function of R and of the water content of the soil. The calculations were made using the ideal slope concept described in section 5.2.

Thawing soils are, however, very different from those soils which are normally the subject of consolidation studies in conventional engineering practice. The compressibility coefficient C_v may be hard to assess for thawing soil. The value of a is dependent on many site factors and is correspondingly difficult. The concept of the thaw consolidation ratio nevertheless provides an instructive theoretical approach, which, with further knowledge of the parameters involved, may allow reasonable analyses for predicting the stability of specific slopes.

The greater the permeability, k, the more rapid the drainage and this tends to hold the excess pore water pressures down, and thus soil strength is retained. A high compressibility m_v, on the other hand, has the opposite effect, as would a high value of a (rapid thawing). The process of thaw and consolidation exerts a control on the pore water pressures which is unlikely to be fully accounted for when the parameters k and m_v are determined. So long as a contiguous ice structure is present it will resist consolidation and

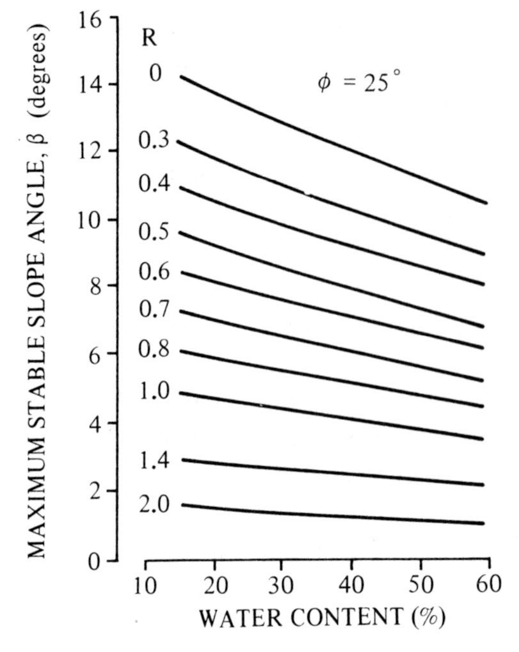

Figure 5.8 Maximum slope angles of slopes for various values of thaw-consolidation ratio R, as a function of water content (modified after McRoberts & Morgenstern 1974).

carry weight which might otherwise be transferred to the water; in addition, the volume decrease of the ice as it melts causes a drop in pore pressure, in so far as the containing matrix cannot contract. It is only when ice has disappeared, or at least is in small, discrete masses that consolidation can proceed in a more or less 'normal' fashion. An additional complication arises because of the suction of the soil water in contact with ice at any temperature below 0 °C, according to the thermodynamic phase relations discussed in Chapter 7. Finally, although the aggregates are over-consolidated, they are collectively loosely packed. Consequently, when there is little or no excess ice, the water liberated on thawing may be barely sufficient to fill the voids, and rather than raised water pressures, the pressures may be less than for the same material prior to freezing. This appears to be the effect noted by Nixon & Morgernstern (1973) and referred to as 'residual stress'.

In summary, the total consolidation on thawing consists of: a volume decrease because of the smaller specific volume of water relative to ice; a possibly very large volume decrease due to melting of ice inclusions and drainage (this component of consolidation can be estimated from the amount of such ice present, and is responsible for most of the settlement of the ground); and a further consolidation in which the heavily consolidated flakes or individual nodules become more closely packed, this depending on the coefficient of consolidation of the particular soil. Both the rate of thaw and the soil properties related to consolidation determine the stability or lack of stability of thawing, frost-heaved materials on slopes, and thus whether rapid slides or flows will occur in addition to the immediate effects of subsidence.

5.5.3 ***Thermokarst***

Terrain which has characteristic forms resulting from the thawing of permafrost with excess ice is called *thermokarst.* The name is derived from the Slav word, karst, which refers to the effects of dissolution of limestone rocks, that is, pits and depressions of various kinds, caverns and other voids, all of which give conspicuous surface relief in typical karst terrain.

Thermokarst can follow simply from a deepening of the active layer but usually thawing continues below this, that is, there is thawing of what has become relict permafrost. Thus there will be a layer of soil above the permafrost, which remains unfrozen year-round. If very large ice bodies are present the thawing of permafrost will inevitably result in extensive subsidences of the ground. The liberated water and attendant loss of strength leads to movements of more soil than that directly involved in the subsidence. The perimeter of the subsidence will slump progressively, thus extending the area affected (Figure 5.9).

Thermokarst depressions are usually water filled and may become small lakes. The lakes may develop an elongate form and uniform orientation as a result of wind-produced water currents which, moving onto the shore, cause additional thawing. Thermokarst occurs in areas of low relief, primarily because the characteristic ponds and lakes could not persist on sharply sloping ground. Some believe mud flows and similar features caused by the thawing of permafrost in slopes, should be considered thermokarst. Perhaps ponded water in the vicinity is necessary to justify using the term. Of importance, however, is that thawing of ice-rich permafrost and the ensuing consolidation is arguably the biggest single effect permafrost has on the form of landscapes. Consequently, it is convenient to use the term thermokarst fairly broadly to cover terrain *conspicuously* affected by 'thaw settlement'.

'Thaw settlement' is of course, ubiquitous, with thermokarst being only the extreme natural expression. It is also the basis of most geotechnical problems in the cold regions even though the *direct* effects of frost heave (rather than of thaw) may be more intractable. Any permafrost with excess

Figure 5.9 Thermokarst depression.

ice may produce thermokarst. Thawing ice masses from a recent glacier may give kettle holes and perhaps extensive associated 'dead ice' topography (Mannerfelt, 1945). This might also be regarded as thermokarst. But it is worth noting that glaciers may extend far beyond the limits of permafrost, and such topography is not necessarily indicative of permafrost more generally.

The amount of relief (the depth of subsidence) developed in thermokarst depends on the amount and distribution of the excess ice and on the thickness of permafrost thawed. It is difficult to predict the thaw and settlements from limited soil sampling.

Moving water, whether in wind-blown lake currents, a river or stream, results in rapid thawing of sediments over which it passes (Chapter 8). Conductive–convective heat flow between the turbulent water and the sediment is effective. Simultaneous erosion of the banks with loss of vegetation and movements of the soil, are also important. The perhaps questionable terms, *thermal erosion* and *thermal abrasion* (the thermal effect precedes the erosion), refer to these processes. Because there is a removal of sediment by moving water, gullies develop from small streams, and as they enlarge the slumping of saturated, thawed material becomes a dominant effect. New layers of frozen ground are exposed to thawing. While some authors refer to an absence of drainage in thermokarst – the ponded water is in sharp contrast to the dry pits and depressions of limestone karst produced by downward infiltration – the effects of surface runoff are a conspicuous element in thermokarst. Even on slightly sloping ground water courses over permafrost tend to enlarge, sometimes dramatically, into gullies and the network of drainage channels is extended.

The nature of thermokarst is modified by the disposition of the excess ice. Thawing of ice wedges in polygons results in a network of ponds or channels with *thermokarst mounds*. The more or less regular arrangement has led to the term 'cemetery mounds' – although they may be larger than the name suggests. In all cases, submergence, even in small ponds, accelerates thermokarst development by its effect on the energy-exchange at the surface.

The thawing giving rise to thermokarst is commonly initiated by microclimatic change, that is, a local change in the nature of the ground surface which leads to warmer temperatures beneath (Chapter 3). A warming of the atmospheric climate (climatic change), on the other hand, can lead to thawing of permafrost over wide areas. Even then, locally (microclimatically) warmer permafrost will thaw first. Much of the permafrost over thousands of square kilometres is presently at temperatures close to 0 °C and is at risk of thawing, if the current small increases in world temperatures (mainly due to rising carbon dioxide concentrations, Davies, 1985) continue as predicted.

Nevertheless, such microclimatic factors as killing or removal of vegetation, or the presence of standing water, can cause a more sudden and striking development of thermokarst. Destruction of vegetation for roads or foundations, or by careless passage of vehicles over natural surfaces, is the dominant cause of anthropogenic thermokarst which develops in the course of a few years. The natural succession of vegetation and the (perhaps cyclic) development of surface landforms as in the development and decay of ice wedges (French, 1976) also produces thermokarst. Perhaps tens or hundreds of years later refreezing occurs, with renewed development of frost-heaved land forms. Pingos form in infilling lakes – lakes which can be of thermokarst origin themselves.

The extent of thermokarst in natural terrain (undisturbed by man) varies greatly. The biggest factor is probably the amount of excess ice. In the Soviet Union those regions having the greatest susceptibility to the thermokarst development following anthropogenic disturbance are in the far northern, colder regions, according to Grave (1983), and this would apply too, to natural thermokarst. Typically, thermokarst would be expected towards the warmer limits of permafrost, but also in Canada very conspicuous thermokarst occurs in cold places (for example, Banks Island, French, 1974). There seem to be several reasons. If the permafrost is close to the ground surface the excess ice will be, too. When it thaws it will do so more rapidly on this account (the rate of penetration of temperature decreases with the square root of depth), and the subsidence is more marked and the effect more immediate at the ground surface. There is also the 'chasing' effect – thawing proceeds deeper as a direct effect of the ground surface itself moving downward (by subsidence). This too is more marked when the initial thawing of permafrost is close to the ground surface. Furthermore, the relatively sparse and fragile vegetation in the very cold regions may mean that ground surface disturbance has a particularly large effect on the ground temperature.

Grave (1983) considers that seven percent of the permafrost region in the Soviet Union is 'extremely sensitive' to thermokarst development. Not only will thermokarst be prominent in this area, producing substantial terrain features, it will cover much, probably well over half, of the land surface. Elsewhere thermokarst will be more occasional and vary in its degree of development. Although it is not clear why, a smaller area of the North American permafrost appears to be in the 'extremely sensitive' category with respect to the development of thermokarst by natural factors. There are, however, many areas tens of square kilometres in extent (for example in the Mackenzie Delta) where melting of ice is producing characteristic thermokarst.

Although thermokarst is the result of thawing of permafrost and thus of a disturbance of the thermal regime, the characteristic landforms are likely to persist, even if there is some reestablishment of cooler conditions. This persistence follows from the nature of the features, and the thermal regime they establish. Water bodies cause the underlying ground to become warmer than elsewhere because of the heat exchange processes associated with them. Other features such as gullies and the associated mass movements also tend to be self-perpetuating – because the soil movements and the absence of vegetation on such features also results in deeper annual thawing. These effects probably explain the extent of thermokarst and the fact that such terrain occurs also in regions where the contemporary climate is cool and not apparently undergoing significant warming.

5.6 The extent and variety of movements on slopes

Deformation of the ground when frozen does not appear to constitute a large element of the sum total of downslope movement, although it is of great geotechnical significance. Small deformations of frozen ground can be responsible for large stresses, for example, on buried pipelines or other sensitive structures. Even more important is the control the deformation properties of frozen ground exert on the process of frost heave and ice segregation. The deformation of frozen ground is so distinct from that of all other ground as to require a special chapter (Chapter 9). But its direct effects are, with a few exceptions, not very apparent in the natural terrain.

Certainly, movements arising from completed cycles of freezing and thawing are far more evident in naturally occurring downslope displacements and instability, and more significant for denudation in a geological sense. The several unique effects of freezing and thawing have their origin in the ice segregation process. The structural changes and microscopic displacements of particles that give rise to the creep of soils exposed to freezing and thawing are distinct from the loss of frictional strength that follows from high pore water pressures. Yet these two effects may occur in one and the same soil mass although not, of course, at the same time. Consequently, some of the surface features, the terrace or lobate forms, that result may have characteristics associated with both processes. The interparticle displacements occur primarily during the freezing of the mass even though the resulting downslope displacements may only be completed on thaw. High pore water pressures on the other hand, are only fully effective on thaw, when the strength of the ice skeleton has been lost, and will be dependent on the amount of ice available to thaw as well as the rates of thaw and the consolidation characteristics. Loss of cohesion on thawing, where particles have been separated by segregation ice, seems inevitable. However, soils subject to repeated freezing and thawing are unlikely to have a large component of cohesional strength at any time, in the unfrozen state. Movements due to high pore water pressures and to loss of cohesion commonly involve displacement along shear planes, although a viscous type of flow also occurs frequently.

The importance of one process compared to another, varies greatly from one terrain to another. In an outstanding study of a small alpine area in Swedish Lappland, Rapp (1960) showed that movements of soil in landslides, mudflows and other forms of relatively rapid movement – which are the result of strength loss due to high pore water pressures and, to some degree, loss of cohesion – exceed the displacements due to solifluction by a ratio of twenty to one. Rapp's study considered the masses of the materials involved and the vertical component of downslope movement per year. Solifluction was far more widespread but the depth involved was relatively slight and, more important, the rate of movement of material was so slow.

In low-lying regions of ice-rich sediments characteristically the site of thermokarst development – a quite different environment – there is usually an even greater preponderance of movements due to high pore pressures. The landslides and mudflows are to be analysed in terms of the thaw and consolidation rates and the instability arising from insufficient rate of drainage. In regions of greater relief and without such large bodies of excess ice, especially if permafrost is absent, the role of landsliding, that is, of rapid, pore pressure induced movements, is comparable to that of analogous effects in many temperate climates, only the origin of the entrapped water being different. But in the thermokarst regions the entire terrain is dominated by mass movements of this kind, abundant because of the water being released from the melting of the large quantities of excess ice (Figures 5.7, 5.9).

The importance of these phenomena in the cold regions can easily lead to other processes of slope denudation being overlooked. Jahn (1975), and Lewkowicz (1983) examined the rates of sediment transport by oversurface water flow. Paucity of vegetation in the high Arctic and restricted infiltration may make the surface susceptible to erosion by raindrops and 'sheet-flow'. Lewkowicz found that the concentration of dissolved materials in water flowing over the surface, although varying significantly, tended to exceed that of suspended sediment by an order of magnitude, and on average, represented a denudation of a centimetre or so per thousand years. This finding supported Rapp's (1960) much-discussed observations of solution being the biggest single mechanism of denudation.

Rapp (1960, 1985) describes well a larger-scale erosion and transport by running water which occurs especially during exceptionally heavy rains. On steep slopes, long narrow tongues of debris extending downslope are characteristic features of the mounts of Lappland, and of Svalbard (Larsson, 1982). Although extreme rainfall is the main agent, elsewhere rapid snowmelt is important. Overland flow is sometimes restricted to saturated materials downslope of late-lying, melting snow. Gullies may be formed at the same time where the water is not already laden with debris. Gulley erosion can occur also in association with rapid degradation of permafrost. It is promoted by loss of vegetation due, for example, to human activity such as highway construction or pipeline burial (Williams, 1986). The surface run-off responsible for the formation of such gulleys, follows from saturation associated with retardation of infiltration due to underlying frozen ground, whether seasonal or permafrost.

Restricted drainage is important too, in *string bogs*. These are sinuous ridges of peat and vegetation (Figure 5.10), a metre or so wide and tens

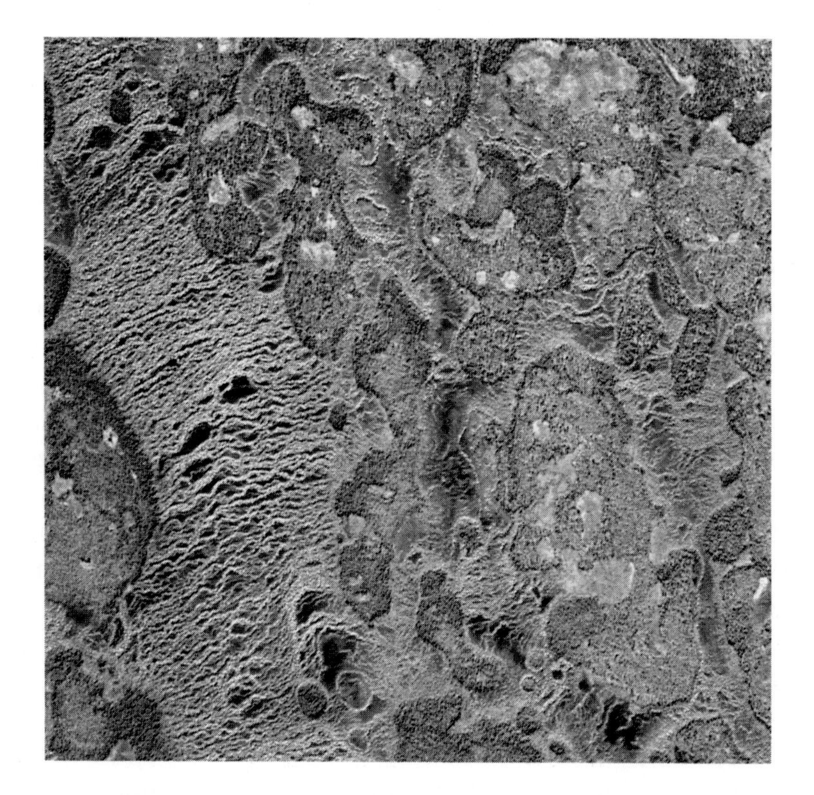

Figure 5.10 The pattern of more or less parallel sinuous threads is characteristic of string bogs, seen from the air (Photograph Crown copyright).

of metres long, separated by a few metres of ponded water. They are common in the extensive mires or peat bogs of Finland, and Ungava-Quebec (such terrain is known as muskeg in Canada). String bogs are restricted to cold regions (although apparently, not necessarily with permafrost) and the strings occur along the contours of slightly sloping surfaces. They are believed to be formed by stresses arising from pressures of the ponded water, but with lateral ice pressures, buoyancy of the peat, and ice segregation also being involved (Seppälä & Koutaniemi, 1985). The natural succession of plant species may be another factor as the drainage conditions of a particular string develop. Measurements by Seppälä & Koutaniemi showed movements sometimes of tens of centimetres per year to occur in various directions, including upslope. Thus there is no steady component of downslope movement comparable to that occurring in solifluction.

In mountainous regions, snow avalanches may carry large amounts of mineral soil and the deposited matter is often conspicuous in the terrain. Avalanche boulder tongues are long, downslope ridges terminating in a boulder-rich and somewhat wider front. Avalanches follow the same path repeatedly. Those occurring in the spring, so-called slush avalanches (Rapp, 1985), are particularly liable to entrain large amounts of soil and rock.

Rockfalls are often seen and heard in mountainous regions. The fall of even small boulders poses problems for highways and railways. Many falls occur during the spring thaw (Bjerrum & Jörstad, 1966) when the cementing effect of ice is lost. Fracture of the rock occurs during winter in such cases. Many falls also occur in the autumn. The classic view is that water trapped in cracks and other openings freezes and the expansive effect causes pressures sufficient to rupture the rock.

The situation is analogous to the bursting of a frozen water pipe and the maximum pressure which can be generated is given by the freezing point equation, equation (7.5). If the rock (or ice) does not yield and the water does not escape, then pressure develops because of the volume increase of water on freezing. The pressure lowers the freezing point in accordance with equation (7.5). Conversely, as the temperature falls the pressure rises, by 1.3×10^4 kPa per °C and when the temperature falls some fraction of a degree, rupture may occur.

Alternatively, the disruptive mechanism may include ice segregation (frost heave) with the rock itself acting as the porous substrate and with a consequent flow of water to the developing ice mass. The heave of bedrock (section 6.3.1) is a result of the same process, which is governed by equations (7.8) and (7.9). This will only occur where rocks are appropriately porous and wet.

There is a greater diversity of features produced by effects of cold climates than can be described in this chapter, which has concentrated instead on the processes causing movement down slopes, particularly those originating in the properties of the earth materials themselves. It may seem that there is a bewildering complexity of processes involved and, indeed, our understanding of them is far from complete. The effects of the various processes are modified by the environment in which they are operating. The variation and distribution of forms of instability is thus a result of the various topographic, climatic, biotic and other environmental factors, operating in the diverse terrains of the cold regions.

6

The forms of the ground surface 2: structures and microtopography of level ground

6.1 Surface characteristics

Explorers of the Arctic tundra regions in the nineteenth century were impressed by strange, orderly geometric formations on the surface of the ground. They came upon expanses of circles free of vegetation, upon fields of half-metre high hummocks 'resembling flocks of resting sheep', and sometimes on vast networks of boulders in interconnected circular or polygonal arrangements. The latter were on a scale that, on occasion, could conceivably have been man-made and sometimes were believed to be so. These structures, and others on a similar scale which were restricted to sloping ground, were striking and unique features of the tundra parts of the cold regions, as conspicuous as the treelessness itself. Many merely descriptive accounts of such *patterned ground* were published even well into this century. Troll (1944) listed some 1500 articles, and, although it was already obvious that freezing and thawing was responsible, the precise mechanisms were often obscure. A variety of theories, sometimes fanciful, had been developed, yet with little detail and little concrete evidence even of rates of formation let alone of the physical and mechanical processes involved.

Washburn (1956) developed a classification laying emphasis on geometric form, and whether sorting occurs to give accumulations of uniform grain size. In his more recent, well-illustrated review Washburn (1979) describes the geomorphological processes which may have a role in formation of patterned ground. There remain complex unanswered questions of mechanics and thermodynamics, however.

In so far as patterned ground is unique to the cold regions, its many forms must originate from freezing and thawing – indeed because superficially similar (but genetically quite different) patterns are found in warm lands, it is desirable to define patterned ground as resulting from freezing and thawing. The term is not usually applied to forms of soil disturbance which, although they may be continuous with specific forms of patterned ground on level surfaces, are modified by the effects of slope. It is clear though that processes important in the formation of patterned ground will often occur on sloping ground – indeed they may provide an embellishment to the surface forms characteristic of downslope movement described in the previous chapter. For example, solifluction terraces sometimes show a sorted surface layer of small stones. Thus many of the effects discussed in the present chapter may contribute to slope-forming processes.

6.2 Features characterised by accumulation of ice

The dimensions of patterned ground features vary greatly; there may be multiple occurrences of patterns repeated on a scale of decimetres, or sometimes each 'unit' is several, even tens, of metres across. Some forms occur individually and, in this category, there are several which, although striking to see, are little more than the surface expression of the accumulation of ground ice. This may be segregation ice or intrusive ice or a combination, as is the case in the pingo. If pingos be considered a form of patterned ground, they are certainly the largest.

6.2.1 Pingos

Pingos often look like small volcanoes and may be 50 metres in height. They are distinct from the often ridge- or mound-shaped hills known as palsas which are rarely more than one to three metres high. Both are the result of perennially accumulating ice masses but the circumstances are different.

Pingos are considered in two groups. The 'Greenland type' (Muller, 1963) are found, characteristically, at the foot of a slope within which water flows. The passage of the water is hindered by a near-surface frozen layer and this leads to artesian pressures. These, in association with the expansive pressures of the growing ice body, are believed responsible for growth of this type of pingo. Liestöl (1977) describes such pingos in Spitsbergen and the numerous associated springs. Much of the trapped water there originates under glaciers and icings are common in the winter at this location.

The 'Mackenzie Delta' type of pingo occurs in low-lying ground on ponds or lakes which have been more or less infilled with sediments or have drained following erosion of an outlet. Mackay (1979) has reviewed this type comprehensively. The original water body must have had a size such that permafrost was absent beneath (see section 4.5). As the pond shallows, the ground beneath cools because of the changed microclimatic conditions. The encroaching frozen ground extends around the unfrozen ground below the lake and if the soil materials are coarse-grained, a displacement of water ahead of the frost line occurs. This follows from there being essentially no frost heave in coarse material. The 9% volume change on freezing is accommodated by extrusion of water from the pores, in such materials, with the water tending to accumulate in pockets or lenses (Figure 6.1).

If there are scattered layers of fine-grained material present as well, these, on the contrary, *will* be sites for ice segregation (frost heave) to occur. Such segregation will be aided by the relatively high pore water pressures of the water extruded from the coarse sediments. The two effects combine to raise the ground surface to the considerable heights found. The pressure exerted by the ice is equal to the elevated pressure of the extruded water plus the increment $P_i - P_w$ by which the ice pressure exceeds the water pressure. The latter depends on temperature (see section 7.6). The artesian pressures observed in 'Mackenzie Delta' pingos (Mackay, 1978a, 1979) are by themselves insufficient to explain the pressures required for uplift of the soil mass and the bending of the frozen surface layer. Even though the deformation occurs slowly and therefore provides relatively little resistance (see Chapter 9), the forces generated by frost heave seem to be essential in most cases. Occasionally however, the surface of a pingo rises and falls ('pulsating pingo' – Mackay, 1977b) with the pressure of internal lenses of water, the water periodically bursting through the sides of the pingo. The top of a pingo can crack ('dilation cracking') with subsequent infilling by ice from surface water. Ice wedge ice may also be present, dating from before the growth of the pingo. The various forms of ground ice can be distinguished by grain size and crystal orientation, and inclusions (Mackay, 1985*a*).

Pingos may grow in height by decimetres or even a metre per year. They are widespread where permafrost is normally present, sometimes widely spaced (see maps in Washburn, 1979) and sometimes close enough that one may be seen from the next. Often the top is eroding. There are many reports of 'collapsed' or fossil pingos (French, 1976). The melting of the ice core may result from the changing microclimate of the fully-grown pingo (the top of a large pingo is often eroded) or, presumably, may follow from general changes of climate.

6.2.2 Palsa

The much smaller palsas have a core of ice-rich frozen soil and are characteristically found in areas of discontinuous permafrost. (Seppälä, 1979, 1986). A palsa* is usually surrounded by permafrost-free ground and also may occur individually or in close proximity with many others. Palsas

* We follow Seppälä's (1972) proposal for palsa as the singular, and palsas as the plural, for the English forms of this Swedish and Lapp word.

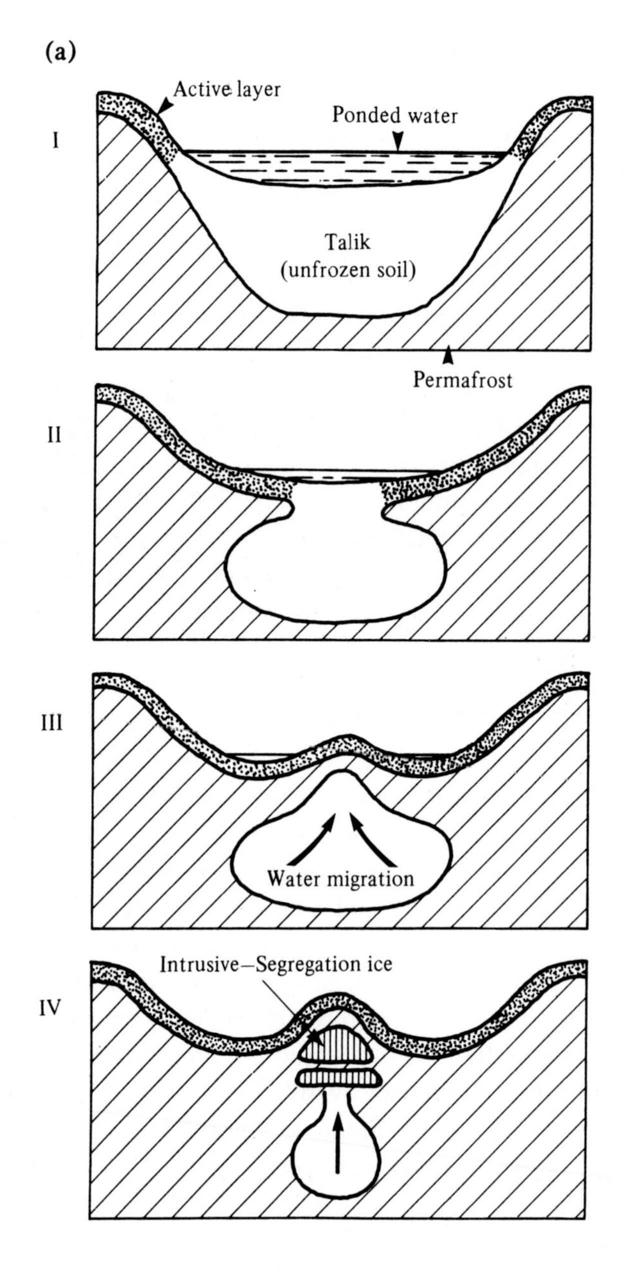

Figure 6.1 (*a*) Stages in development of pingo (after Kudriavtsev, 1978). As the body of unfrozen soil below the shallowing lake freezes, the rising water pressures within it feed a core of segregation ice. (*b*) Ice core exposed in eroded pingo. The ice (which has apparently been tilted after formation) shows bands which, according to Ross Mackay (personal communication), who took the photograph, probably represent annual growth layers.

Figure 6.1 (*cont.*) (*b*)

vary considerably in form and size, frequently being only 1–2 metres in height but occasionally several times that, and more or less round, or elongated with a ridge-like or mound form (Figure 6.2). Sometimes they extend laterally to produce a peat plateau which may cover hundreds of metres. They are usually composed largely of peat but examples occur with mainly mineral soil, with or without a substantial surface organic layer. The elevated surface has a thinner snow cover, allowing greater winter cooling, while in summer the surface material (especially if organic) will dry out and provide thermal insulation. Thus the interior temperature is lower than that of adjacent ground. In fact, the core of the palsa is an isolated, small body of permafrost. It contains segregation ice, the heave being responsible for the height of the palsa. Even though the palsa may appear to be composed of peat, sufficient mineral particles (a small layer of silt for example) must be present, as ice segregation does not occur in pure peat. In fact, pure peat is used as a foundation material in Scandinavian railways because it does not show frost heave (Skaven-Haug 1959).

The initial development of the palsa is probably due to an unusually thin cover of snow. Experiments by Seppälä (1982) showed that the ground below an artificially cleared patch, froze nearly twice as deep and remained frozen through the following summer with a persistent 'bump' of some 10 cm due to heave. The heave doubled in the next year. It seems that the 'unusual' cooling (which could occur naturally with lower snow fall,

Figure 6.2 A palsa – in Suttisjoki region. Finnish Lapland. Photo: Matti Seppälä.

unusual drifting or other effects) has to be sufficient to give a large-enough initial bump. Changes in surface moisture and vegetation will then be such as to tend to preserve the newly-formed permafrost. Seppälä's experiments showed the *Carex* sedge and *Eriophorum* died and Sphagnum mosses began to do so, during the first season. These are replaced by the shrubs and lichens which distinguish the palsa surface.

The process of formation of palsa is an interesting demonstration of the importance of surface conditions in determining ground temperature. The extensive distribution of palsas also illustrates how large areas of the earth's surface are locally prone to formation of permafrost, or equally, its disappearance. Palsas may disintegrate by thawing of the frozen core as a result of microclimatic changes. These may be modification of the vegetation or snow cover, and sometimes cracks develop across the palsa, allowing warming of the interior by sunshine, or by percolating meltwater (Seppälä 1982).

6.2.3 *Hydrolaccoliths*

A range of features which often resemble palsa but which occur where permafrost is continuous are formed by growth of intrusion ice. They are known under the general name 'hydrolaccolith' but in North America are often called 'frost blisters'. The water, of recent meteoric origin (Pollard & French, 1984) is confined under pressure in the still unfrozen part of the active layer just above the permafrost during winter freeze-back. Presumably the water is expelled ahead of the frozen ground as it spreads downward and then moves laterally to the blister, where the overlying frozen layer yields sometimes with an explosion. The large volume of water arriving at the blister suggests that it comes from a considerable volume of soil extending some distance away. Pollard & French (1985) found four layers of clear ice in the core of a frost blister, the layers each having a distinct crystallography and bubble structure. Ice crystals were arranged parallel and essentially vertical, although slightly and differently tilted in each layer. The ice core may form and melt away each winter while several layers may sometimes reflect a succession of cold years with much of the ice surviving through the summers. More often they are due to 'pulses' of water within a winter. A water-filled opening is sometimes observed below the ice in early summer. Assuming the accumulation of the water and its subsequent freezing is responsible for the winter-time uplift of the blister, the process is similar to that in the formation of the Greenland-type of pingo. The main difference is that the generally smaller-sized and shorter-lived frost blisters occur on relatively level ground while the Greenland-type of pingo occurs characteristically at the foot of substantial slopes where a sufficient quantity and head of water occur.

6.3 Differential freeze–thaw effects

Second in importance only to the frostheave process itself in explaining the characteristic nature of the earth's surface in cold regions, is the diversity of the ground surface and soil conditions upon which the cold climate is imposed.

The amount of frost heave is extremely sensitive to the lithology and hydrological regime of natural soils and the heaved surface is thus very irregular. The heat and mass exchanges through the ground surface are regulated by the nature of the surface and its plant and snow cover, and these are a further control on the pattern of ground freezing.

The larger surface features already discussed reflect this diversity of behaviour, but there are many smaller 'microtopographic' forms of cold regions, striking in their symmetry in patterned ground, which are the product of local patterns of freezing and thawing and which require a strictly local and detailed analysis. The microclimatic (surface) conditions and soil variations thus assume special importance.

6.3.1 *Uplift of stones*

Even in temperate regions with substantial winter freezing farmers are aware that, through the years, stones and boulders rise to the surface. Such 'growing stones' must have been a particular nuisance to early generations striving to remove the stones from their land. In tundra it is common to see accumulations of large or small stones, often as part of patterned ground.

A number of explanations have been proposed (reviewed by Washburn 1979) but there are relatively few experimental studies which have demonstrated unequivocably a particular mechanism. Any object extending through sufficient depth may be raised as the soil around it (and to which it is frozen) is lifted by frost heave. Fence posts, utility poles and similar insertions in the active layer are often affected, being raised incrementally by up to many centimetres per year. The adhesive forces of the frozen ground are much greater than of the unfrozen ground so a post is pulled out of underlying unfrozen soil as soon as winter freezing extends sufficiently around the upper part. When thawing occurs, from the surface downwards, the soil settles around the post which is held up by the last remaining seasonally frozen soil. After that has thawed, there is insufficient dragging weight to overcome the adhesion of the lower, unfrozen soil. A post may often be hammered down subsequently although the cavity beneath tends to become blocked.

If these effects can be referred to as 'frost pull', perhaps the 'frost push' process is equally important – although more uncertain. Frost push describes a stone or boulder being lifted relative to the surrounding material because of frost heave below it. An essential point when the freezing occurs downwards, appears to be that there is a more rapid assumption of freezing temperatures at the base of the boulder or stone. The isotherm penetrates more rapidly because the stone has no latent heat of fusion and a higher thermal conductivity, and ice forms at the base before the stone is completely surrounded by freezing soil. The heaving pressure generated by this ice presumably displaces soil *downwards and laterally*. Upwards movement of the stone is resisted by the frozen material above. This suggests that frost push in a strict sense will be limited to where a stone has only a thin cover of frozen soil.

A layer of ice does not necessarily imply 'pushing' – whether upwards or downwards. The cavity produced below an object subject to 'frost pull' tends to become filled with ice simply because the cavity is a region of low pressure and this favours the segregation of ice. Van Vliet-Lanoe *et al.* (1984) discusses the ice segregations that occupy dessication cracks, themselves caused by migration of water to the freezing layer. Stones, being impermeable, prevent flow of water along temperature gradients in a frozen fringe (section 8.5) and this also should result in ice accumulations under them. Obviously the process of stone uplift is complex: an additional element may be the movement of water downwards to the growing ice layer, from beside the stone. This can result in a consolidation of the soil losing the water. Such consolidation constitutes a movement of the soil downwards relative to the stone (which has thus risen).

More easily understood is the uplift of bedrock, or 'bedrock heave'. Dyke (1981) describes large blocks lifted five centimetres annually by segregation ice. Layers of soil provided the conditions for ice segregation, in some cases, while elsewhere the weathered, or unweathered rock appears to have been sufficient. Experiments by Mackay (1984*a*) showed that the porous surface of rock can be sufficient to produce ice segregation and heave. Confined water under high pressure may increase the heaving pressure in a manner somewhat comparable to the effect in pingos.

Water passing through a porous stone may feed an ice layer on the cold side of the stone thus pushing it into unfrozen soil.

Mackay's experiment also demonstrated that stones may be moved upwards when the frost line is advancing *upwards* from the permafrost. Obviously the stone is then pushed into unfrozen soil, so that this form of uplift, which appears to have been first noted by Mackay, is relatively effective. However, significant amounts of upward freezing occur only above rather cold permafrost.

The movement of stones, or larger particles, relative to finer particles is referred to as sorting, and is an important process in the formation of patterned ground. It also occurs on a microscopic scale in the soil (section 2.2.8).

In addition to differential heaving, sorting can occur by the action of wind (removing fine-grained material) and by the action of water. On sloping ground the washing away of fine material can leave elongated accumulations of stones and boulders, sometimes known as stone stripes. These processes occurring in combination with the effects of freezing and thawing are responsible for the characteristic forms of certain patterned ground.

6.3.2 Soil hummocks

The surface of the ground is commonly very irregular and bumpy in tundra regions so that walking is difficult; artificially levelled surfaces rapidly become uneven. In forested periglacial regions too, there is much soil disturbance even though roots may provide restraint. Indeed the extent of 'undisturbed' soil and soil profiles is very limited, perhaps to uniformly coarse-grained well-drained soils lacking any tendency to ice segregation and frost heave.

Quite often the disturbance takes the form of well-defined and regularly spaced mounds, usually known as hummocks, and perhaps 50 cm high (Figure 6.3). The common, larger forms of hummocks require a fairly deep active layer. They can occur independently of permafrost, especially in temperate climates. Tufnell (1975) reports that hummocks grew in 20 years on the English Pennines. They seem less widespread and less developed in continental climates, where in the absence of permafrost, deep frost

Figure 6.3 Hummocks, Dovrefjell, Norway. (a) Summer (b) Winter. Note snow is absent from tops causing greater winter cooling than in the interhummock spaces.

Figure 6.3 (*cont.*)

penetration is often followed by much thawing from below due to seasonally stored summer heat (section 1.3.2). Hummocks may appear in highly regular form, but the term is also applied to extremely irregular surfaces. There are often conspicuous differences between the vegetation cover of the elevated parts and the troughs. The hummocks may also have bare tops.

Excavation normally reveals a much disturbed soil profile, often with irregular streaks of organic matter or other colorations suggesting fluidity at some time past. The disturbance, a form of *cryoturbation* (section 2.2.8) often extends to a depth roughly equal to the hummocks' height.

Various ideas have been put forward to explain their formation. According to the 'cryostatic pressure' hypothesis (which has been invoked to explain various forms of patterned ground), the freezing from the surface downwards during the winter occurs in such a way as to entrap bodies of still-unfrozen soil above the permafrost. The entrapped soil would be under increasing pressure as the frozen layer encroached and could, according to the hypothesis, be extruded through points of weakness. There is only limited evidence of this happening and extensive investigations of hummocks by Mackay (1980*b*) found rather firm soils and an absence of the high pressures implied. In coarse (non-frost susceptible) soil only, ice forming in the pores causes the extrusion of pore water with an associated pore water pressure increase. While this appears to be important in pingos and hydrolaccoliths, there is little evidence for it giving rise to flows of soil (rather than water).

If cryostatic pressures were to occur capable of moving unfrozen material to above the original surface, they must exceed the stresses due to overburden weight and the resistance of the frozen material. Frost heave can produce pressures of such magnitude although it is not clear how either the heave displacement or the pressure generated by the freezing soil would be transmitted so as to cause the flows of unfrozen material envisaged.

In fact, when high pressures have been observed (100–200 kPa) they were probably *within* frozen ground (Pissart, 1973, see also section 7.6), and would thus deform the frozen ground itself. Indeed, it appears that much cryoturbation is simply differential frost heave, that is, uneven displacements of the frozen material resulting from grain size variations, or differences in temperature or moisture conditions (see Figure 8.12). Such pressures and displacements of the frozen soil are not, of course, 'cryostatic' as that term is generally understood.

Strength is lost in unfrozen soil, and the soil may even be more or less fluid, if water is trapped in pores and voids in excess of the normal saturation moisture content. This is more likely to occur at thaw than during freezing, in materials that are not coarse-grained. The excess water arises from the melting of ice segregations. Its pressure rises due to the weight of overlying material and as a result the effective stress is zero. Deformation and displacement is likely to occur in such newly thawed material, particularly on sloping ground (cf. section 5.2). These movements will tend to be downslope and such soil structures or patterns as may be developed would have a downslope orientation.

Cryoturbation includes various flow-like forms often found in soil profiles on level as well as sloping ground (even where hummocks are not well developed). The involutions and turbulent patterns have also been ascribed to soil flow initiated by differences of density. This seems no more likely than the cryostatic effects. Unless the soil were extremely weak the forces (weight) due to the greater density would usually be insufficient for penetration of one body of soil into another. Mackay (1980*b*) did however observe such a weak layer directly above frozen ground, and in a detailed study develops the concept of a circulation cell (Figure 6.4*a*). Essential

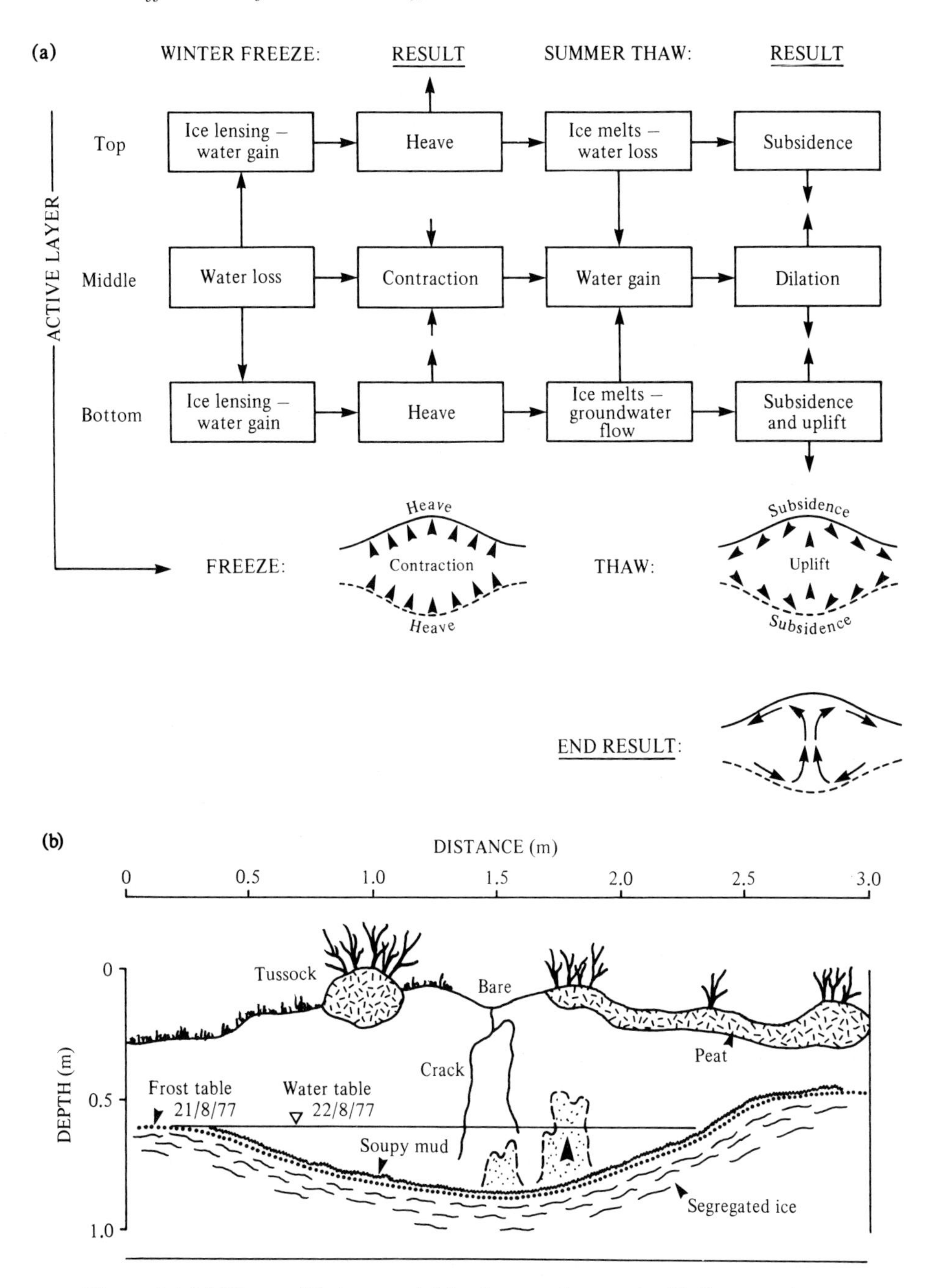

Figure 6.4 (*a*) The equilibrium model for hummock growth. (*b*) Cross-section of a hummock in late-summer showing tongues of soil which are inferred to ascend from the bowl-shaped surface of the frozen ground. The thawing of ice lenses in the ice-rich material below this surface, and the permeability of the newly-thawed soil are thought important in this respect. Both figures are after Mackay (1980b).

features of this are the heave in the upper part of the active layer during autumn refreezing and, in early summer, lens formation and heave as water migrates downwards to refreeze within the still-frozen basal part of the active layer (section 8.6.2). Between these two levels of ice accumulation the soil has a reduced water content and consolidates due to the suction induced, but expands by absorbing meltwater in the summer thaw. Thawing of the upper and lower heaved layers results in a radially outward movement from the centre near the surface of a hummock, and a radially inward movement in the basal part of the active layer below the hummock. The underlying frozen ground has a bowl-shaped depression. The sequence of events, according to Mackay, also explains fairly well the tendency of material to move up in the centre of a hummock (Figure 6.4(b)). Where permafrost is absent the active layer extends deeper below a hummock (giving the bowl-shaped frozen surface), if the hummock surface and soil are such as to give a more rapid cooling (Figure 6.3(b)) or if the annual energy exchange is modified so as to give a lower mean annual temperature. In permafrost areas, deeper active layers occur with warmer mean temperatures. Mackay (1980*b*) also demonstrated that a kaolin slurry in a (bowl-shaped) circular pan assumed a hummock form (replacing the initially flat surface) after some 20 freeze-thaw cycles. The forces involved, in Mackay's view, are, additional to those following from the frost heave, essentially gravitational, and thus smaller than those hypothesised (but not demonstrated) for the cryostatic pressure.

The diversity of hummocks implies that there are several processes of formation. Certain plant species (especially the cotton grass, *Eriophorum*) promote hummock formation additional to the processes associated with differential freezing and thawing.

Differential frost heaving, including the sorting of materials at the microscopic scale and the role of cracks due to dessication are all discussed by Van Vliet-Lanoe *et al.* (1984), and Van Vliet-Lanoe (1985). She also demonstrated the development of cryoturbations, including features due to injection of one kind of material into another, in both field and laboratory experiments. An interesting idea discussed, is that of the effects of gradients of frost heave. This refers to situations where frost heave is greatest in the layers at the surface and decreases downwards (positive gradient) or the reverse, where frost heave increases with depth (negative gradient). If vertical cracks form (extending down from the surface) there may be a displacement of soil towards each crack and upwards adjacent to it (because of the lower resistance and perhaps more rapid cooling of the crack). Thus soils of different heaving abilities will come to lie side by side.

The effect will be reinforced by an uneven penetration of the isotherms and of the frost line, because of the different heat capacities of the soils according to their ice content. Substantial differential heaves may thus occur with an orientation determined by that of the cracks. When thaw occurs, displacements due to differential heave will be at most only partially reversible. Although the formation of hummocks is still not entirely understood, differential frost-heave can explain many of the structures within them. Furthermore, the various processes are also important in the development of other forms of patterned ground of similar size.

6.3.3 Earth circles

A rather common feature of tundra regions with or without permafrost are clusters of more or less circular, vegetation-free patches (Figure 6.5). They are also known as mudboils, non-sorted circles, medallions and by other terms. The surface of each 'circle' (they are a metre or two in diameter, sometimes less, sometimes more) is usually noticeably flat even though often covered in stones. However, there are also forms in which the surface is raised into a shallow dome. The surrounding vegetation is often dominated by lichens.

Shilts (1978), discussing mudboils, emphasises the weakness of the sub-surface layer, compared to the surface which tends to be drier and firmer, and the consequent flow of mud through the surface if appropriate stresses occur. He suggests the latter might be cryostatic pressure or that sloping ground could be sufficient. Mud is sometimes seen to break through the surface and flow in a tongue downslope. Shilts' observations were in a permafrost area.

An important factor appears to be the nature of the sub-surface material, which Shilts (1978) reports as having a low liquid limit and low plasticity index (these two geotechnical properties are described in, for example, Lambe & Whitman, 1979). The significance seems to be that, as the material is generally saturated, only a small amount of additional ('excess') water at thaw would be liable to render the material liquid. The lack of strength of the liquid material would mean that only small stresses (perhaps only the weight of overlying, denser, material) could cause it to break through the surface.

Stony earth circles can be initiated by removing the vegetation, which results, in a year or two, in an accumulation of stones on the surface and disturbance (Figure 6.6) due to frost heave (Williams, 1958). The element of soil or mud flow does not then seem to be important, and permafrost is not necessary. These circles are restricted, however, to windblown, and thus

Figure 6.5 Stony earth circles, Trollheimen mountains, Norway.

substantially snow-free exposed ground. Presumably wind can strip the shallow lichen vegetation; Seppälä (personal communication) observed reindeer eating the lichens through holes in the snow. The absence of snow means that the first few centimetres of the soil will be subject to numerous freeze–thaw cycles thus accelerating the sorting process. The accumulation of stones on the surface is probably aided by removal of fine grains by the wind. During early summer, liquefaction can be produced merely by trampling, with water and mud appearing at the surface.

The latter behaviour suggests an affinity with that described by Shilts. Rieger (1983) describes the liquefaction of Arctic soils as thixotropic. The water chemically bound to particle surfaces (perhaps by aluminum and iron hydroxides and organometallic complexes) is released by the disturbance. There is then a state of no effective stress, the weight of particles being carried by the water in the pores. Tumel & Mudrov (1973) consider such thixotropic material to be a loess-like weathering product, arising from freezing and thawing, and maintain that it is widely responsible for various characteristics of solifluction and patterned ground.

6.3.4 Stone polygons and rings

Characteristic stone polygons are assemblages of boulders in a roughly polygonal network (Figure 6.7). Each polygon can be several metres

Figure 6.6 (a) Patch of 55 cm diameter cleared of above-ground vegetation (b) same patch two years later

Figure 6.7 Stone polygons. The reindeer antlers are about one metre across.

across but they occur in a range of size down to 'micropolygons' only a decimetre or so across. In fact, the 'polygons' vary considerably in appearance and sometimes are best described as rings or circles (Figure 6.8), having roughly circular accumulations of finer material lying in a mass of boulders. Such polygonal and circular features may also occur without the significant sorting of stones and boulders (Figure 1.11). The latter may, in fact, be more or less absent, with the polygonal form being demarcated by vegetation around the perimeters, and sometimes by cracking, or by updoming of the centres. Thus, these patterned ground features could be part of a continuum from such forms as hummocks and stony circles through to the typical stone polygons. The size of sorted polygons and circles usually corresponds with the dominant sizes of boulders or stones in the surface materials, so that, for example, small circles or polygons tend to occur where large boulders are absent.

The origin of polygons is generally cracking by dessication or thermal contraction. The cracks are vertical and due to horizontal (lateral) contraction of the surface layer. Sorting often occurs by movement of coarse material towards and into the cracks. Circles which have fine-grained centres surrounded by irregular arrangements of stones and boulders

Figure 6.8 Fully-developed stone rings (each ring is 1 to 2 metres diameter). Spritsbergen (photo: Bernard Hallet)

probably originate similarly. They are distinct from the stony earth circles described in the previous section, the origin of which does not seem to involve the cracking.

Experiments by Pissart (1974) on small stone polygons showed that small stones accumulate on the surface within two years of the ground being artificially levelled and mixed. He also showed that the stones which have reached the surface move towards and eventually fall into thermal contraction cracks. This additional material causes compressive stress when the ground warms somewhat and expands. Fine-grained material below the polygon as well as on its periphery is pushed upwards and inwards. Stones rise to the surface of the progressively raised, fine-grained material, and ultimately fall into the bordering cracks. The compressive effects are augmented and the process continues.

Some authors have referred to movements of this type as 'convection' but this seems inappropriate. Convection is the density controlled flows of a liquid being warmed. Although open vertical cracks may accelerate cooling at the sides of polygons, the density differences resulting from such relatively small temperature inequalities would produce forces far too small to cause differential movements of soil and rock material. The considerations relative to cryoturbation apply: the density differences give rise to gravitation (weight) induced forces but these are normally far smaller than the strength of the materials being penetrated.

Ballantyne & Matthews (1982) describe sorted circles a metre or more across: these resemble stone polygons, with borders of boulders somewhat depressed relative to the finer-grained centres. They have formed on ground exposed from a glacier more than 250 years ago. They believe the circles formed in a few decades and were favoured by the moisture and temperature conditions adjacent to the glacier. Small, sorted polygons developed in 35 years adjacent to a retreating glacier according to Ballantyne & Matthews (1983). A succession from simple dessication cracking to fully developed sorting of stones into the cracks could be observed.

The cryostatic pressure hypothesis has also been invoked to explain stone polygons, but experimental observations do not seem to have been reported. The sorting of material laterally adjacent to a crack was demonstrated by Corte (1962). The crack causes a lateral heatflow, and is a region of low resistance, towards which stones and possibly boulders are moved by the differential heaving considered in section 6.3.1.

In some situations fine material (silt size or finer) in a narrow range of grain size may be separated from the soil matrix. Corte proposed that fine particles would be pushed ahead of ice lenses and through the larger pores. Certainly, growing ice layers reject particles, and this has been observed in several experiments (e.g. Römkens & Miller, 1973). It seems clear that freezing induces sorting in several ways. Sorting by wind of surface materials disturbed by frost heave was observed by Rissing & Thorn (1985). The same authors suggest chemical weathering is modified below the vegetation-free circles.

Norwithstanding the role of size of stone or boulder in determining polygon size, the properties of the soil more generally determine the frequency of thermal contraction cracking, and thus the spacing of cracks and the size of the patterned ground features which result. The enormous variety of patterned ground, particularly hummocks, circles, polygons and other 'multiple' forms must be due to a wide range of processes and effects (Washburn, 1979). The process of cracking, especially that due to thermal contraction, is in several respects unique in frozen ground. It appears to have a role second only to the frost heave process itself in explaining the local topography of cold regions. Cracking of frozen ground is also important in geotechnical considerations. Understanding contraction cracking and its most obvious result – ice wedge formation in polygonal patterns – draws attention to the importance of rheological properties (considered in Chapter 9).

6.4 Thermal contraction and cracking

Saturated soil must expand on freezing by an amount depending on the amount of water frozen (Figure 1.9). At some temperature which may be several degrees below 0 °C, freezing of most of the water is complete and thermal contraction of the soil then dominates when cooling occurs. The temperature at which the soil contracts rather than expands depends on the unfrozen water content and its relation to temperature. Fine-grained soils, in which significant ice-formation (with the associated expansion) occurs at several degrees below 0 °C, may contract only at temperatures colder than −5 °C or −6 °C. Sands, in which almost all water freezes near 0 °C, may show contraction on cooling to only −0.5 °C.

The tendency to contract results in a cracking of the ground surface and ultimately a network (in plan view Figure 6.9) of more or less vertical cracks, commonly with a mesh of ten metres or more but which may be much smaller, down to perhaps 20 centimetres. The latter much smaller and often more local cracking may sometimes be confused with that due to drying. The large cracks, which at their coldest may open several millimetres, extend vertically downwards some tens of centimetres to a metre or more, and become filled with ice or soil. In the former case, many years' accumulation gives *ice wedges*, a widespread, and remarkable form of ice in permafrost having great geotechnical significance. Mackay (1970) estimates that 2.5 km^2 of terrain in the Mackenzie Delta can have 160 linear km of the wedges. *Soil wedges* are formed if the cracks are filled with sediment directly. Such wedges are not restricted to ground underlain by permafrost. Soil wedges may also arise by infilling as ice wedges melt (Black, 1976).

The process of thermal contraction cracking is well described by Lachenbruch (1963) although details are still uncertain. Consider the layer subject to contraction as extending over a large area, and also that the layer is not able to move (contract) freely because of its continuity. The restraints on contraction then result in a stretching and tension on cooling. That is, according to the temperature the layer ought to be smaller but its strength against rupture is such that it maintains, initially at least, its lateral extent. Thermal contraction of a material is described by its coefficient of thermal expansion and for frozen soil the value is about $10^{-5}\,°C^{-1}$. This is a strain:

$$\frac{\Delta l}{l}, \text{ linear, or } \frac{\Delta V}{V}, \text{ volumetric}$$

per degree C of temperature change. It corresponds to a shortening of 10 m of ground by about 1 mm for a 10 °C fall of temperature. Considering the layer of ground being cooled, the stretching is a strain of this magnitude but opposite sign. It is due to the tensile stress induced. This strain is the amount by which the lateral extent exceeds what it 'ought' to be according

Figure 6.9 Ice wedge polygons seen from several hundred metres height (photos: D. Lawson).

to the temperature, expressed as a ratio. Knowledge of the stress–strain relationship (see Chapter 9) for the frozen ground allows the tensile stress to be calculated. Note that thermal contraction of a body does not by itself involve development of stress but if there is a resistance to the contraction, this produces a tensile stress in association with the stretching (the strain).

The tensile stress so developed can lead to the cracking of the ground, which occurs suddenly when the tensile strength is exceeded. This might seem simple enough, but the tendency of frozen ground to creep (see Chapters 5 and 9) complicates the matter. The temperature of frozen ground does not drop suddenly to that giving the stress necessary for cracking. On the contrary the temperature falls gradually (the more gradually, the deeper in the ground), and during the time involved the ground creeps in response to the tensile stresses developing. The creep is a deformation, a tendency to spreading of the material, which reduces the stresses.

Here it is important to make a distinction between elasticity, which involves a reversible process, and plasticity which involves a non-reversible deformation. The stretching that occurs immediately on cooling is largely elastic. When cracking occurs, the frozen ground contracts due to the elasticity and the tensile stress is lost. The nature of creep, however, represents a rearrangement of the material that allows the stress to fall, or relax, without the contraction. Thus if sufficient creep occurs with a sufficiently low rate of stress development, there will be no cracking. An analogy would be the stretching of a piece of elastic fabric. It might rupture and spring apart. On the other hand, if it remained stretched a long time it might simply loose its elasticity and thus its tension.

Not only must the temperature of the ground fall sufficiently low (to −6 °C or so it appears) to produce cracking, it must do so sufficiently rapidly, before significant stress relaxation by creep occurs. The temperature conditions required will vary because the strength and creep properties of frozen soils vary greatly (Chapter 9). The reopening of the cracks, year by year, rather than formation of new ones, follows from the relative weakness of the ice fillings, or, in the case of sediment-filled cracks, from the inherent weakness along the cracks, compared to the undisturbed frozen ground. Furthermore, this implies that cracks progress upwards to the surface from their points of origin in the top of the ice wedge (the top of the permafrost) each winter (Mackay, 1984*b*).

The spacing of cracks and their geometrical relationship are also described by Lachenbruch (1963). The developing stresses during cooling will have a generally lateral (horizontal) orientation, since this is the direction in which contraction is constrained. Cracks, therefore, tend to be vertical.

The formation of a crack releases the tensile stresses that were normal to the crack, that is, those that caused it. Subsequent cracks in proximity will therefore not be parallel but, rather, normal to the initial one, if viewed from above. If the initial crack should be somewhat curved, the subsequent cracks will still be essentially orthogonal but tending towards an, ultimately, polygonal network. Cracking patterns may also be orderly rectangles – this being so especially where there is an initial asymmetry, such as the shore of a lake, which significantly affects the pattern of stress in the near-surface layers. The frequency of cracks and the depths to which they extend is compared with ground temperature conditions by Romanovskij (1973). Continental climates produce greater thermal gradients and the depths to which sufficiently low soil temperatures and sufficiently sharp temperatures changes occur will also depend on mean ground temperature and surface conditions, especially snow cover. (Mackay, 1978b, 1984b). Deep, widely spaced cracks (usually those with ice wedges) are said to require mean ground temperatures of −5 °C or colder. Presumably such mean temperatures are necesssary because the winter cooling is much reduced at depth. Romanovskij points out that smaller cracks, more closely spaced, may occur even where mean ground temperatures are slightly above 0 °C (and thus in the absence of permafrost). Isolated cracks occur occasionally in seasonally freezing ground far from permafrost, presumably because of extreme weather conditions with sharply falling temperatures.

6.4.1 Soil and ice wedges

So-called primary soil wedges developed by infilling of thermal contraction cracks in permafrost occur in the dry regions of Antarctica, where the sediments may be eolian (Black, 1973). Much more widespread are soil wedges which have formed above ice wedges. Indeed, because the ice wedge cannot extend into the active layer (it would melt), some kind of infilling by soil is bound to occur if the active layer also cracks in the winter. According to Romanovskij (1973), sediments may be carried down into cracks by percolating flood or meltwater and produce soil wedges, in a seasonally freezing layer.

Thermal contraction cracks normally reopen and close annually. Ice wedges develop by accumulation of hoar or melt water, in the open cracks. Mackay's (1975b) observations demonstrated this and that the additional ice did not fill more than about 20% of the maximum winter opening. In some winters, late-lying snow prevented water entering cracks before they reclosed. The amount of ice added by sublimation is probably small, because when the cracks are fully opened, the near-surface layers are the coldest and vapour pressure gradients would be towards the ground surface, not downwards.

Multiple wedges – one wedge appearing on the top of another – are observed fairly frequently, and these follow from a rise of the upper surface of the permafrost. The cause may be climatic or microclimatic or, alternatively, the accumulation of sediment on the surface can lead to a rise in the level of the top of the permafrost. The fact that a younger wedge may lie *directly* above an older one also shows that cracking is initiated at some depth in the ground, at the top of the pre-existing wedge for example (Mackay, 1974a, 1984b). If cracks started at the ground surface they would not, presumably, occur immediately above the existing wedge.

As the ice wedge thickens, in the permafrost, year by year, there is a progressive deformation of the adjacent ground when it expands during the warm period. The restricted space results in the soil being pushed up into ridges above and to either side of the top of the ice wedge. It is these ridges which are observed on the surface, as the characteristic polygonal or orthogonal patterning.

7

Thermodynamic behaviour of frozen soils

7.1 Soil: a porous system

Those features of ground and terrain that are unique to cold climates arise mainly because of the special conditions which soils and other porous media impose on the water freezing within them. These conditions in soils have only been described in detail in the last three decades, yet are largely what would be expected from a basic knowledge of physical chemistry or thermodynamics. Applying the fundamental sciences to the interpretation of field situations requires careful attention to identifying the terms and concepts used by physicists and chemists with those used by geologists, engineers, soil scientists and others. In the following sections, the physics, chemistry, and thermodynamics of freezing soils are examined further.

The coexistence of ice and water in the soil pores is the most fundamental attribute of frozen soils. Consequently, the study of frozen soils requires consideration of phase change and freezing points.

7.1.1 'Freezing points' and latent heat

The freezing point of a substance is generally determined by observation of an abrupt interruption in cooling (Figure 7.1) caused by the release of latent heat of fusion as the liquid turns to solid. The temperature first reaches a minimum, just below 0 °C in the case of water, but this involves an unstable *supercooling*.* Freezing occurs following nucleation (the abrupt formation of a stable ice crystal) and the liberated latent heat immediately causes the observed temperature to rise to, or even somewhat above, the *equilibrium freezing point*.* For pure water at atmospheric pressure, this is 0 °C. It will be less than 0 °C ('depressed') if, for example, there are dissolved salts. Many years ago, such freezing points for soils were

* For definitions of this and other terms in general science, see science texts or dictionaries of science (e.g. Gray & Issacs, 1975).

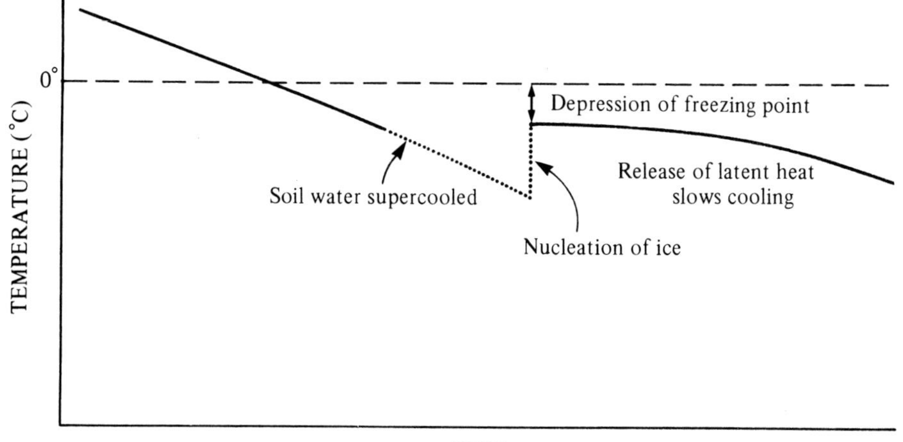

Figure 7.1 Cooling of sample of soil under constant rate of heat extraction. The dotted line represents supercooling, the cooling of the soil water below its equilibrium freezing point.

observed to be a tenth of a degree or so below 0 °C (Bouyoucos & McCool, 1916). However, in such experiments, a constant rate of cooling of the soil samples is not resumed even after the temperature has fallen quite substantially (Figure 7.1). This should have suggested that freezing and the liberation of latent heat in soils is not limited to a single temperature (or freezing point). However, it was only in the early fifties that Nersessova (Inst. Merzlot. 1953–57) reported observations, made with a simple calorimeter, that large quantities of heat were involved in the warming of soils that clearly remained frozen. This was latent heat associated with thawing of some of the ice at temperatures below the 'normal' melting point of 0 °C. Evidently the freezing of the water, or thawing of ice, in soils actually occurs over a range of freezing 'points': that is, the equilibrium freezing point must be changing as the unfrozen water content is decreasing or increasing. It is unlikely that supercooling – the existence of water because of absence of ice nuclei – is responsible for much of the unfrozen water in frozen soils. This is in contrast to plants, for example, where supercooling is common and important (Franks, 1980).

7.1.2 Proportions of ice and water

Measuring the quantities of heat added to or removed from a frozen soil to change its temperature allows calculation of the accumulated amount of ice, or of water (Figure 1.4). The heat required for merely warming or cooling – that is, the heat capacities of the mineral or organic

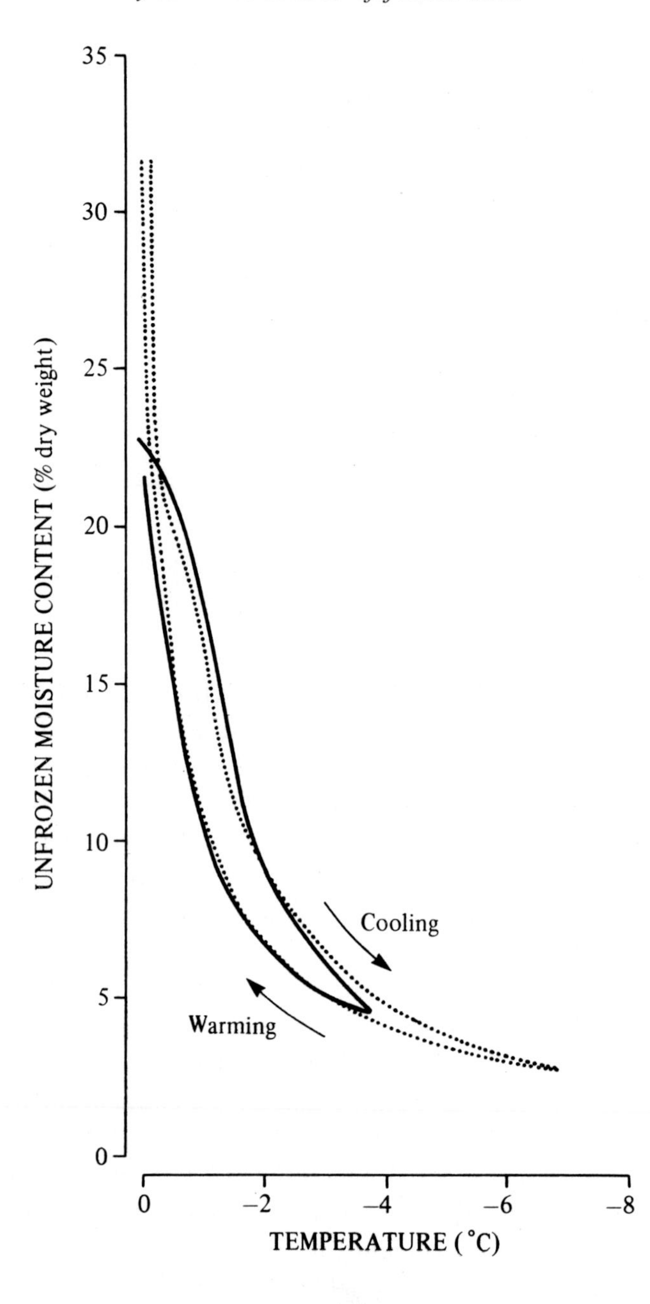

Figure 7.2 The unfrozen water content of a frozen chlorite–illite clay (Leda Clay, Ottawa, Canada), as a function of temperature. The values were obtained for two samples, one of which had a total moisture content of 32% dry wt., (which was cooled to −7 °C), and the other, 23% (cooled to −4 °C). After Williams (1964b).

soil material, water and ice – is small by comparison with the quantity of latent heat of fusion: to cool 1 g of water 1 °C involves the removal of 4.2 joules, while 334 J g^{-1} must be removed to freeze it.

After appropriate deduction for heat capacities, the heat, in joules, involved in a change of temperature of a frozen soil divided by 334 gives the amount of ice formed or melted. Below about −4.5 °C the latent heat of fusion is less. Low *et al.* (1968) suggest it is about 328 J g^{-1} at −5 °C and 317 J g^{-1} at −12 °C. If the frozen soil is saturated – that is, if there are no gas-filled pores or cavities – then the changing proportions of ice and water can be observed through the changes of volume of a sample (ice having a specific volume, in m^3 kg^{-1}, about 9% greater than water). Such observations can be made with a dilatometer (Williams, 1976) and, although technically demanding, are probably the most direct method of determination.

When the temperature of a frozen soil is raised or lowered, ice thaws or water freezes as indicated in Figure 1.4, but there is not a single, unique curve for a particular soil. The curve is modified depending on the thermal history of the soil. Figure 7.2 shows the unfrozen water contents (as a function of temperature) for temperatures reached by cooling and by warming. This is hysteresis, or partial irreversibility, a phenomenon known from the wetting and drying of soils (discussed later in this chapter). It is ascribed to differences in water behaviour on entering and on leaving soil pores, and on being adsorbed or released from particle surfaces. Similar phenomena are involved in the freezing or thawing of the soil water. The 'warming' curve, furthermore, is modified by the lowest temperature reached (consider the implication in the temperature cycle illustrated of cooling to, say, only −2 °C). The 'cooling' curve is also modified by the extent of any previous cycle of freezing and thawing, even though this may have been long ago. The freezing of soil water causes irreversibile modifications to the fundamental structure of the soil. The effects are of scientific interest and were considered in section 2.2.8; for many practical purposes, however, the variation between different soils (Figure 1.4) is more important.

Observations (e.g. Williams, 1964b, 1967) show that if the water content* of the frozen soil is expressed as % dry weight (i.e., weight of water contained/weight of sample after drying), it is essentially independent of the ice content. It follows that the water content is independent of the total moisture content ('moisture' referring to the sum of water and ice). If the moisture content is low, then the curves will be truncated but not otherwise

* In this chapter 'water' refers exclusively to the liquid phase, not to the ice or vapour.

changed (Figure 7.2). Schofield & Botelho da Costa (1938) measured the temperature at which freezing *commenced* (the 'freezing point') in samples, of a single soil, dried to various water contents. Put together, such results produce a similar curve.

The reasons are as follows: the presence of the water in frozen soil is due to its modification by confinement in pores and proximity to mineral surfaces. In a saturated soil, large or small amounts of ice are present in masses or layers larger than pore size. Just how much such ice there is has no effect on the water in association with the mineral particles nor, therefore, on the (liquid) *water* content expressed as *% dry weight*. If the water content is expressed *volumetrically* (volume of water per unit of volume of frozen soil), the value will change according to the volume of the ice inclusions. The volumetric water content is in this respect a somewhat arbitrary quantity, although often required in calculations for practical purposes.

More recently it has been noted that the unfrozen water content (% dry weight) may not be absolutely independent of total moisture content. The main reason is the effect of dissolved salts. When the ice content is high, the salts (ejected from the ice on freezing) may significantly affect the freezing 'points' (Banin & Anderson, 1974). Using nuclear magnetic resonance, Tice *et al.* (1982) showed that progressive removal of ice did not significantly affect the water content.

The effect of externally applied pressures (the 'confining' pressure) on the water content of frozen soil samples is generally small (Figure 7.3), thus the composition of the soil and its temperature determine the (unfrozen) water contents. As would be expected, the finer-grained and smaller-pored soils generally have larger water contents. Anderson & Tice (1972) showed that the unfrozen water content at any temperature could be predicted from the specific surface area of the soil. Based on data for a wide range of soils, they determined a general phase composition equation as follows:

$$\ln \theta_u = 0.2618 + 0.5519 \ln S - 1.449\, S^{-0.264} \ln T \qquad (7.1)^*$$

where

θ_u = unfrozen water content (g g^{-1})

S = specific surface area (m^2 g^{-1})

T = temperature in degrees below 0 °C

Measurement of specific surface area requires specialised techniques. Meas-

* The term 1.449 S was misprinted in Anderson & Tice (1972) as 1.4495

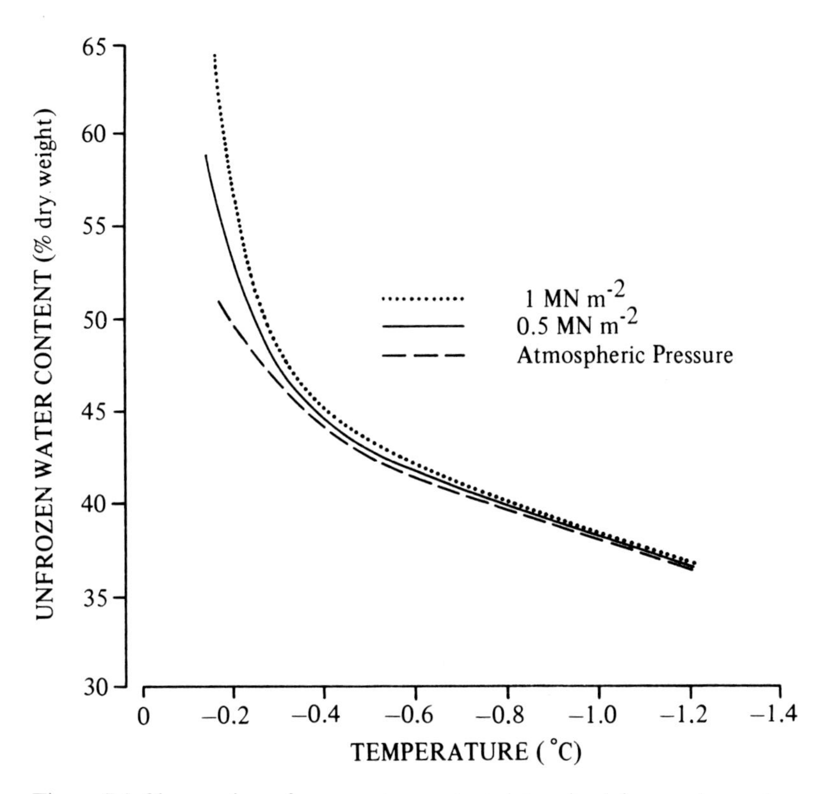

Figure 7.3 Changes in unfrozen water content determined from volume changes of frozen Tanzania clay, due to externally-applied pressure (from Williams, 1976).

urement of the energy status, or potential, or soil water, on the other hand, provides a relatively simple means of predicting the amounts and behaviour of water in frozen soils. As we shall see, the relations involved are fundamental (section 7.2).

7.1.3 Water contents and thermal properties

The progressive freezing of soil water through a range of temperatures determines the thermal properties of soils at those temperatures (Chapter 4). The apparent heat capacities are largely composed of the liberated latent heat and thus depend on the amount of ice formed. The curve of unfrozen water content (Figure 1.3 and 7.2) reflects the progressive formation of ice and allows us to calculate the apparent heat capacities for a range of temperatures (Williams, 1964*a*). Although the heat capacities are expressed per degree Celsius, the values are, of course, not constant through any particular one degree (Figure 7.4).

Changes in total moisture content affect the heat capacity in a rather

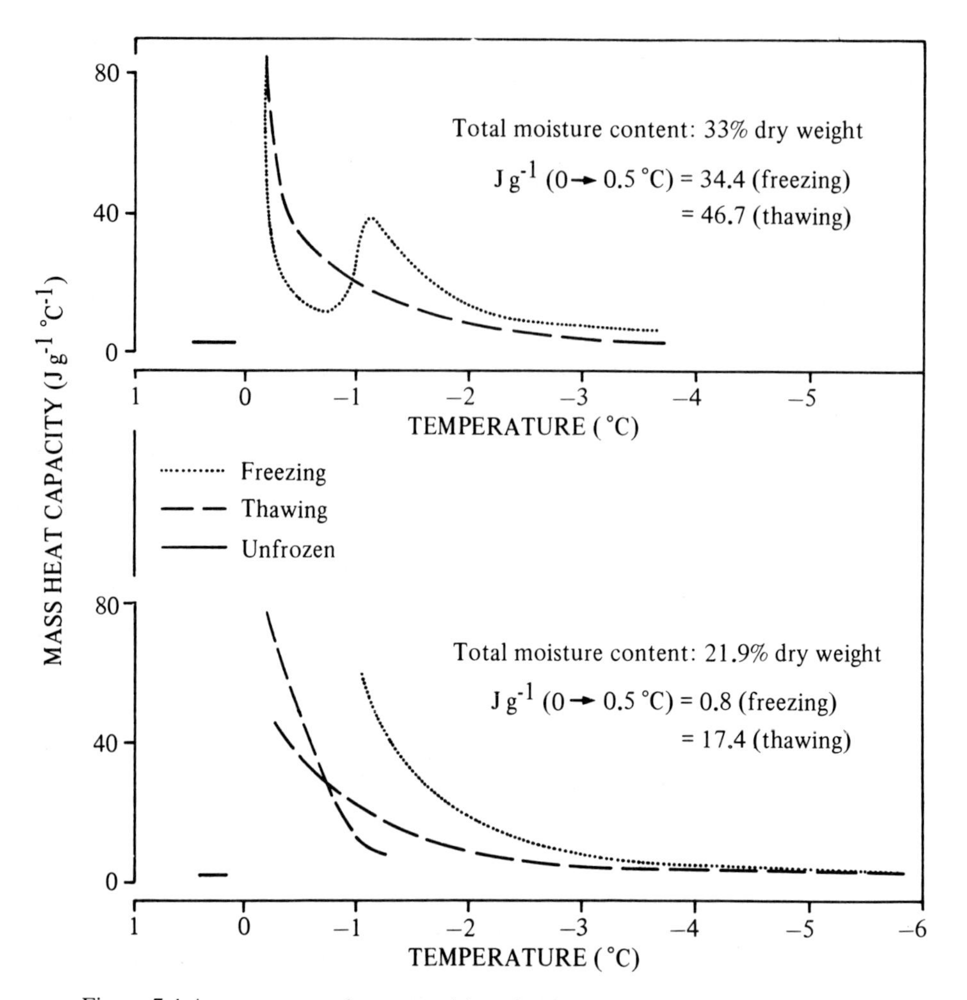

Figure 7.4 Apparent mass heat capacities of samples of Leda clay, in the frozen state, as a function of temperature. The values are primarily determined by the latent heat component and the apparently anomalous peak for one sample merely reflects the slope of the curve for unfrozen water content (Figure 7.2.) After Williams (1964a).

complex way. Additional moisture in a saturated soil freezes at temperatures very near to 0 °C, as explained in the previous section, so that the apparent heat capacity will be augmented (usually greatly) for those temperatures. In Figure 7.4 the amount of heat actually involved in the temperature change 0 °C to − 0.5 °C is also shown. For lower temperatures, the curves of unfrozen water content, expressed gravimetrically as a percentage of mineral material, are independent of total moisture content, as described in the previous section. However, the mineral matter will be more dispersed, as will the remaining water, because of the additional ice present. As a consequence for these lower temperatures, the heat capacities (whether for unit mass or unit volume) will be lower if the moisture content is greater (at least for all temperatures where there is still significant freezing).

Apparent mass heat capacity, c, $J\,kg^{-1}\,°C^{-1}$, is shown for several samples of Leda clay, in Figure 7.4. These values could be converted to volumetric heat capacity, C, $J\,m^{-3}\,°C^{-1}$, by multiplying by ρ_s, the bulk density of the soil. However, the assessment of the moisture content and the bulk density is often difficult in practice. It is necessary to take into consideration the volume changes associated with freezing as well as with the moisture content itself. Under natural conditions, variations in the moisture content of a soil with location or through time, make the assessment of thermal properties more difficult than does the characteristic temperature dependence of the thermal property itself.

The peak shown in one of the curves is not anomalous, although, perhaps, unusual. It is explained by the curve of the unfrozen water content for the same sample (Figure 7.2), where it is seen that the decrease in unfrozen water content with temperature is greater at $-1.1\,°C$ than at $-0.7\,°C$ or so.

The thermal conductivity ($W\,m^{-1}\,K^{-1}$) of frozen soils is, of course, also affected by the changes in ice and water contents discussed. The assessment of thermal conductivities involves a number of considerations additional to those for heat capacity, and these were reviewed in section 4.3.1.

7.2 Energy status, or potential, of soil water

The concept of potential of soil water is especially developed in agricultural soil science and in hydrology. Understanding of the potential concept as applied to soils is generally a prerequisite to understanding the water relations of freezing soils.

Numerous forces act on soil water, such as those associated with the attraction of soil particle surfaces and with osmotic effects: more generally there are forces due to the earth's gravitational field and the weight of overlying masses. The sum of these effects at a point constitutes the *total potential* of the soil water. It is differences in potential which give rise to movement of water, from higher to lower potential.

Measurement of potentials, and of the permeability or hydraulic conductivity coefficient of the soil, are required for application of the basic equation for flow, which is represented in the form of Darcy's law:

$$q = kA\frac{\Delta\psi}{\Delta Z} \qquad (7.2)$$

where

q = flow $m^3 s^{-1}$

k = hydraulic conductivity (permeability*) $m s^{-1}$

A = cross-sectional area of flow path m^2

$\frac{\Delta\psi}{\Delta Z}$ = gradient of potential $m m^{-1}$

The units of potential, ψ, are joules per kilogram but often other quantities, more easily observed, such as pressure (Pascals, $= N m^{-2}$), heights (m) of water columns (as in equation 7.2), etc. can be used as a measure instead. The potential which arises due to a soil's capillarity and particle surface adsorption is often referred to as soil water *suction* (having units of pressure). The joule being the unit of energy, differences in potential are differences in energy status, which in turn reflect the difference in the forces acting on the water.

The principle in the determination of the potential of soil water is comparison with a reference body of water whose potential is easily and clearly defined. Thus, for example, if a sample of moist but not saturated soil is brought into contact with water, the tendency of the water to move into the sample will be overcome if a sufficiently reduced pressure can be applied to the water (an experimental arrangement for this is shown in Figure 7.5, discussed further below). The water then has the same potential as that in the soil and is in mechanical equilibrium with it. Normally, pure, free water (i.e. water not confined in small spaces such as pores or drops) at the same temperature and elevation is the reference (*Int. Soc. Soil Sci.*, 1963). It is logical to ascribe a potential of zero to the pure water when it has atmospheric pressure. Then the pressure reduction required to bring such water to a state of equilibrium with that in the soil is a measure of the potential of the soil water.

The definition of the potential is that it is the work expended (joules) in transferring water from such a pool to the soil water at the point in question. In the case considered (unsaturated soil), the potential has a negative sign. It is a consequence of the porous and particulate nature of the soil and is called matric potential. The potential associated with depth of water is also considered a matric potential. It has a positive sign, increases with the depth, and can be easily measured directly with a

* Engineers use the term permeability for this quantity. Hydrologists, on the other hand, refer to it as hydraulic conductivity, while using 'permeability' in a somewhat different sense.

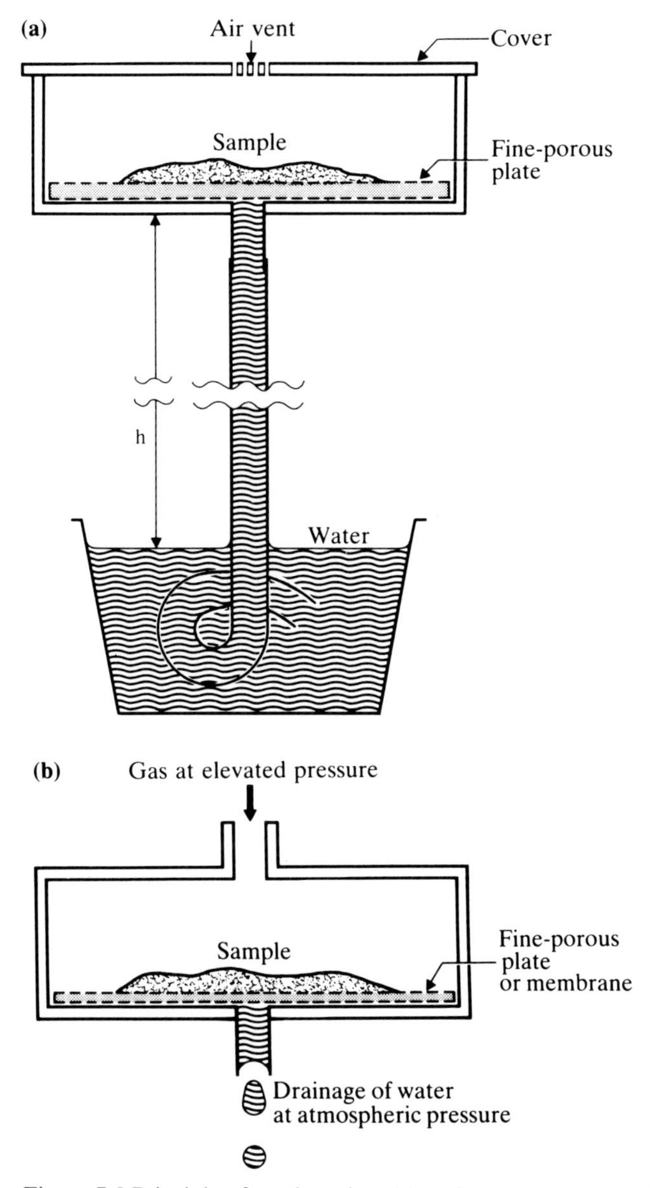

Figure 7.5 Principle of suction plate (*a*) and pressure plate and pressure membrane apparatus (*b*). In (*a*) the suction applied corresponds to *h*, and is the difference between atmospheric pressure and the water pressure at the base of the sample. In (*b*) the water pressure is atmospheric and the air pressure raised.

pressure gauge (pore water pressure meter or piezometer). Another type of potential is that due to an elevation difference, and is known as a gravitational potential. This is often called 'head' and measured as a height. Accounts of the principles of moisture movement and potentials in soils are

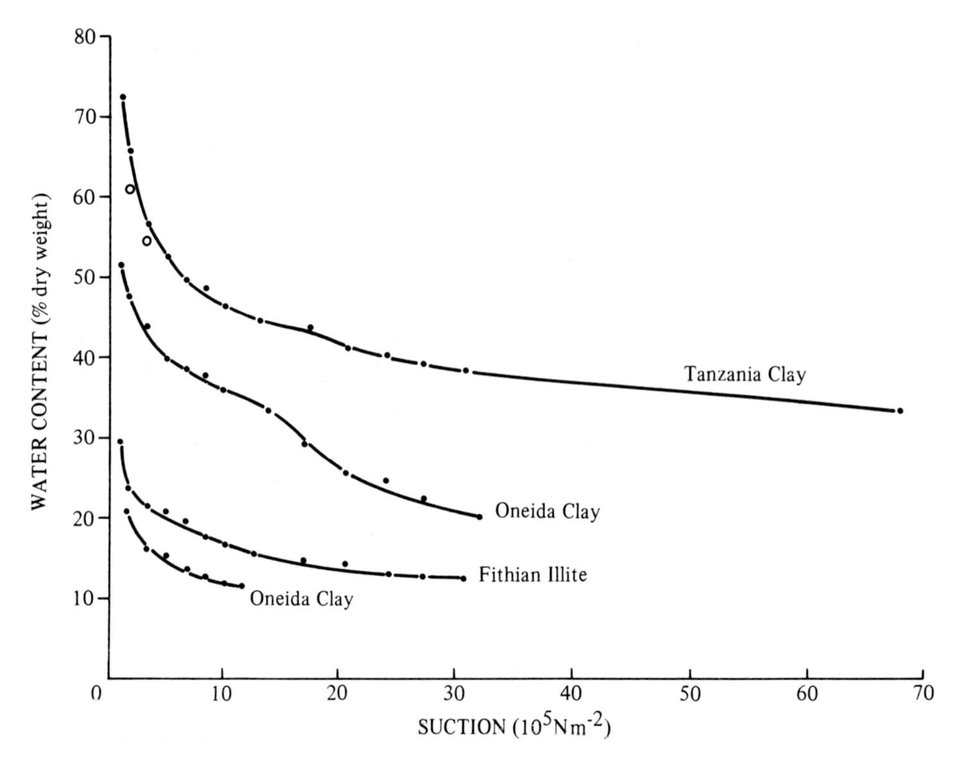

Figure 7.6 Examples of the relation of water content to soil water suction (matric potential). Curves of this kind are known as soil moisture characteristic curves. In these examples, each point was reached by loss of water. If the samples had been progressively wetted, the curves would have been slightly different due to hysteresis.

given in Davidson (1978), Day, Bolt & Anderson (1967), and Williams (1982).

It is unfortunate that different terms are used in different fields of science and technology for what is often essentially the same quantity: 'suction' in soil science, 'pore water pressure' in soils engineering, 'potential' in soil science and hydrology, 'Gibbs free energy' in physical chemistry, and others. The study of frozen ground draws on all these fields, and it is desirable to recognise these terms and their various applications. In freezing soils it is the matric potential which is particularly affected by the freezing process.

If the suction, value '*h*', in the arrangement in Figure 7.5(*a*) is increased, the water content of the soil sample will be less at equilibrium. The water content of the soil, plotted on a graph as a function of the suction, or matric potential, constitutes the *soil moisture characteristic*, or suction–moisture content relation, of that soil (Figure 7.6). More complex experimental arrangements than that illustrated in Figure 7.5(*a*) are required if the soil water characteristic is to include (as in Figure 7.6) low moisture contents where the matric potential is low (that is, the suction is large). If the column of water (Figure 7.5(*a*)) is lengthened more than a few metres with the intent of applying a greater suction, *h*, the column will break and the water runs out. Instead, the water column may be removed and the suction effect created by raising the air pressure around the sample. The water (Figure 7.5(*b*)) is then at atmospheric pressure but, as before, the suction is given by $P_a - P_w$, where P_a is the gas pressure and P_w the water pressure. Numerically, $P_a - P_w$ is equal to the applied gas pressure in this experimental arrangement, which is called a *pressure plate* or *pressure membrane apparatus* (cf. Williams, 1982). In the simpler arrangement (Figure 7.5(*a*)) P_a is, of course, atmospheric pressure.

Movement of water in freezing soils leading to the accumulation of moisture (giving frost heave), or indeed any movement of water within the ground, is analysed following these principles. The transference of water to ice and vice versa changes the matric potential of the water in a manner analogous to the change of potential when the water content is changed by drying or wetting of the soil. It is also this modification of the state of the soil water (measured as the matric potential) which explains the existence of the water at temperatures below the normal freezing point of 0 °C.

7.2.1 *Freezing point depression, potential, and free energy*

The concept of soil water potential has limitations in explaining the effects of a *difference* in temperature. For example, there may be a tendency for water to move from a point in the soil to another at a different temperature, but we cannot analyse the effect of temperature on the potential from first principles. For one thing, in definitions of potential it is specified that the system be isothermal. Clearly the reference body of water could not be 'isothermal' with respect to two temperatures.

As Edlefsen & Anderson (1943) wrote: 'Potential . . . was invented to deal with changes of mechanical energy . . . in studies on soil moisture . . . the criteria . . . for energy changes and for equilibria . . . set up by potential, are sometimes too limited. Potential, for example, takes no explicit account of the effect of temperature on the total energy change of a system. Fortunately another function, called 'free energy', was invented many years ago in the field of thermodynamics.' They might have added, however, that the ease of definition and of measurement of potential makes it a more suitable and easily utilised quantity in many studies of soils.

Gibbs free energy is defined as:

$$G = E + PV - Ts \tag{7.3}$$

where

E = internal energy

P = pressure

V = volume

T = temperature

s = entropy

The differential form, as defined by the equation

$$\mathrm{d}G = -s\mathrm{d}T + V\mathrm{d}P \tag{7.4}$$

is widely used in physical chemistry in studies of equilibria in chemical reactions, melting points and their modification, and other applications (Nash, 1970). $\mathrm{d}G$ can be expressed per mole, or as $\mathrm{J\,g^{-1}}$, and is often referred to as the *chemical potential*. The Gibbs free energies for ice and water are represented in Figure 1.3. The free energy of water in soils is modified by capillarity and adsorption such that it equals that of ice at temperatures below 0 °C. Such temperatures are freezing 'points', and the phenomenon is that of equilibrium freezing point depression. The modification is indicated in Figure 1.3.

Capillarity and adsorption are responsible for the matric potential. Indeed, the matric potential is numerically equivalent to the difference in Gibbs free energy of the soil water (relative to that of 'ordinary' water at the same temperature) arising because of capillarity and adsorption. In the previous section it was seen that the matric potential could be expressed as a pressure, ΔP, and from equation (7.4) it is apparent that, if there is no temperature change, ΔPV is equal to ΔG. The modification of the free energy of the soil water by dissolved salts is measured as the osmotic potential.

By recognising the similarities between soil water potential and free energy it becomes possible to bring the experimental methods and knowledge gained in studies of soil moisture into association with fundamental studies and relationships in physical chemistry and chemical thermodynamics. Theoretical studies applying fundamental thermodynamics to the effects of temperature in frozen soils include Everett (1961) and Low, Anderson & Hoekstra (1968). For many purposes it is sufficient to be aware that there is such a well-established and fundamental basis for understanding the properties of freezing soils. The concept of potential remains the most useful and, with due attention to definitions, the least likely to be misunderstood of the approaches to the general description of moisture in freezing soils.

Neither the free energy nor potential tell us about the physical situation specific to the interior of the soil. Expressing the matric potential as an equivalent pressure (or suction) does not mean that such pressures necessarily occur in the water within the soil, at the microscopic level. For example, the adsorption forces of mineral surfaces responsible for much of the suction at low soil water contents can be regarded as pulling the water molecules onto the surfaces, and onto each other – a situation which might be thought to be associated with elevated pressure within the adsorbed films. All we know is that water in bulk, with those pressures, would be in equilibrium with the soil water. In reality, the microscopic behaviour and the nature of soil water is not well understood. Nevertheless, the behaviour of the soil can often be considered as though the soil water simply had pressure as the cause of its potential which in any case is commonly measured in units of pressure.

7.2.2 ***Potential of water in freezing soils***

Schofield (1935) observed a relationship between the freezing points and the suction (the matric potential) in soil at various moisture contents. He pointed out that the depressed freezing points were to be expected because of the correspondence of matric potential to free energy.

Koopmans & Miller (1966) and Williams (1964b, 1967, 1976) also demonstrated the correlation between the matric potential and temperature in saturated frozen soils. The soil water characteristic curve, which relates suction and water content (e.g. Figure 7.6), was compared with that relating (unfrozen) water content to freezing temperature for the same soil (e.g. Figure 7.2). For each temperature, a suction corresponding to the water content was plotted. The correlation (Figure 7.7) was in general agreement with the point demonstrated by Schofield.

Regardless of the type of soil, the water alongside the ice has a matric potential that is temperature-dependent. These findings are the basis for a method of predicting unfrozen water content versus temperature curves for frozen soils using soil water characteristic curves. They also demonstrate that freezing (involving reduction of water content by transfer to ice) and drying (involving drainage or evaporation) have similar effects on the remaining water. The curve relating water content to temperature is sometimes known as the soil-freezing characteristic.

Further, direct evidence as to the state of the water in the frozen soil was

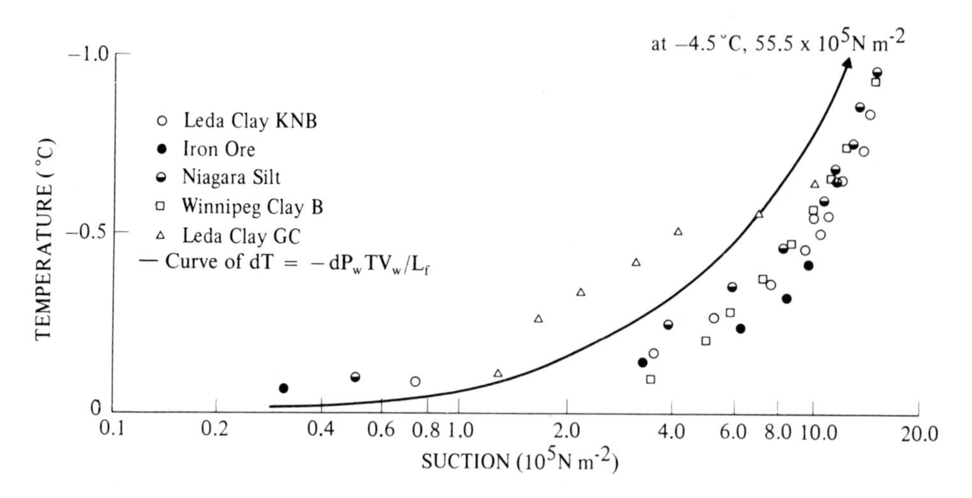

Figure 7.7 Experimental determinations of relation between temperature and suction. The continuous line is the relationship that can be predicted from Schofield's (1935) work. The samples were under atmospheric pressure.

provided by an experiment demonstrating the consolidation that occurs on freezing (Williams, 1966*a*). The suction developed pulls together the particles to produce thc denser soil. It is commonly observed that clay frozen and thawed for the first time exhibits a shaley texturc. The individual, hard flakes (which lay between discrete ice lenses (Figure 1.1) have been exposed to the effective stress produced by the suction of the water (Williams, 1967; Chamberlain, 1980). This consolidation (the term is used in the soil mechanics, not geological sense: see Williams, 1982), or volume decrease, is comparable to that observed in soils on drying. It is partly irreversible in that the soil does not expand to its original volume when the effective stress is relaxed (section 5.5.2). Note that, unlike water in an ordinary container, water in small spaces (such as soil pores) can apparently carry tensile stresses: that is, very large negative pressures. The term 'suction' does not imply any limitation to −1 atmosphere gauge pressure.

There are thus several lines of evidence concerning the value of the matric potential of the water in a frozen soil and, for temperatures near 0 °C, the relationships are clearly established. Water migration in frozen soils is to be expected along the gradients of such potentials. At temperatures below −5 °C the calorimetric methods for determining water content are hampered by uncertainty over the latent heat of fusion. Knowledge is also lacking on the matric potential, although it can be argued that it would be governed by the same general equations. The amount of water at such temperatures is so small that the dilatometric method is also difficult to apply. However, Tice *et al.* (1982) have recently shown that the unfrozen water content can be determined down to quite low temperatures (−15 °C and colder) by pulsed nuclear magnetic resonance.

In a drying soil there are interfaces between the water and air, while in a freezing soil there are interfaces between water and ice. Associated with interfaces is *surface tension* or *interfacial energy*. As will be described in the next section, surface tension influences the pressure and thus the potential of soil water. Its effects are also seen in many common situations: for example, in droplets in clouds, in thin tubes, and in soap bubbles. The interfacial energy has a value of about 30 $mN\,m^{-1}$ for water–ice interfaces (Hesstvedt, 1964) and about 72 $mN\,m^{-1}$ for water–air interfaces (Weast, 1979).

It appears that the suction for a particular water content in the frozen soil and that for the same water content in the unfrozen (partly dried) soil are often in the same proportion: 30 to 72. The reason for this is best understood by further consideration of the freezing process.

7.3 Pressure and temperature relations

Pressure changes the freezing point of water. This well-known fact is usually illustrated by the equation

$$\frac{dT}{dP} = \frac{(V_w - V_i)\,T}{L_f} \tag{7.5}$$

V_w = specific volume of water $m^3\,kg^{-1}$

V_i = specific volume of ice $m^3\,kg^{-1}$

L_f = heat of fusion of ice $J\,kg^{-1}$

T = temperature K

where dP is the change of pressure; dT is the change of freezing point, − °C; and T is the normal freezing point, 0 °C = 273.15 K, when the pressure is atmospheric.

It is often known as the Clausius–Clapeyron equation, although representing only a particular variant of that equation. Equation (7.5) tells us that the freezing point of water falls by 0.074 °C per MPa: if air is dissolved in the ice, it lowers the freezing point further (Harrison, 1972) – by 0.0028 °C for air dissolved to saturation under one atmosphere pressure. Equation (7.5) is relevant if the pressure increment applies *equally* to both phases, ice and water – which is normally assumed. Nevertheless, it is not the usual circumstance in the freezing of water in soils.

The pressure of ice in layers in soil such as those seen in Figure 1.1 is likely to be at least as great as the pressure acting on the body of soil as a whole. In the case of a free-standing soil sample, the latter is simply the atmospheric pressure (normally referred to as zero). At the same time, abundant evidence, already described, shows that the *water* in the frozen soil is *not* at atmospheric pressure.

At a freezing point, as noted in Chapter 1 ΔG of the two phases must be equal: that is, from equation (7.4):

$$\Delta G_i = \Delta G_w = -s_i \Delta T + V_i \Delta P_i \tag{7.6}$$

$$= -s_w \Delta T + V_w \Delta P_w$$

As before, the subscripts i and w refer to ice and water respectively. The difference in entropies of two phases at the temperature of equilibrium (the freezing or melting point) relates to the latent heat of fusion:

$$s_i - s_w = \frac{L_f}{T} \tag{7.7}$$

Thus equation (7.5) can be derived from equation (7.6). The condition for equation (7.5) is that $\Delta P_i = \Delta P_w$. Other forms of the Clausius–Clapeyron equation can be derived for the cases where ΔP_i is *not* equal to ΔP_w (Edlefsen & Anderson 1943). For example:

$$\frac{dP_w}{dT} = \frac{L_f}{TV_w} \tag{7.8}$$

describes the change of freezing point when the pressure on the water is lowered, while the ice remains at constant pressure. This equation is also quite general. Equations of this type have a widespread application in studies of crystallisation, in physical chemistry and atmospheric meteorology, for example.

Equation (7.8) corresponds well with the experimental observations in Figure 7.7. It shows that for ice and the water in the soil $P_i - P_w$ (the suction) increases by approximately 1.2 MPa per °C below 0 °C.

If the ice (in bodies larger than pore size) has a pressure *higher* than atmospheric (for example, due to the weight of overlying material) the situation is not described fully by equation (7.8). However, equations (7.5) and (7.8) may be used in sequence. Firstly, the lowering of the freezing point below 0 °C that occurs because of this elevated pressure, assumed to be a 'total' or 'system' pressure applied to the ice *and* to the water, is determined from equation (7.5). Then, the effect of the *difference* $P_i - P_w$

which gives a further, much larger, component of the freezing point depression, is calculated using equation (7.8). The matric potential of the water can thus be calculated as a function of temperature and of overburden or other 'resisting' or other 'confining' pressures. The concept of the development of the two pressures, that of the ice and of the water, represented by $P_i - P_w$, helps greatly in understanding the special behaviour and pressures generated by soils on freezing, as will be demonstrated subsequently. Equation (7.8) may be combined with equation (7.5) to give:

$$T - T_0 = \frac{(P_w V_w - P_i V_i)\, T}{L_f} \tag{7.9}$$

In this equation, T_0 is the 'normal' freezing point of water (273.15 K = 0 °C). The pressures P_w, P_i are expressed conventionally, with positive values if above atmospheric and negative below. As noted before, the water in the soil does not necessarily have such a hydrostatic pressure, P_w, in the strict sense but, considered macroscopically, the soil–water system behaves as though it does.

The equations demonstrate the relations between temperature, water potential (expressed as pressure of water), and the pressure of the ice – the latter being the source of the expansive forces in a frost-heaving soil.

The capillary theory of frost heave is based on analogy with the capillary model for soil moisture in unfrozen soils. The theory provides an explanation, albeit a qualified one, for the occurrence of distinct ice and water pressures in frozen soils. Suction of the soil water in unfrozen soils is, for quite moist or wet soils, an effect of the confinement of the air–water

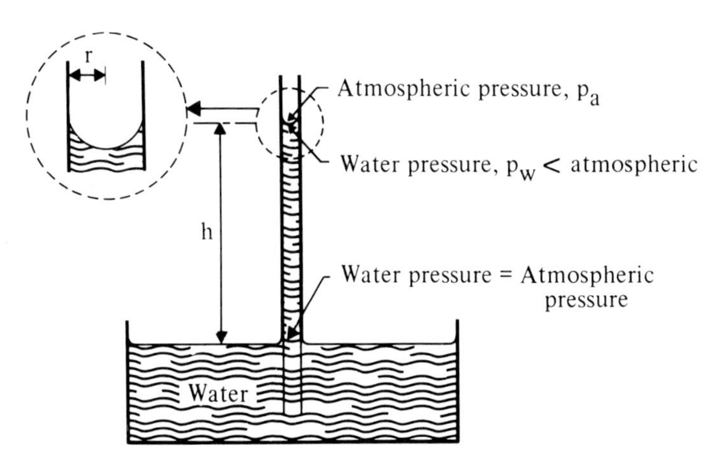

Figure 7.8 Water rises up a capillary tube to a height *h* that depends inversely on the radius of the tube.

interfaces in soil pores. It is well known that water rises up narrow bore glass tubes by an amount inversely proportional to the radius (Figure 7.8). The pressure of the water at the meniscus is lower than that of the air by an amount

$$P_a - P_w = 2\frac{\sigma_{aw}}{r_{aw}} \tag{7.10}$$

where

σ_{aw} = surface tension, air–water

r_{aw} = radius of meniscus

Equation (7.10) can be applied to the situation in a sample of soil. The r_{aw} is the 'radius' of the pores in which air–water interfaces are situated. With lower water content in the sample these interfaces will lie in smaller pores, and the suction, in the water, $P_a - P_w$, is correspondingly greater.

Assuming that ice–water interfaces in the frozen soil behave similarly, the pressure of the water is lower than that of the ice by:

$$P_i - P_w = \frac{2\sigma_{iw}}{r_{iw}} \tag{7.11}$$

where

σ_{iw} = interfacial tension, ice to water

r_{iw} = radius of interface between ice and water

Combining this equation and equation (7.8) gives:

$$T - T_0 = \frac{V_w 2\sigma_{iw} T_0}{r_{iw} L_f} \tag{7.12}$$

This equation is an expression of the fact that ice penetrates into smaller and smaller pores as the temperature falls.

The similarity of equation (7.11) to (7.10) leads to the conclusion (Koopmans & Miller 1966, Williams 1967) that there should be an analogous suction in the water when the soil contains ice in part of the pore space, to that when the equivalent space is filled with air. For *the same water contents*, the suction in a frozen soil should be 0.42 times that in similar material when unfrozen, the value 0.42 being the ratio σ_{iw}/σ_{aw}.

If, however, the soil is compressible (clay-rich) and ice segregation is occurring, there will be no such difference. This is because the ice-free layers between the ice segregations will be undergoing consolidation and the water content–suction relation will effectively be independent of the presence or absence of adjacent ice. These circumstances apply only over the few tenths of a degree, where ice has not penetrated the pores and is restricted to segregations.

The findings illustrated in Figure 7.7 are compatible with, but do not prove, equations (7.11) and (7.12). In fact, equations (7.11) and (7.12) are probably meaningful only for temperatures within a degree or so of 0 °C. With small pores the layers of water adsorbed on the particles occupy the greater part of the pore space, and this water has modified properties. It is this water that freezes at temperatures below about − 1.5 °C. In larger pores at such temperatures, the adsorbed water cannot extend across the pore which will be ice-filled, and the only concave menisci will be at re-entrants where particle surfaces approach each other. In addition, hemispherical menisci, found with air–water interfaces, would not be expected in the case of an ice–water interface. The form of the interface is presumably modified by the crystal structure of the ice. The regularity of the curves of unfrozen water content (Figures 1.4, 7.2) suggest there is no abrupt transition from a 'capillary-controlled' to an 'adsorption-controlled' freezing. Clearly, too, the simple capillary model is not directly applicable when the soil being frozen is not saturated: that is, when part of the pore space is filled with air. The application of capillary theory to porous media is considered generally by Defay & Prigogine (1951).

7.4 Origin of frost heave and frost heave pressures

Frost-heave refers to the raising of the ground surface as a result of ice accumulation of the kind shown in Figure 1.1 The accumulation occurs by migration of water to the freezing layer, and the ice is segregation ice (section 2.2.2). Sometimes the term 'frost heave' is used to refer to the ice segregation itself, but this is undesirable. Ice segregation may also occur by transfer of water from immediately adjacent pores, without overall increase of moisture content. The 9% expansion always associated with change of water to ice is not normally regarded as frost heave. The heave may be equal to half or more of the thickness of soil frozen: in the extreme, freezing may occur solely by enlargement of an ice layer. The pressure exerted by the soil as it tends to expand with the accumulation of ice is the frost heave pressure, and it has its origin in the pressure in the ice segregations. However, for the soil to heave the ice must first overcome the resistance to its expansion represented by the strength of frozen soil in which it is contained (Wood & Williams 1985b). Accordingly, the pressure in an ice lens is likely to be greater than the pressure exerted on an external surface.

Beskow's remarkable work (Beskow, 1935) established the general nature of frost heave, and the fact that the occurrence and amount of heave depended on ground conditions: soil type and the weight of overburden, freezing temperatures and the availability of water. In pioneering experiments, he froze soil samples under varying conditions of loading, and with access to water at various suctions. His results clearly demonstrated the importance of these factors.

Experience in regions of seasonal ground freezing has shown that frost heave tends to be greatest in silty soils, and absent in exclusively coarse-grained materials such as coarse sand and gravel (Linnell & Kaplar, 1959, Chamberlain, 1981). Frost heave will be greater, other conditions being equal, in moist or wet situations. Under natural conditions heave often varies greatly from point to point: this gives an uneven, bumpy ground surface. Coarse-grained 'non-frost susceptible' materials are in demand for foundations of highways or elsewhere where heaving must be avoided.

Frost heave depends upon the lower pressure in the *water* adjacent to the ice (in the following we shall continue to refer to pressure, rather than potential, when convenient: the relationship of the terms (see section 7.2) can be borne in mind). This lower pressure causes the flow of water to the freezing layer. The heave results from the *increment* of pressure in the ice, relative to that of the soil water, great enough to separate and lift the overlying material. The two effects combine to give the pressure difference $P_i - P_w$, discussed in the previous sections. The difference in pressures of the two phases provides the basis for understanding the frost heave process, at least in saturated soils. Consideration of unsaturated soils, where air occupies much of the pore space, is deferred to section 8.4.5.

Normally, at a point in the ground, in the absence of freezing the amount of water present and its potential depend on the hydrological environment. If the static (equilibrium) situation is disturbed, flows occur involving exchange of water with that of neighbouring ground, until a new static situation is reached. The situation during freezing and heaving is indicated in idealised form in Figure 7.9. When the gradient towards the freezing ground becomes zero – that is, when the potential of the water adjacent to the ice is again in equilibrium with the ground water in general – ice segregation ceases. The frost-heave pressure, the pressure of the segregating ice, P_i, will have reached its maximum.

If $P_i - P_w$ is defined and if P_w is given the value of the pressure of the water ultimately available to form ice, then P_i is also defined. It follows that

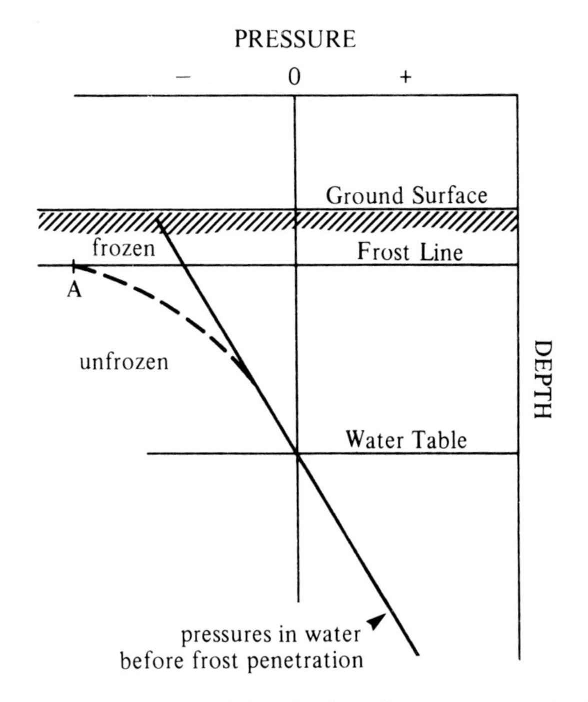

Figure 7.9 Idealised distribution of water pressure in the ground under static conditions in the absence of freezing (continuous line) and (dashed line) during penetration of frost. Under static conditions the linear gradient merely reflects the depth effect and does not imply movement.

where the hydrological environment gives high water pressures (wet places), the ice pressure, P_i, which with some approximation we equate with frost heave pressure, correspondingly, tends to be high. In less wet situations the frost heve pressure will be less. This presupposes that the heaving soil is resisted by weight of overburden, a structure, or other source of force of reaction. If such is not present, the pressure, P_i, cannot rise and instead P_w falls to satisfy the temperature-determined value of $P_i - P_w$ (equation (7.9)). Frost heave is then more rapid. It is well documented that wet conditions favour heave and that applied loads reduce the rate of heave. Indeed, if sufficient pressure is applied, frost heaving can be stopped and further freezing results in water being displaced ahead of the ice and into the unfrozen soil. A pressure of MegaPascals is required for fine-grained soils.

While the pre-existing moisture regime of the ground is of basic importance, equally so is the magnitude of the relevant $P_i - P_w$ generated by the freezing of the soil in question. Temperature is not normally uniform throughout a freezing soil, and the question arises as to where the ice segregation occurs. Before the significance of the *water* in frozen soil was understood, it was tactily assumed that ice segregation occurred exclusively at the boundary between the frozen and unfrozen layers. As this 'frost line' penetrated the ground, heave would occur in each increment as it froze. The penetrating frost line is of course the location of the highest (freezing) temperature (0 °C or very slightly below): ice segregation is often concentrated in its vicinity. In fine-grained, and consequently finely porous soils, the ice within pores is particularly confined and, in accordance with equation (7.12), the temperature at the frost line is slightly lower than in coarse-grained soils. It follows that the value of $P_i - P_w$ at a penetrating frost line will tend to be greater the more fine-grained a soil is. Many experiments have demonstrated this (e.g. Beskow, 1935; Williams, 1967).

While the tendency for heave might therefore be thought greatest in the most fine-grained soils, the largest heaves experienced in engineering practice (other factors being equal) are usually associated with soils with silt-size particles. This is because the migration of water to the freezing layers is restricted by the low permeability of clays.

The rate of flow of heat is also important with respect to the distribution of ice. High rates (rapid cooling) are generally associated with thinner ice lenses. On the other hand, in soils which are quite permeable the water may migrate to the freezing zone so fast that virtually all the heat being removed is the latent heat of fusion. Ice lenses then grow to a large size or will be very abundant (see section (8.5.2)).

Most laboratory experimental studies of frost heave involve measurements made over a few days at most, and with relatively rapid penetration of frost. Under these conditions most ice formation occurs at or near the frost line, and this largely determines the volume changes and pressure effects being measured. But the continuity of the liquid phase from the underlying unfrozen soil, and into the frozen layer, suggests the migration of water to some point within the frozen layer. Because the value $P_i - P_w$ increases as the temperature is lower (equation (7.9)), there is then the possibility of ice segregation occurring with much higher heaving pressures. However, such migration would be restricted by the low permeability of the frozen soil to water. Indeed, when the frost penetrates the ground at rates normally associated with winter freezing, it seems that ice segregations mainly develop very near the frost line. Heaving associated with such freezing may be called 'primary' heaving to distinguish it from the slower 'secondary' heaving that occurs *within* frozen layers at various temperatures (the terms 'primary' and 'secondary' heaving are defined somewhat differently by Miller, 1972, 1978). The importance of 'secondary' heaving lies in the very large heaving pressures that may develop over time, and the slow and therefore not immediately detectable heaving process. The phenomenon appears particularly important to geotechnical engineering in permafrost regions, or elsewhere that frozen ground persists over many years (such as around cold storage plants). This topic is discussed further in Chapter 8.

Frost heave pressure, which is of great practical importance, seems easily defined. But its prediction, and that of the rates of the ice segregation itself, is difficult and even the subject of heated debate. For example, frost heave pressure is a serious and controversial problem in the design of gas pipelines for cold regions (Williams, 1986).

If the pressure is exerted against a structure or earth mass which can yield to some degree before developing the full reaction force, then the pressures may be less. This will be so if the frost heave necessary to take up the displacement does not occur in the time available. Frost heave pressures of 100–300 kPa probably occur in association with quite large heaves, developing relatively rapidly. In particular situations values several times this may occur with significant consequences, although perhaps only after months or years. If soil is rigidly confined, extremely high values develop according to equation (7.5).

7.5 Permeability of frozen soils

The movement of water through soils and rocks depends upon the gradient of potential in the water, and on the ease with which water passes through the material. The latter property of the material is called the permeability (or alternatively, hydraulic conductivity) and has the units $m\,s^{-1}$. By an experiment in which an ice lens was observed to increase in size although situated a few millimetres *within* already frozen material, Hoekstra (1969) demonstrated that water moves through frozen soil. He also calculated (Hoekstra, 1966) a rate of water migration in frozen, partly saturated soil on the basis of changes in moisture content with time. An experiment using various frozen soils (Burt & Williams, 1976) showed that the permeability decreased with temperature as well as being related to soil type (Figure 7.10). The experiment was performed under isothermal conditions and involved a permeameter in which water in two reservoirs was prevented from freezing by dissolved sugar (Figure 7.11). Cary & Mayland (1972) proposed that the permeability of a frozen soil to water should be similar to that of the soil at above-freezing temperaures when unsaturated – that is, whether a part of the pore space is occupied by air or by ice,

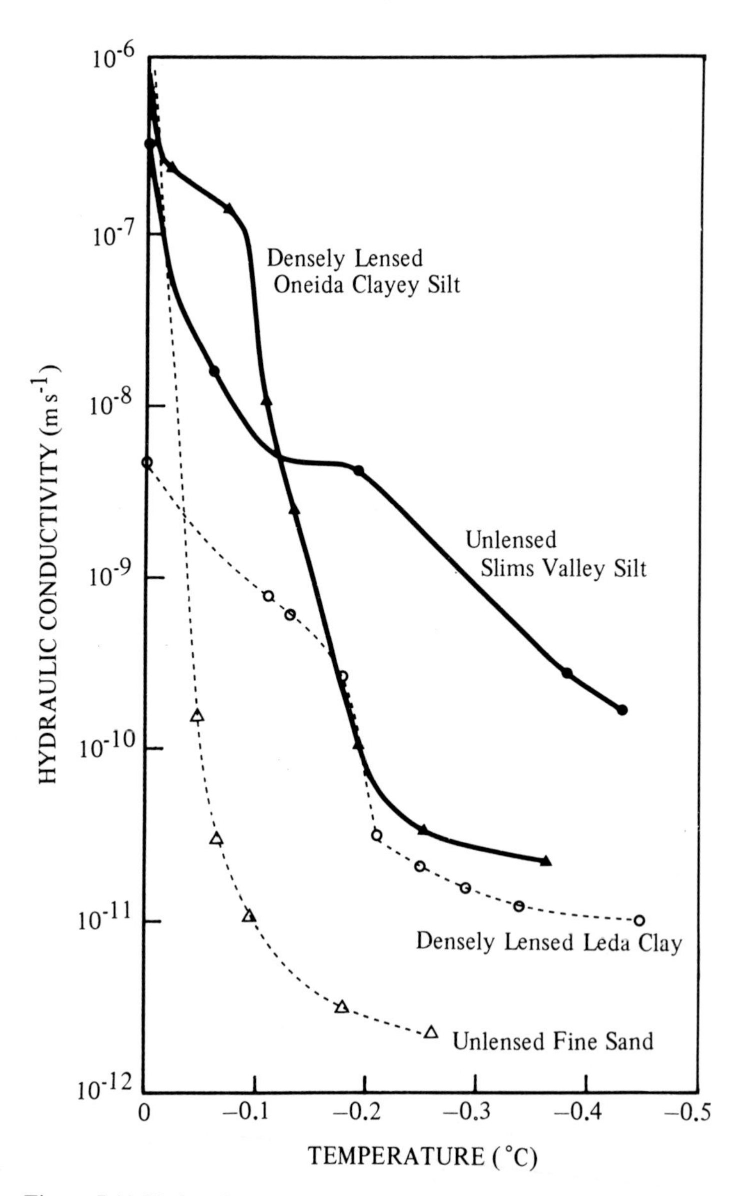

Figure 7.10 Hydraulic conductivity of various frozen soils, as a function of temperature. After Burt & Williams 1976.

the permeability would be the same at the same water content. However, the presence of ice segregations in the frozen soil makes the analogy imperfect.

In an ingenious experiment with a layer of ice between semi-permeable membranes, Miller (1970) demonstrated transport of molecules from a solution on one side of the ice and into the water on the other, on application of pressure to the one reservoir. This involved regelation, with ac-

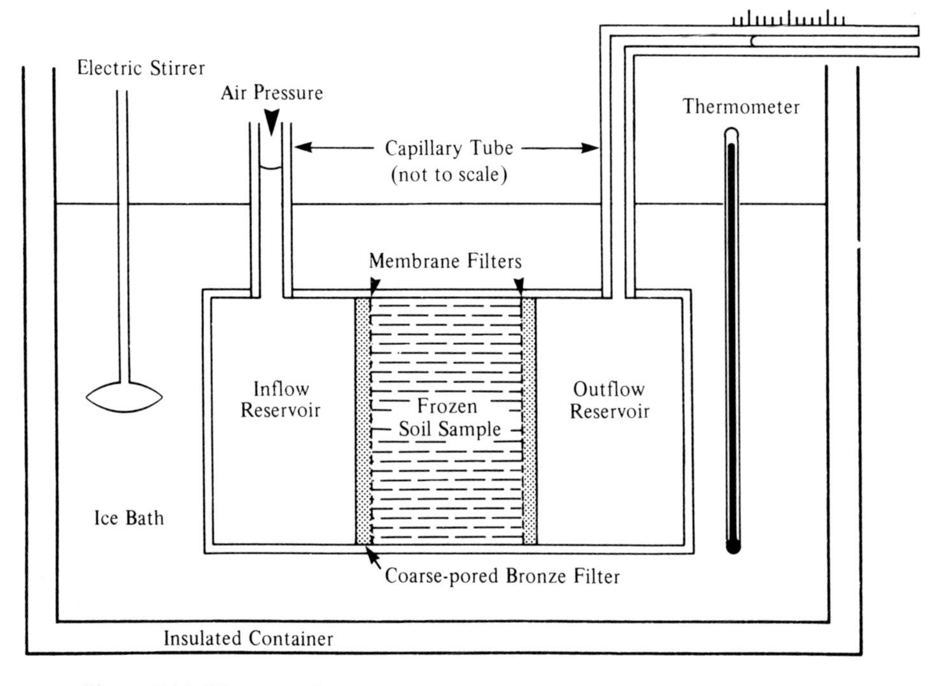

Figure 7.11 Diagram of apparatus used to measure hydraulic conductivity of frozen soil under isothermal conditions.

cretion of ice to one face of the ice layer and thawing at the other (section 8.4.3). The experiment suggested that ice masses in a frozen soil do not represent barriers that would greatly limit the soil permeability. The regelation mechanism means that lenses participate in water migration in frozen soil, indeed, molecules of ice move steadily, even though a lens will apparently remain stationary. Miller's experiment has been repeated using supercooled water on both sides of the ice (Horiguchi & Miller, 1980).

Work by Nye & Frank (1973) and Osterkamp (1975) shows that ice at temperatures within a few tenths of a degree of its melting point has numerous water-filled inter-crystalline boundaries. These also allow the passage of water through ice, but the process is distinct from that envisaged by Miller and co-workers, and is only significant in very 'warm' ice.

With the exception of Hoekstra's study (1966), most consideration has been given to permeability of frozen soils which are saturated (that is, without significant air content) or supersaturated (that is, with excess ice). If soil is not saturated and there are continuous air-filled passages, then migration of water vapour and an evaporation–sublimation process occur in addition to the movement in the liquid and perhaps ice phase.

Although permeabilities for the frozen state are generally low, some soils have permeabilities at temperatures down to −1 °C or −2 °C which are similar to those found in very dense clays in the unfrozen state. Consequently, significant migrations of water are probable, if appropriate gradients of potential (notably those associated with temperature) persist over a period of years. A temperature gradient of 1.0 °C m^{-1} and a permeability of 10^{-11} m s^{-1} (a rough value for silty soils at temperatures of −0.5 °C and somewhat lower) gives a migration of water (following Darcy's law) equivalent to the accumulation of a 1 m-thick ice layer in thirty years. But there are uncertainties in this calculation, and the migration might be an order of magnitude more or less.

While significant migrations of moisture occur within the 'frozen fringe' of soil (Penner & Walton, 1978), where temperatures are within a few tenths of °C of the freezing 'point' (i.e. the initial freezing temperature) less is known about lower temperatures. Mageau & Morgenstern (1980) reported laboratory observations of significant water movements at temperatures of 0 °C to −2 °C. Field observations suggest larger values. These are considered further in Chapter 8.

7.6 Frost heave as a thermodynamic-rheologic process

In the ground, a layer of frozen soil of considerable thickness may be significantly affected by movement of moisture within it. Though such movements may be limited to soils in a narrow range of temperature, such temperatures occur over a considerable distance under the low temperature gradients often prevailing. The magnitude of heaving pressures observed is the result of ice accumulation occurring where the temperature is below 0 °C, rather than solely at the boundary of freezing ground.

The enlargement of an ice body that is surrounded by frozen material requires that the confining material yield. This will only occur if the thermodynamically determined pressure of the ice, as discussed in previous sections, is great enough. If yielding does occur, it will involve creep (deformation) of the frozen ground. The rate of creep will depend on the pressure the ice exerts (as discussed in Chapter 9). Conversely, the rate of creep will exert a control on the rate of accretion of ice and thus on the frost heave. Consequently frost heave occurs at a rate which depends in a complex manner on the thermodynamic conditions of temperature and water and ice pressures – conditions that are modified by the creep properties of the particular frozen soil.

Experiments using microtransducers inserted in frozen soil samples subjected to a temperature gradient have shown pressures of more than 100 kPa locally in the sample. The observed pressures are related to temperature and also undergo changes with time in association with moisture movements in the sample (Williams & Wood, 1985; Wood & Williams, 1985b). In these experiments there was a source of water at atmospheric pressure contiguous with the sample. The fluid nature of water and the relative ease of its movement means that there is a strong tendency towards equalisation of water pressures even within the frozen sample. Thus the pressure the ice exerts will (if the water pressure is taken as atmospheric) be greater than atmospheric, corresponding to the difference $P_i - P_w$, according to the temperature, as given by equation (7.9).

Under natural conditions, the pressure of the water may well be less than atmospheric depending on the hydrological regime, and this would reduce the value of the pressure attainable by the ice, according to equation (7.9). The creep rate of the frozen soil would be less and thus the frost heave rate. This is in accordance with the general observation that frost heave is reduced by drier conditions.

The complexity of the frost heave process and our limited knowledge does not yet provide a basis for quantitative prediction of heave. However both the thermodynamic and rheologic aspects are considered further, in Chapters 8 and 9.

8

Hydrology of frozen ground

8.1 Introduction

The thermal conditions of cold regions are important to hydrology in a number of ways. At a macroscale, the various elements of the hydrologic cycle are modulated in intensity, magnitude and significance when compared to mid-latitude regions. For example, winter snow storage is the single most important feature of the hydrological cycle in cold regions, and each year, for six months or more, arctic water bodies are affected by ice. As a consequence, streamflow is intensely seasonal or even ephemeral. The spring melting of snow and ice, when soils are frozen and infiltration is limited, accentuates peak flows. On the Colville River in Alaska, the break-up period accounts for about 40% of the annual flow (Arnborg *et al.*, 1967). On smaller rivers, the spring flood may account for as much as 90% of the discharge (e.g. McCann *et al.*, 1972). In summer, the runoff response is modulated by the thawing of the active layer.

The groundwater regime is profoundly affected by the presence of permafrost, which limits infiltration and water movement. Where permafrost is continuous, the groundwater contribution to runoff is probably less than 10%, but can increase to 20–40% in regions of discontinuous permafrost (MacKay & Loken, 1974). In the fall, freezeback of the active layer can restrict subsurface flow, forcing it to the surface and resulting in unique features such as surface icings and frost blisters. The local configuration of permafrost itself is affected by the presence of water bodies (see Chapter 4), and unique hydrologic features, such as thermokarst lakes and beaded drainage, can result from the melting of ground ice. A brief, but comprehensive, review of arctic hydrology is provided by MacKay & Loken (1974).

In a quite different respect, freezing temperatures create unique *hydrodynamic* conditions within the ground. This truly distinctive aspect of hydrology in cold regions springs from the fact that at temperatures below 0 °C thermal gradients control water movement in the ground (Chapter 7). The hydrological regime of frozen ground, therefore, is intimately linked to the thermal regime; gravitational and hydrostatic forces, in comparison, are generally negligible. Such processes are not generally treated in the literature dealing with the groundwater or watershed hydrology of permafrost regions, where the frost table is viewed simply as an 'impermeable' barrier and permafrost as an aquiclude or aquitard (e.g. Brandon, 1965, 1966; J.R. Williams, 1970; Slaughter *et al.*, 1983).

In keeping with the focus of this book, this chapter deals mainly with the unique coupling of thermal and hydrological conditions that exists in the frozen ground environment, as opposed to describing the operation of conventional hydrological processes within a permafrost setting. Before discussing these special features, however, some interesting aspects of surface and groundwater hydrology will be summarised briefly.

8.2 Surface hydrology

It has been suggested that the cold climate of high latitudes, which results in prominent frost action and restricts geomorphological processes dominant in other climates, has produced a distinctive periglacial landscape. Indeed, the details of this landscape (Chapters 5 and 6) certainly do lend an individuality to cold regions. Perhaps because of this, there is a point of view that since streams in periglacial regions flow for only a few months of the year, fluvial processes play a minor role in landscape development in cold regions. However, according to Church (1972), the large-scale organisation of arctic landscapes is not unlike that of other regions, and a well integrated drainage network exists. Though arctic rivers flow only during a few summer months, and there is little or no movement of surface water, or even near-surface groundwater, for much of the year, flowing water is still capable of great erosional and transporting activity.

Church (1974) has provided a summary of the surface hydrology of permafrost regions, while Ryden (1981) and Woo (1986), for example, have reviewed the broader operation of hydrologic processes in the permafrost environment. In general, streamflow response to snowmelt or rainfall is rapid (e.g. Dingman, 1973; Cogley & McCann, 1976; Lewkowicz & French, 1982) and the pattern of subsurface drainage is controlled by spatial and temporal variations in the active layer (Woo & Steer, 1983). As mentioned, groundwater contribution to streamflow in areas of continuous permafrost is negligible leading to a characteristically ephemeral regime (e.g. Dingman, 1973; Onesti & Walti, 1983). Streams are frozen over in winter, and are often frozen solid. Only the very large north-flowing rivers maintain a reduced flow beneath the ice (MacKay & Loken, 1974).

Overall, however, these features are really not very different from those of southern regions, where seasonal frost can lead to rapid spring runoff, or where an impermeable horizon lies close to the surface.

8.2.1 Bank stability in permafrost regions

Block slumping is the main feature of river bank (and coastal) retreat in permafrost regions, especially in semi-cohesive fine-grained sediments (Lawson 1983). These frequently contain excess ice, and when thawed they become very weak. In combination with undercutting, and the consequent increase in shear stresses, this leads to collapse of soil in large blocks. Often, there is plastic deformation and flowage of mud (see Chapter 5).

The presence of ice wedges exerts a strong control over the pattern of failure; fractures along ice wedges occur both parallel and perpendicular to the bank (Figure 8.1). After collapse, degradation of the blocks and bank proceeds as exposed permafrost thaws and material is removed by river action. Wave action and sloughing are important auxiliary processes. Rates of recession of 10 metres per year have been reported in the Colville Delta,

Figure 8.1 Block slumping controlled by ice wedges. Photo: P. Sellman.

Alaska (Walker & Arnborg, 1963) and Outhet (1974) reported retreat of up to 30 metres in a single year at individual locations in the Mackenzie Delta.

Bank erosion is initiated in the spring, when water contacts the frozen bank, and thawing begins. With increases in flow, undercutting of the bank proceeds, resulting in a so-called *thermo-erosional niche.* Undercutting is continued by peak flows due to summer rains, or backwater effects in coastal reaches. In semi-cohesive, fine-grained sediments, niches can develop rapidly, for strength is largely provided only by the frozen condition of the material. Niches up to 8 metres deep were reported by Walker & Arnborg (1963).

Whilst not particularly spectacular, block slumping can continue regularly for many years; Smith (1976) determined an average rate of bank recession of about one metre per year over a 30 year period in the Mackenzie Delta. Similarly, Henoch (1960) reported that the Peel Channel at Aklavik, NWT, retreated 25 metres in 20 years.

On a larger scale, entire valley sides of ice-rich material may slump or flow when river erosion exposes ground ice to melting (e.g. McRoberts & Morgenstern, 1974; Burn, 1982). McDonald & Lewis (1973) reported that the headwall of one active thaw slump retreated more than 10 metres in one year, delivering large amounts of sediment to the river. Although related to fluvial action, the phenomenon is ultimately one of soil mechanics (see Chapter 5).

8.3 Groundwater

The traditional view of groundwater in permafrost regions is that permafrost acts as an impermeable barrier to water movement. In this view, groundwater movement is restricted by the presence of both perennially and seasonally frozen ground, and groundwater exchanges are mainly accomplished through unfrozen zones (taliks) that penetrate the frozen ground. Thus the local configuration of permafrost has a significant influence on the pattern of groundwater circulation. Because of the high heat capacity of water, the movements of groundwater constitute a mechanism for the transfer of large quantities of heat. However, relatively little is known about the effects of groundwater movement on the local configuration of permafrost (but see, for example, Dostovalov & Kudriavtsev, 1967).

Kane & Stein (1983*a*) point out that, with respect to groundwater, a distinction must be made between areas of continuous and discontinuous permafrost. Where permafrost is continuous, there is essentially no hydraulic connection between groundwater in the active layer and sub-permafrost groundwater. In areas of discontinuous permafrost, hydraulic connections exist through taliks which perforate the permafrost. The extent of frozen ground in a river basin is one of the most important factors determining the base flow of streams, and groundwater flow becomes progressively more important as one moves southward and permafrost becomes less continuous.

Overall, the major part of groundwater recharge occurs during spring snowmelt. Kane & Slaughter (1973) demonstrated the connection between the surface and sub-permafrost aquifers via taliks beneath lakes, which act as conduits for recharge or discharge. However, because of the generally low precipitation in arctic regions, the long periods of sub-freezing temperatures, and the restricting influence of frozen ground, rates of groundwater recharge are orders of magnitude lower than in non-permafrost regions (van Everdingen, 1974).

8.3.1 Icings and frost blisters

A distinctive hydrological circumstance can arise in permafrost regions as the thaw season comes to an end and the ground freezes back. Surface effusions of groundwater in the form of icings, frost blisters or frost mounds are prevalent in some areas (e.g. van Everdingen, 1978, 1982). These features originate from the expulsion (injection) of free water confined by downward freezing in the active layer (Figure 8.2). As water flow is restricted in a *narrowing cryoconduit* (a residual unfrozen passage through the active layer), a sufficient hydraulic potential may develop to deform or rupture the overlying frozen material (Pollard & French, 1984). In the case of blisters and mounds, a water-filled chamber or reservoir is formed and fed by continuing groundwater discharge. This can be considerable where the cryoconduit is linked to a perennial spring source. For example, van Everdingen (1982) recorded vertical growth rates of up to $0.55\,m\,day^{-1}$ in frost blisters.

A continuing discharge of groundwater onto the surface in winter results in the formation of *icings* (section 2.2.4). These consist of horizontal layers of ice from a few centimetres up to many metres in thickness. They may form as a result of the constriction of interflow or river flow by seasonal frost (as above), or from the discharge of a natural groundwater spring. The largest icings are of the latter type.

Kane *et al.* (1973) measured groundwater pore pressures adjacent to a stream throughout the winter. The pressure head increased throughout the period with the advance of seasonal frost (Figure 8.3). The interspersed periods of sharp decline in head occurred when water was released onto the

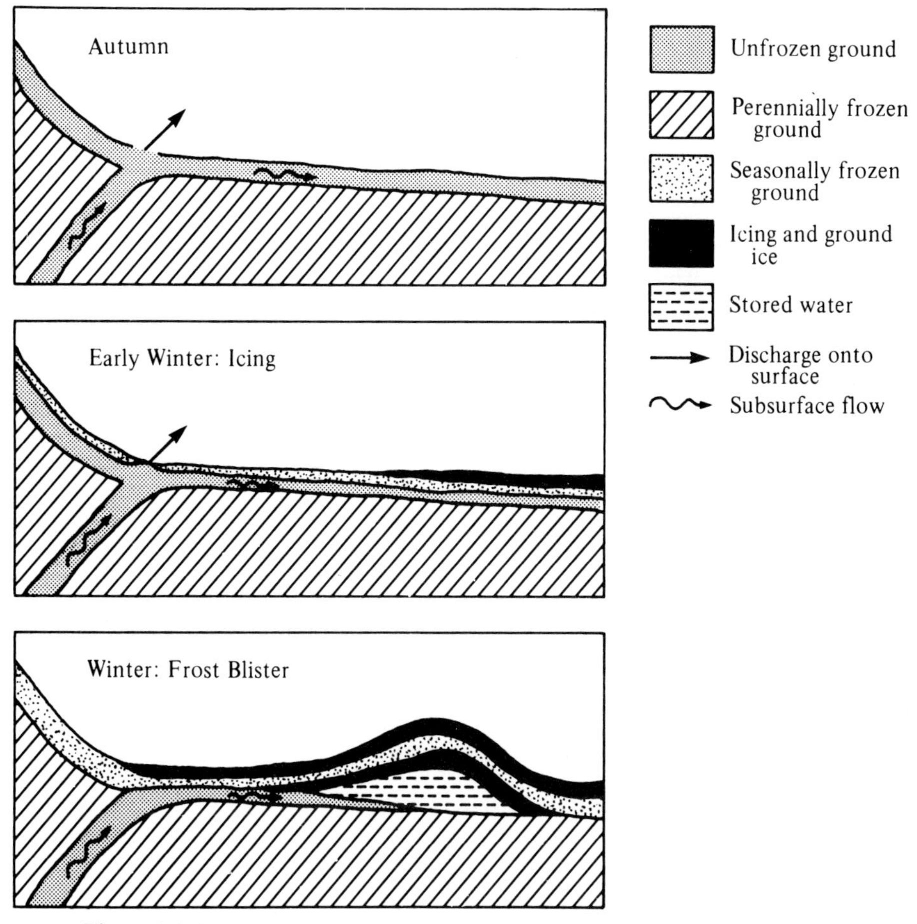

Figure 8.2 Sequence of events in the formation of a frost blister (from van Everdingen, 1978).

icing surface. Before water can discharge, the pressure head must exceed the icing surface elevation. The increasing duration between peaks probably resulted from a decline in the water available to the stream. By winter's end, the icing completely filled the creek.

Very long icings are found in narrow river valleys, while in wider valleys they form into extensive tabular masses. Round or oval icings occur at the foot of slopes and on alluvial fans. The largest reported icing is 80 km^2, with a volume of 200 million m^3 (Tolstikhin & Tolstikhin, 1974). The total volume of icings in the Sagavanirktok River basin, Alaska, has been estimated at 123 million m^3 (J.R. Williams & van Everdingen, 1973). This represents an average discharge from all sources of about 6 m^3 s^{-1}, and the icings add about 23 m^3 s^{-1} to streamflow over a two-month melt period.

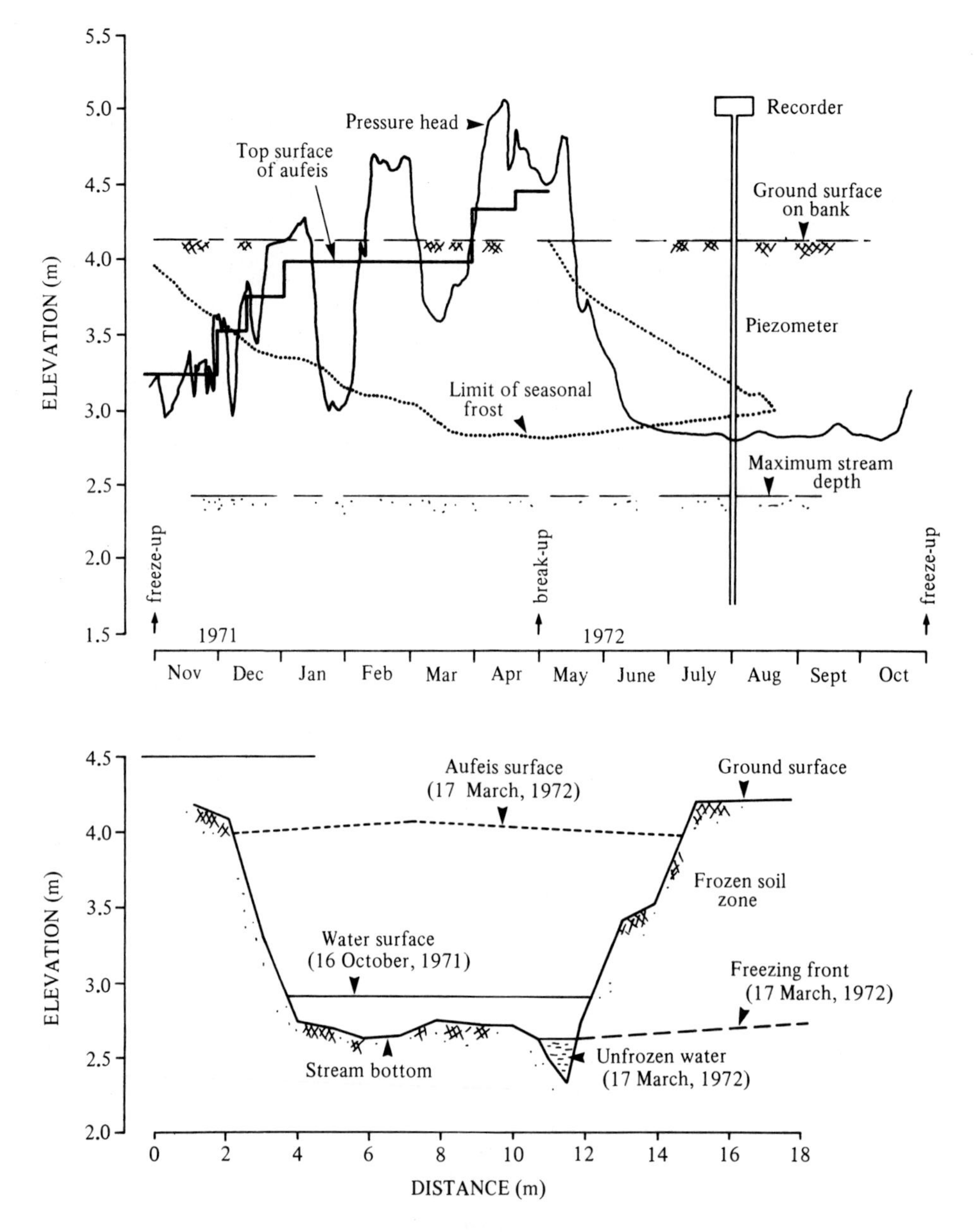

Figure 8.3 Pressure head variation and aufeis formation, Goldstream Creek, Alaska (from Kane *et al.*, 1973).

8.3.2 Water supply

The problems of developing water supplies in the North relate partly to the presence of permafrost and partly to the severity of the climate. While permafrost limits the extent of aquifers, winter freezing affects aquifers in the active layer and in taliks by reducing and occasionally cutting off the supply of water. This can lead to seasonal reductions of flow from wells, and even cessation of flow. Surface sources, such as lakes and rivers, although used in some local instances, suffer from problems such as winter freezedown and the concentration of impurities (Sherman, 1973).

Although groundwater is largely confined in sub-permafrost aquifers, substantial volumes of unfrozen material may be present beneath rivers and lakes. Shallow aquifers in taliks beneath lakes and rivers are the only known sources of potable groundwater, other than springs, along the arctic coast of Alaska (J.R. Williams, 1970). In this region, deep sub-permafrost water is frequently saline, but the presence of permafrost acts as a barrier to protect shallow aquifers from seawater intrusion. Alluvial deposits in river valleys are the sources of the largest and most economic supplies of groundwater in Alaska. The extent of frozen ground within the alluvium affects the volume of available groundwater (Figure 8.4). In January, the

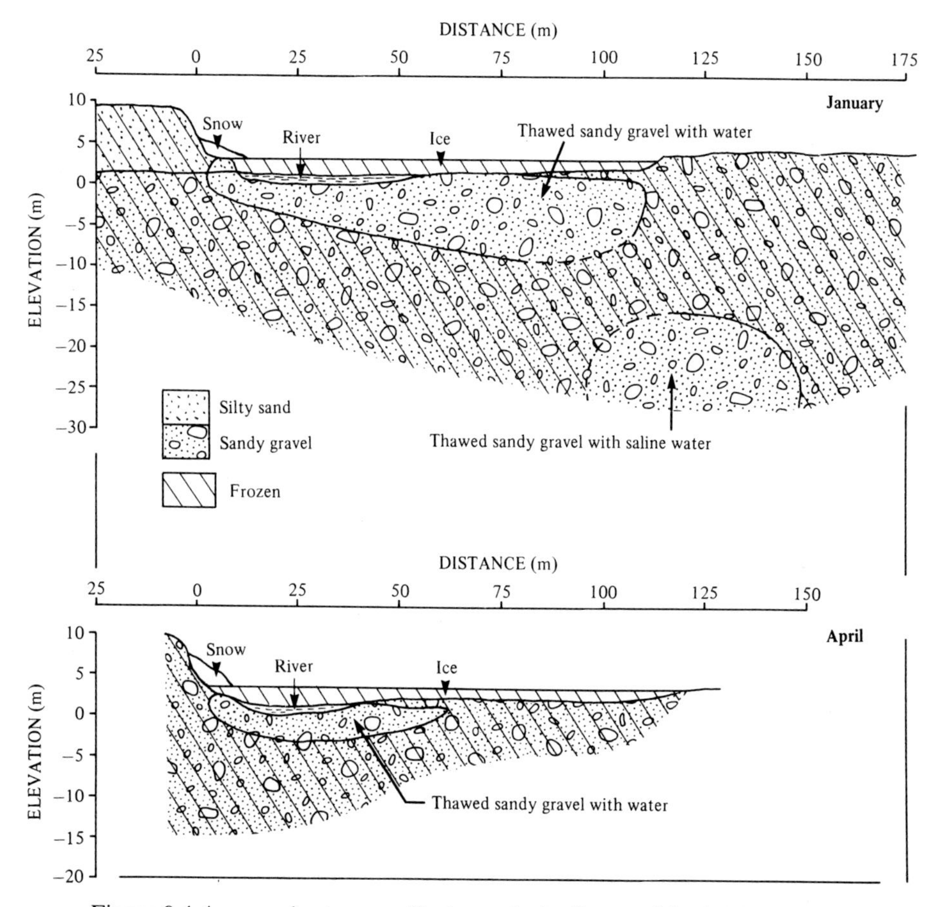

Figure 8.4 A groundwater acquifer beneath the Sagavanirktok River, Alaska (from Sherman, 1973).

unfrozen zone had a cross-sectional area of 760 m^2 with a potential yield of 50 000 m^3 day^{-1} (Sherman, 1973). By April, which is the time of maximum freeze down, the unfrozen zone was reduced to 170 m^3, with a daily yield of only 11 000 m^3.

8.4 Hydrodynamics of frozen ground

Although it is generally assumed that subsurface movement of water in permafrost regions occurs only through unfrozen zones that perforate the frozen ground (taliks, cryoconduits), water movement within frozen ground cannot be excluded, particularly over geomorphic time scales (see Harlan, 1974). From a hydrodynamic point of view, frozen ground does not constitute a barrier to all water movement, but rather should be regarded as a confining medium of low, but finite, permeability (an aquitard). Further, this should be viewed in relationship to the large gradients of water potential which are developed within frozen soils. Previously, a lack of quantitative information on the hydraulic (and thermal) properties of frozen ground precluded a full appreciation of its hydrological significance, but this is now beginning to emerge.

8.4.1 Unfrozen water content

Significant amounts of liquid water can exist in soil–water systems in equilibrium with ice at temperatures considerably below 0 °C (see section 7.1.2). This water exists in small capillaries and as films adsorbed on the surfaces of soil particles. Burt & Williams (1976) demonstrated that this unfrozen water is mobile, but that as the temperature falls, the permeability decreases due to the freezing of capillary water and the decrease in thickness of the water films around soil grains. It does not fall to zero, however, and some permeability to water remains, even at quite low temperatures.

8.4.2 Gradients of water potential due to freezing temperatures

Water in frozen soil is subject to several forces that lower its potential (pore pressure), creating a suction relative to bulk water (sections 7.2 and 7.3). The matric potential, created by the capillary and adsorptive forces of the soil matrix, is the major component of soil water potential in frozen soil, and, regardless of the soil type, the water alongside the ice has a matric potential ($P_i - P_w$) that is proportional to the temperature depression below 0 °C. The relationship is described by the Clausius–Clapeyron equation (section 7.3).

This temperature–pressure relationship translates into enormous suction gradients given the temperature gradients commonly found in natural soils.

For example, a temperature gradient of only $1\,\mathrm{K\,m^{-1}}$ can give rise to a gradient of hydraulic potential of about $1.2\,\mathrm{MPa\,m^{-1}}$ or 120 metres of water head per metre of soil. The significance of this is that a temperature gradient in frozen soil will cause heat and soil moisture to flow interdependently in the same direction – i.e. water is drawn to zones of lower temperature and, as it accumulates there, ice lenses grow (section 8.5). This ice segregation process causes the total moisture content of the frozen soil to increase. Whereas soil freezing can produce very considerable volumetric expansion, thawing of frozen soil is often accompanied by weakening and consolidation. *Frost action* phenomena are related to the strains which result from these processes. In addition, irreversible changes in soil structure occur during repeated freeze–thaw cycles, which affect soil permeability and mechanical properties (e.g. Chamberlain & Gow, 1979; Yong *et al.*, 1985).

Another implication of the Clausius–Clapeyron relationship is that as the temperature of an unsaturated soil drops below 0 °C, water will not freeze until the appropriate temperature depression is reached for soil water under the suction ($P_i - P_w$). Thus the unfrozen water content curve (section 7.1.2) not only describes the relationship between the unfrozen water content of saturated soils and temperature, but also indicates the temperature at which pore water freezing begins in unsaturated soil.

Furthermore, because of the temperature–pressure dependency, an equilibrium temperature gradient can be defined such that liquid, vapour and solid exist in a state of hydrostatic equilibrium. Using a form of the Clausius–Clapeyron equation, Harlan (1974) calculated that hydrostatic equilibrium would occur with a vertical temperature gradient of $0.00816\,\mathrm{K\,m^{-1}}$. This is much smaller than the gradients usually found in nature – for example, for the temperature profiles shown in Figure 4.4, gradients range from 0.016 to $0.038\,\mathrm{K\,m^{-1}}$. Thus, given an adequate supply of water, we may expect a constant migration into the base of permafrost, and in addition a steady upward redistribution of moisture toward the ground surface. However, where the ground is cold (and the unfrozen water content is small), the permeability will be quite low and the rate of upward redistribution will generally be small. Closer to 0 °C, though, these effects may be extremely important and overall the process may contribute to the widespread occurrence of ice-rich ground in the upper layers of permafrost (but see section 8.6.2). (The effect of the overburden pressure in controlling the ice pressure at depth must also be taken into account, of course, since this will reduce the gradient of potential driving water towards the upper layers of permafrost.)

8.4.3 Moisture transfer mechanisms

While the movement of water through unfrozen soil towards a freezing front is well-known, frozen ground itself was for long considered to be hydrologically inactive. In recent years, however, studies have demonstrated that a mobile liquid phase persists in frozen soils at temperatures significantly below 0 °C (e.g. Burt & Williams, 1976; Horiguchi & Miller, 1983; Yoneyama *et al.*, 1983; Ohrai & Yamamoto, 1985; Smith, 1985). In addition, Miller *et al.* (1975) argued that a continuous liquid phase would not only allow transport via water films, but also transport from regelation of pore ice. Regelation is driven by alterations to the equilibrium ice–water pressure difference; the rate, therefore, depends on the rate at which liquid water arrives at or departs from the ice mass. Finally, vapour diffusion may occur in unsaturated soils.

LIQUID TRANSPORT

Burt & Williams (1976) measured water flow through frozen soils, using an applied pressure gradient, although Miller (1978) argued that they measured the sum of liquid flow and regelation transport in their permeameter (see below). An apparent conductivity was calculated as the ratio of the pressure gradient to the observed flow, assuming Darcy's Law to apply. They showed that the conductivity decreased with temperature, as well as being related to soil type (section 7.5). Subsequently, other researchers e.g. Loch & Kay, 1978; Horiguchi & Miller, 1980; Mageau & Morgenstern, 1980; Perfect & Williams, 1980; Oliphant *et al.*, 1983, measured the movement of water in frozen soil in response to a temperature gradient. In these cases, it was assumed that the temperature gradient led to a gradient of potential according to the relationship in equation (7.8).

Determinations of the hydraulic conductivity of fine-grained frozen soils give values of about 10^{-12} m s^{-1} at -0.5 °C to 3×10^{-13} at -2 °C (e.g. Oliphant *et al.*, 1983; Ohrai & Yamamoto, 1985; Smith, 1985). These values are similar to those found in dense clays in the unfrozen state.

REGELATION

Burt & Williams (1976) reported that the presence of an ice lens across a sample did not prevent water movement through the sample. In an ingenious experiment using a frozen permeameter, Miller (1970) had also demonstrated that ice masses do not represent barriers to mass flow, and that molecules are transported across the ice. This movement involves pressure-induced freezing of water molecules at the upstream (warm) side of the ice mass and melting at the downstream (cold) side. Miller *et al.* (1975) described this process of *regelation* in thermodynamic terms, paraphrased here as follows.

Recall equation (7.9):

$$T - T_0 = \frac{(P_w V_w - P_i V_i)T}{L_f}$$

where T_0 is taken to be 273.15 K (i.e. 0 °C). This indicates that when $T < 273.15$ K, P_i must exceed P_w. If on one side of an ice mass, P_w is increased somewhat (by water inflow, under an applied pressure for example), the equilibrium temperature, T, rises above ambient. Since the ambient temperature is below the equilibrium temperature, ice will form on that side and the ice pressure rises. As this increase in P_i is felt at the other side of the ice mass, where P_w has not changed (i.e. $P_i - P_w$ increases), the equilibrium temperature falls below ambient, favouring melting. The net result is a translational movement of ice. In the case of ice in soil pores, water would freeze at the warm (inflow) end, with latent heat release, and melt at the cold (outflow) end, with absorption of latent heat. Thus the process involves heat flow as well as water flow.

Some have suggested that the flow observed in this experiment is due in part or in whole to liquid flow along ice grain boundaries. However, in another experiment, Horiguchi & Miller (1980) showed that this was negligible compared to regelation transport. Further, they reported that the 'permeability' of ice lenses at temperatures between −0.05 °C to −0.15 °C was higher than that of frozen soil (10^{-10} m s^{-1} compared to about 10^{-11} m s^{-1}), and that it decreased more slowly with temperature. Additional results were obtained by Wood & Williams (1985*a*) at temperatures down to −0.3 °C.

While regelation in the ice sandwich permeameter was induced by a pressure gradient, Romkens & Miller (1973) determined that regelation could also be thermally-induced. They observed that mineral particles embedded in ice subjected to a *temperature gradient* moved from the cold to the warm side, at a rate that was temperature-dependent. As the corollary of this, they assume that where soil particles remain stationary with respect to some external frame of reference, pore ice will move through the soil under a temperature gradient from warm to cold, at a rate that is temperature-dependent. It is unlikely that the ice actually creeps through the pores, but rather the water molecules move in a sequence of freezing and melting steps (but see section 8.5.3).

The picture of water movement that emerges from these considerations is of a series–parallel process, with series flow (regelation) through pore ice and ice lenses, and parallel transport in unfrozen water films surrounding soil particles (see also section 8.5.3 and Figure 8.9).

VAPOUR DIFFUSION

Vapour diffusion in soils is driven by a gradient in soil water vapour pressure induced by the temperature gradient. In frozen soils, the appropriate vapour pressure is that over ice, since the unfrozen water is in thermodynamic equilibrium with the ice phase. Furthermore, where soils contain any excess ice (or, at least, ice larger than pore size) as well as unfrozen water, saturation vapour pressures may be used to calculate the vapour flux.

The equation governing the flow of vapour in soils is given as (Giddings & LaChapelle, 1962):

$$q_v = -D_v \frac{d(e/RT)}{dz} \tag{8.1}$$

where q_v is the mass flux per unit area ($kg\,m^{-2}\,s^{-1}$), D_v is the diffusion coefficient for water vapour in soil ($m^2\,s^{-1}$), which depends on the water content, e is the saturation vapour pressure over ice (Pa), R is the gas constant for water vapour ($0.46\,kJ\,kg^{-1}\,K^{-1}$), T is the temperature (K) and z is depth (m). Since the saturation vapour pressure is temperature dependent, we obtain from (8.1):

$$q_v = \frac{-D_v}{RT}\frac{de}{dT}\frac{dT}{dz} \tag{8.2}$$

for temperatures within a few degrees of 0 °C. The negative sign in the equation indicates that the flow is in the direction of decreasing temperature.

Several studies have determined the outward flux of moisture from frozen soil occurring throughout the winter, using some form of 'vapour trap' placed on the soil surface (e.g. Santeford, 1978; Woo, 1982; Smith & Burn, 1987). This flux is partially responsible for the development of depth hoar within arctic snow packs, although Woo (1982) indicated that in a high arctic environment its contribution may be very small. In addition, Parmuzina (1978) and Cheng (1983) observed that the surface layer of soil experienced some desiccation over the winter period. Both Santeford (1978) and Woo (1982) interpreted the evaporation from the ground as a continuation of vapour diffusion through the soil, in accordance with equation (8.2), with the soil temperature gradient driving water vapour towards lower surface temperatures. However, Smith & Burn (1987) argue that the vapour trapped at the surface migrates through the soil in the liquid phase, to evaporate from the surface into the trap. (This is analogous to the situation described by Nakano *et al.* (1984), who separated two soil samples by an air gap but continued to observe moisture transfer between them.)

RELATIVE IMPORTANCE OF MOVEMENT IN THE DIFFERENT PHASES

We can expect the relative importance of the transport mechanisms in frozen soil – i.e. liquid flow (water in water), regelation transport (water in ice) and vapour diffusion – to *vary with the total moisture content of the soil and temperature* (since the phase composition of soil moisture varies with temperature). Both liquid and vapour flow will occur in ice-free soils; in unsaturated frozen soils with ice present, we may anticipate liquid flow (as determined by the Clausius–Clapeyron equation), vapour flow (governed by the saturation vapour pressure gradient over ice), and some regelation transport of pore ice, or across ice lenses.

8.4.4 Moisture transport in saturated soils

Burt & Williams (1976) observed that the presence of ice lenses in saturated samples of silt reduced the permeability, k_f, by two orders of magnitude, for the temperature range 0 °C to −0.1 °C. Both Loch & Kay (1978) and Horiguchi & Miller (1983) obtained values for k_f of between 10^{-8} and 10^{-11} m s^{-1}, for various fine-grained soils between 0 °C and −0.1 °C. In the warmer half of this range, the transport coefficient of ice (10^{-10} m s^{-1}, Wood & Williams, 1985*a*) is less than that of frozen soil (10^{-8} to 10^{-10} m s^{-1}), and therefore ice impedes the transport of moisture. Liquid transport in the unfrozen pore water and in the adsorbed water films around soil particles is at its maximum at these temperatures.

At lower temperatures, however, measurements of the transport coefficient of the ice phase alone (10^{-11} m s^{-1}, Wood & Williams 1985a) are up to two orders of magnitude higher than the hydraulic conductivity of frozen soil. From this, one may conclude, therefore, that ice layers do not inhibit moisture flow below about −0.1 °C. Instead, the relatively slow transport via liquid films away from the cold side of the lens leads to a reduction in the pressure difference ($P_i - P_w$) which sustains regelation transport. Thus the critical phase determining k_f is the liquid phase.

8.4.5 Moisture transport in unsaturated soils

In unsaturated (ice-free) soil, moisture moves by both vapour diffusion and by liquid flow in adsorbed water films. It is difficult to isolate these components experimentally, so that moisture movement in unsaturated soils is analysed using an equation of the form (Nakano *et al.*, 1984):

$$q = -D_\theta \rho_d \frac{\delta \theta}{\delta z} \text{ kg m}^{-2} \text{ s}^{-1} \tag{8.3}$$

where θ is the volumetric water content ($m^3 m^{-3}$)*, D_θ is a bulk soil-moisture diffusivity ($m^2 s^{-1}$), and ρ_d is the dry density ($kg m^{-3}$). D_θ increases with θ, and is related to the hydraulic conductivity, k, by:

$$D_\theta = k \frac{d\psi}{d\theta} \tag{8.4}$$

(Hillel, 1971) where ψ is the matric potential. The term $d\psi/d\theta$ corresponds to the slope of the suction–moisture content curve, which is related to the temperature–unfrozen water content curve for frozen soil (see section 7.2).

Recent experiments reported by Nakano *et al.* (1982, 1983*a*, 1983*b*, 1984) provide estimates of D for soils at -1 °C, at water contents below 0.05 kg kg^{-1}. Their values are comparable to those obtained by Jackson (1965) (10^{-9} $m^2 s^{-1}$) and so they concluded that, as might be expected, the transport mechanism of moisture in ice-free soil is essentially the same as that in unfrozen soil containing a similar amount of water (Nakano *et al.* 1983*a*, p. 892). Following Jackson (1965), they suggested that vapour diffusion may account for up to 60% of the total moisture transport observed in their experiments. Since the maximum values for D_θ obtained by them were only on the order of 6×10^{-10} $m^2 s^{-1}$ (Nakano *et al.*, 1982), this suggests that vapour diffusivity is very low in (frozen) soils ($< 4 \times 10^{-10}$ $m^2 s^{-1}$).

Both Jackson (1965) and Hillel (1971) indicate that while the overall diffusivity increases sharply with water content above 0.04 kg kg^{-1}, this is due entirely to the increase in liquid diffusivity, and, in fact, vapour diffusion diminishes to insignificance. In frozen soils, one can expect, therefore, that as the pore ice content increases (i.e. as soil voids are filled with ice) the paths for vapour transport are closed off. However, the component of flow in the liquid phase will remain constant, since the unfrozen water content is essentially independent of the ice content.

* The volumetric moisture content θ is equal to the gravimetric moisture content w times (ρ_d/ρ_w).

A quantitative estimate of the relative importance of liquid and vapour flow in frozen soil can be obtained when results from equation (8.2) are compared to estimates of flow rates in the liquid phase using Darcy's Law (see Smith & Burn, 1987). For given soil temperature conditions, the flow in the liquid phase, q_l is given by (ignoring signs):

$$q_l = k_f \frac{d\psi}{dT} \frac{dT}{dz} \tag{8.5}$$

For a silty clay soil at $-1\,°C$, for example, k_f may be approximately $10^{-12}\,m\,s^{-1}$ (Smith, 1985), and $d\psi/dT$ is approximately 120 m water head $°C^{-1}$. Thus:

$$q_l = 10^{-12}\,m\,s^{-1} \times 120\,m\,°C^{-1} \times \frac{dT}{dz}\,°C\,m^{-1}$$

$$= 1.2 \times 10^{-10} \frac{dT}{dz}\,m\,s^{-1}$$

Under the same conditions, the mass vapour flow, q_v may be estimated by (again ignoring signs):

$$q_v = \frac{D_v}{RT} \frac{de}{dT} \frac{dT}{dz} \tag{8.2, above}$$

If D with respect to vapour is $4 \times 10^{-10}\,m^2\,s^{-1}$ (see above) and de/dT is the slope of the saturation vapour pressure curve (Figure 2.5) at $-1.0\,°C$ ($46.7\,Pa\,°C^{-1}$, List 1968), we have, under optimum conditions:

$$q_v = \frac{4.0 \times 10^{-10}}{1.2 \times 10^5} \frac{m^2\,s^{-1}}{J\,kg^{-1}} \times 46.7\,Pa\,°C^{-1} \times \frac{dT}{dz}\,°C\,m^{-1}$$

$$= 1.4 \times 10^{-13} \frac{dT}{dz}\,kg\,m^{-2}\,s^{-1}$$

$$= 1.4 \times 10^{-16} \frac{dT}{dz}\,m\,s^{-1}$$

Clearly, the vapour flow is several orders of magnitude less than liquid flow, even if there is much vapour-filled pore space, and may be neglected for most practical purposes (see also Dirksen & Miller, 1966).

8.5 Ice segregation and frost heaving

Frost heaving is the increase in pressure and/or volume displacement when a soil freezes; the particular manifestation depends on the confining stresses on the soil. Where these stresses are sufficient to resist expansion, then (very) large heaving pressures are generated. Where the load to be displaced is simply the small weight of some thin overlying layer, however, then frozen fine-grained soil typically exhibits a highly segregated structure with discrete layers of ice (ice lenses). A redistribution of moisture induced by the freezing process causes the total water content of the frozen soil to increase (Figure 8.5). The resulting frost heave is responsible for a variety of natural features in cold regions such as pingos, palsa and

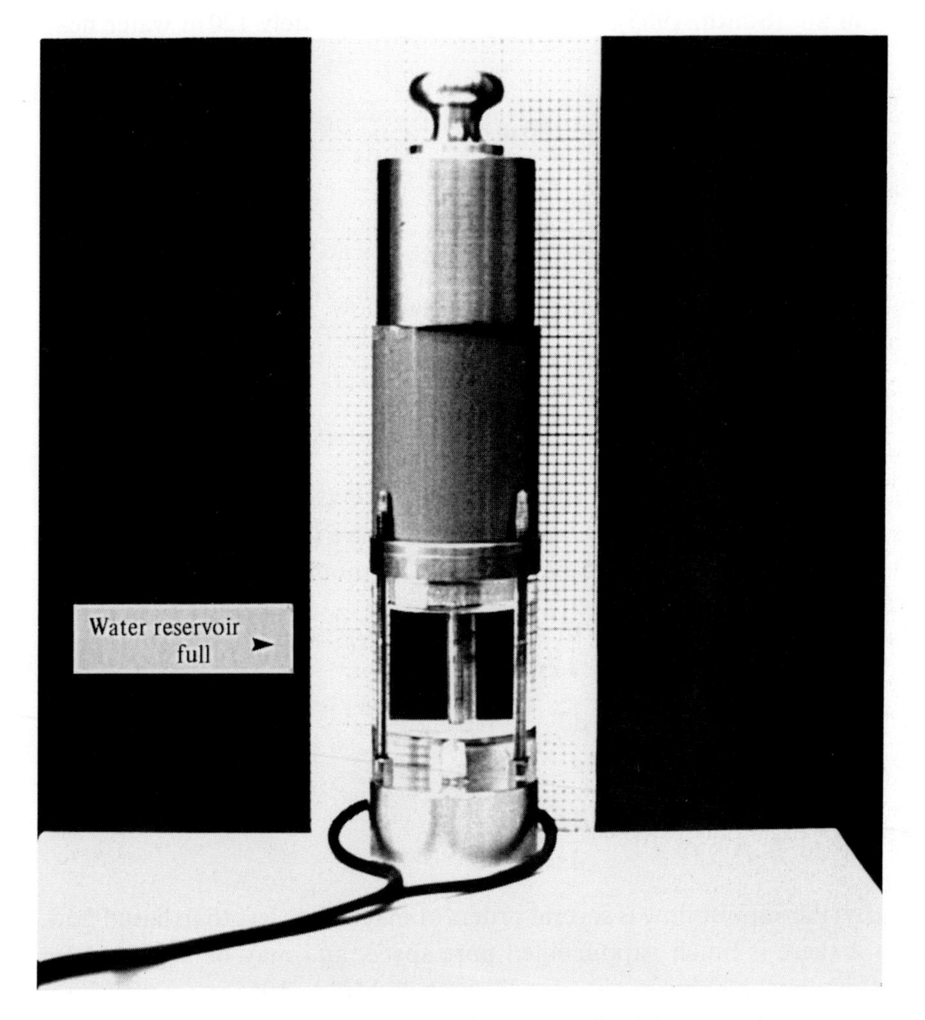

Figure 8.5 Frost heaving. The soil sample draws water from the reservoir on freezing. The expansion of the soil with the formation of ice lenses lifts the weight. Photos: R.D. Miller.

hummocks (Chapter 6), and leads to unique construction and engineering problems. It has been reported that frost heave damages or destroys property worth billions of dollars every year (O'Neill & Miller, 1985).

Frost-heaving has been the subject of innumerable investigations over many decades, although most of our knowledge of this process still derives from laboratory tests conducted on small, sometimes very small, samples of soil. More often than not, these tests have been carried out under high rates of freezing, involving large gradients of temperature, and have run for

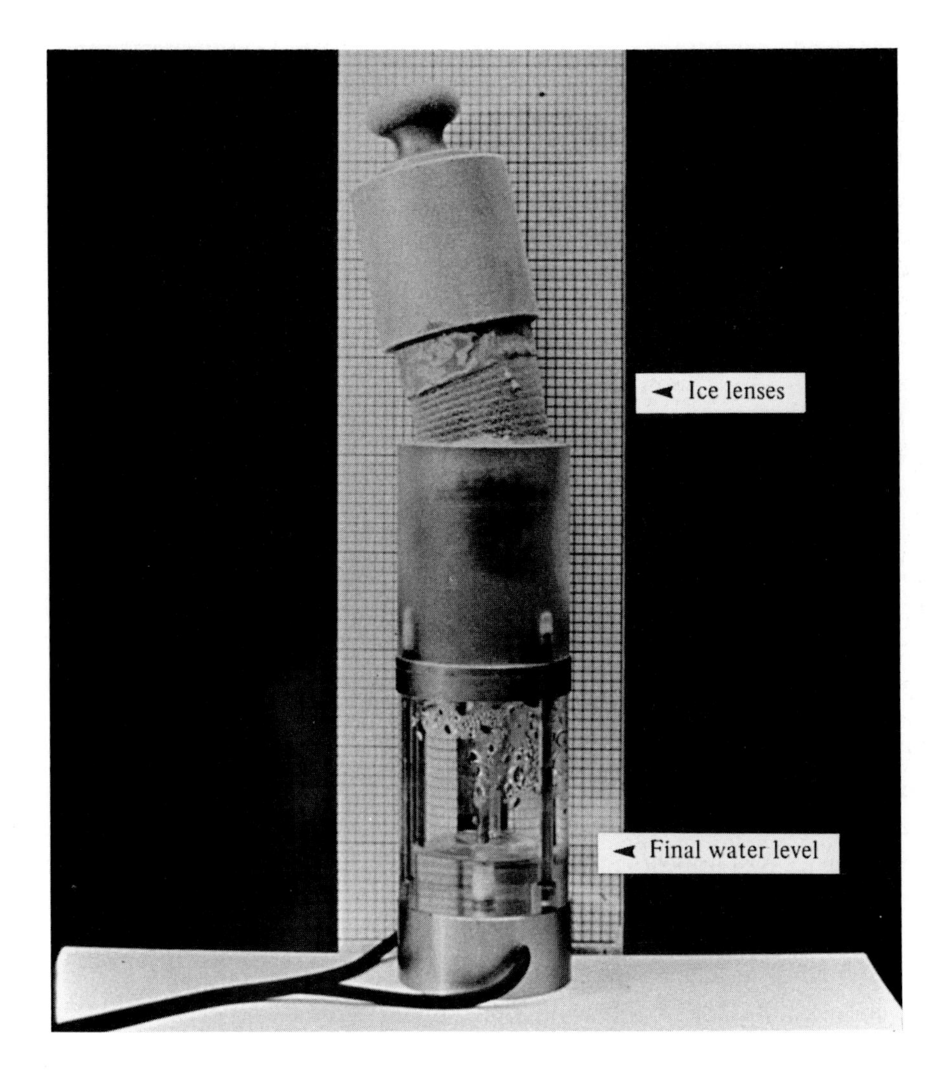

Figure 8.5 (*cont.*)

brief periods of time (a few days at most). In reality, frost-heave involves large masses of soil, slow rates of freezing, small temperature gradients and often extensive periods of time. Thus, the extent to which laboratory observations, and the models consequent to this, apply to frost-heaving in nature eventually has to be questioned.

There are different approaches that can be taken in frost heave modelling. Laboratory experimental data may be generalised into relationships between heave rate and frost penetration or heat flux (e.g. Penner & Walton, 1978; Konrad & Morgenstern, 1980; Nixon *et al.*, 1983). Resulting equations may be used with ground thermal models to predict frost heave under field conditions. Apart from the concerns mentioned above, however, such relationships may lack an adequate physical basis.

As an alternative, one can apply thermodynamic principles, together with the conservation equations for heat and mass, to construct a detailed physical model of the frost heaving process. This approach relates more closely to the theme of this book. However, while the physical approach provides considerable insights into the process, it must be acknowledged that the resultant models are not yet suitable for prediction purposes, since they require a knowledge of more properties of the soil than is reasonably known.

Beginning with the classic works of Taber (1918, 1929, 1930) and Beskow (1935), it was shown that when soils freeze, pressure develops in the direction of ice crystal growth. It was also determined that application of a surcharge would reduce, but not eliminate, frost heave. The energy required to overcome the resistance of the soil (i.e. the heaving pressure) derives from the difference in the pressures of the ice (P_i) and water (P_w) phases which co-exist in soil at temperatures below 0 °C. The capillary model (see section 7.3) was the first attempt to provide an explanation of distinct ice and water pressures in frozen soils.

In the '*ice intrusion*' model, the difference in pressure ($P_i - P_w$) was considered as that required *for ice to proliferate* through the soil pores. Thus the heaving pressure was assumed to originate at the freezing front – i.e. where the ice first appears in the soil pores. By equating this concept of 'ice intrusion' to the well-known intrusion of air in unfrozen soil, it was deduced that the heaving pressure is determined by the largest continuous pore openings in the soil (Williams, 1967). However, by the early 1970s it became clear that the ice intrusion model under-estimated (perhaps greatly) actual values of heaving pressure.

Penner (1967) first noted a discrepancy in the heaving pressure predicted from a simple capillary model and that he observed in laboratory tests, and Hoekstra (1969) also noted that attempts to find agreement between calculated and experimental heaving pressures had failed. The computed values were always too low. Later studies by Sutherland & Gaskin (1973) and Loch & Miller (1975) added further confirmation to the discrepancy.

A logical explanation for the disparity is that ice segregation occurs at a location (temperature) where the difference in pressure between the ice and the water is greater than that at the first entry of ice into the soil pores. The relationship between heaving pressure and temperature (from equation (7.9)) implies that the ice lens must therefore form in the partially frozen soil some distance *behind* the freezing front. From laboratory experiments, Dirksen (1964) and Hoekstra (1966) had both inferred that ice lenses were formed somewhere behind the freezing front, but Miller (1972) first introduced the concept of a *frozen fringe* (Figure 8.6) into a frost heave model. The fact was subsequently confirmed in experiments by Loch & Miller (1975), Loch & Kay (1978) and E. Penner & Goodrich (1980).

The partially frozen layer that exists between the freezing front and a growing ice lens must be important somehow to the formation of the lenses themselves, and two models have arisen to explain ice segregation in this context. These are the *hydrodynamic* model, introduced by Harlan (1973), and what may be termed the *rigid ice* model of Miller (1978, 1980). Both models consider the coupling of heat and moisture flow in the freezing soil.

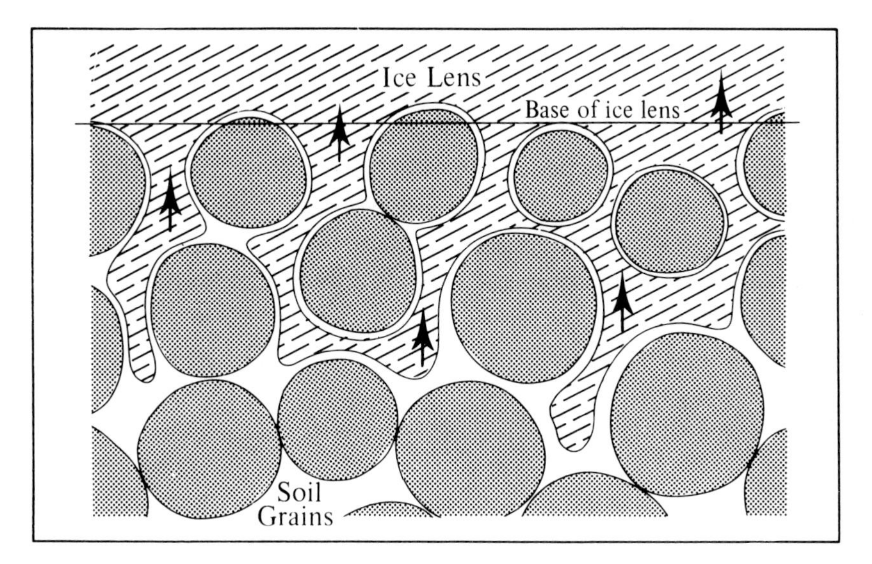

Figure 8.6 Schematic diagram of the frozen fringe (from O'Neill & Miller, 1985).

8.5.1 *Coupled heat and moisture flow at freezing temperatures*

In unfrozen soil, the rate of water flow depends on the gradient of potential (which depends on θ) and the hydraulic conductivity $k(\theta)$. In frozen soil, since the liquid water content is mainly a function of temperature, a temperature gradient is analogous to a water content gradient in an unsaturated homogeneous soil. In addition, the hydraulic conductivity of frozen soil is also a function of the temperature (section 7.5). This implies that a temperature gradient will induce water movement in the direction of lower temperatures, but the water flow will encounter an increasing impedance. Where water accumulates in the colder parts of the soil it must freeze, since the amount of unfrozen water that can exist is limited by the temperature (i.e. the unfrozen water content relationship for the particular soil). The freezing of water results in the release of latent heat, which must then be dissipated or the temperature will rise. Thus, the movement of water in frozen soil depends not only on the hydrological circumstances but also on the thermal conditions in the soil.

8.5.2 *The hydrodynamic model*

This interdependence was expressed in terms of a set of coupled heat and moisture flow equations by Harlan (1973), who proposed a simple analogy with unfrozen soil to obtain an expression for mass transport in a freezing soil. He assumed that pore ice, being a solid, is inherently immobile so that the mass transport involves only liquid flow (vapour flow being minor as discussed in 8.4.5). He stated that if the continuity of unfrozen water films is broken – for example through the formation of an ice lens – 'water migration toward the cold side will be affected significantly' (Harlan, 1973, p. 1315). He further assumed that the pore ice has the same Gibbs free energy as bulk ice at the same temperature and at atmospheric pressure. Thus, through the use of Schofield's (1935) equation (see Figure 7.7) for the freezing point depression of water in moist soil, the driving force for mass flow is assumed to be directly proportional to the temperature gradient.

For heat flow, we have:

$$\frac{\partial}{\partial z}\left(K\frac{\partial T}{\partial z}\right) + C_{w}q\frac{\partial T}{\partial z} + \rho_{i}L_{f}\frac{\partial \theta_{i}}{\partial t} = C_{s}\frac{\partial T}{\partial t} \qquad (8.6)$$

where C_w is the volumetric heat capacity of water, q is the flux of water, θ_i is the volumetric ice content, ρ_i is the density of ice and C_s is the volumetric heat capacity of the soil (see section 4.3.2). On the left-hand side, the first term accounts for heat conduction due to the temperature gradient, the second term is the mass heat transfer associated with the flow of water, and the third term represents the heat source (or sink) related to phase transition. The second term is typically only 0.001 to 0.01 of the first and is usually ignored (Taylor & Luthin, 1978). Equation (8.6) then becomes:

$$\frac{\partial}{\partial z}\left(K\frac{\partial T}{\partial z}\right) + \rho_i L_f \frac{\partial \theta_i}{\partial t} = C_s \frac{\partial T}{\partial t} \tag{8.7}$$

For water flow, we have:

$$\frac{\partial}{\partial z}\left(k\frac{\partial P_w}{\partial z}\right) = \frac{\partial \theta_w}{\partial t} = \frac{\partial \theta_u}{\partial t} + \frac{\rho_i}{\rho_w}\frac{\partial \theta_i}{\partial t} \tag{8.8}$$

where k is the hydraulic conductivity, P_w is the pressure head, and θ and θ_u are the volumetric contents of total moisture and unfrozen water respectively. The unfrozen water content–temperature relationship dictates the maximum amount of unfrozen water at any temperature for the soil. Therefore, we can substitute:

$$\frac{\partial \theta_u}{\partial t} = \frac{\partial \theta_u}{\partial T}\frac{\partial T}{\partial t} \tag{8.9}$$

in equation (8.8) so that it becomes:

$$\frac{\partial}{\partial z}\left(k\frac{\partial P_w}{\partial z}\right) = \frac{\partial \theta_u}{\partial T}\frac{\partial T}{\partial t} + \frac{\rho_i}{\rho_w}\frac{\partial \theta_i}{\partial t} \tag{8.10}$$

Solving for, and equating, $(\rho_i \partial\theta_i/\partial t)$ from (8.7) and (8.10), we obtain:

$$\frac{1}{L_f}\left[C_s\frac{\partial T}{\partial t} - \frac{\partial}{\partial z}\left(K\frac{\partial T}{\partial z}\right)\right] = \rho_w\left[\frac{\partial}{\partial z}\left(k\frac{\partial P_w}{\partial z}\right) - \frac{\partial \theta_u}{\partial T}\frac{\partial T}{\partial t}\right] \tag{8.11}$$

or,

$$\frac{\partial}{\partial z}\left(K\frac{\partial T}{\partial z}\right) + \rho_w L_f \frac{\partial}{\partial z}\left(k\frac{\partial P_w}{\partial z}\right) = \left(C_s + \rho_w L_f \frac{\partial \theta_u}{\partial T}\right)\frac{\partial T}{\partial t}$$

The term

$$C_s + \rho_w L_f \frac{\partial \theta_u}{\partial T}$$

is known as the apparent heat capacity, C_a (see equation 4.9), and therefore we can write:

$$\frac{\partial}{\partial z}\left(K\frac{\partial T}{\partial z}\right) + \rho_w L_f \frac{\partial}{\partial z}\left(k\frac{\partial P_w}{\partial z}\right) = C_a \frac{\partial T}{\partial t} \tag{8.12}$$

Since P_i is assumed to be zero (relative to atmospheric pressure), we can substitute:

$$\frac{\partial P_w}{\partial z} = \frac{\partial P_w}{\partial T}\frac{\partial T}{\partial z}$$

where $\partial P_w/\partial T$ is determined from equation (7.8).

In the case of no water flow, equation (8.12) reduces to equation (4.5) in chapter 4. On the left-hand side of equation (8.12), the second term represents the latent heat released by the formation of ice as a result of water inflow (migration), while the right-hand side includes the latent heat produced as interstitial water freezes *in situ*.

If we imagine water being drawn towards the freezing soil, we encounter a very sharp decrease in hydraulic conductivity within the frozen fringe (see Figure 7.10). Because of this, the inferred mass flux (the product of hydraulic gradient and conductivity) exhibits negative divergence (water 'piles up') – i.e. since:

$$k_f \ll k_u$$

then,

$$q_f \ll q_u$$

where f and u refer to frozen and unfrozen zones respectively or, altenratively, colder and warmer parts of the freezing soil. (We can assume, for simplicity, that the temperature gradient is linear.) Thus, because flow is more rapid through the unfrozen material than the frozen, water accumulates (as ice) in the region of the frozen fringe. The location of an ice lens is assumed to correspond to the point of maximum flow divergence – i.e. where dk_f/dT is maximum (Figure 8.7). From Figure 7.10 we see that this occurs typically in the temperature range $-0.1°$ to -0.2°C or so for fine-grained soils. The latent heat evolved during this process produces a positive divergence of heat flow, completing the coupling between mass and heat transport.

The heat balance at the freezing front may be expressed as:

$$\frac{\partial}{\partial z}\left(K\frac{\partial T}{\partial z}\right) = K_f\left(\frac{\partial T}{\partial z}\right)_f - K_u\left(\frac{\partial T}{\partial z}\right)_u \tag{8.13}$$

so that (8.12) becomes:

$$K_f\left(\frac{\partial T}{\partial z}\right)_f - K_u\left(\frac{\partial T}{\partial z}\right)_u = C_a\frac{\partial T}{\partial t} - \rho_w L_f\frac{\partial}{\partial z}\left(k\frac{\partial P_w}{\partial z}\right) \tag{8.14}$$

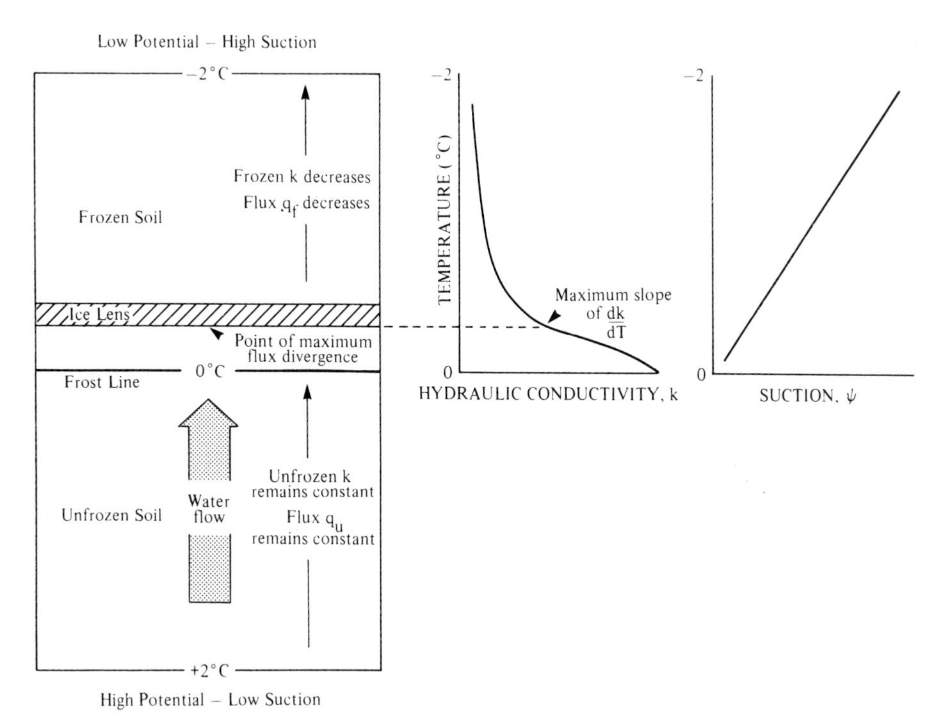

Figure 8.7 Ice lensing according to the hydrodynamic model.

In other words, a balance must be observed between the net conduction of heat away from the freezing front on the one hand, and the freezing of interstitial water (the first term on the right-hand side) together with the release of latent heat of fusion as any transported water freezes (the second term). The first term on the right-hand side also includes changes in the sensible heat content of the soil (see the discussion in section 4.3), but this is negligible compared to the latent heat at temperatures close to zero. If the net heat flow is greater than the term $(\rho_w L_f \partial/\partial z\,(k\,\partial P_w/\partial z))$, because water cannot migrate to the freezing front at a sufficient rate, then some interstitial water will freeze and the frost line will advance into the soil. In this case, the ice lens would cease to grow.

A similar circumstance results if the unfrozen soil itself has a low hydraulic conductivity, which limits water migration to the freezing front. For example, E. Penner (1986) studied ice lensing in a layered soil system and found that lenses formed at the boundary between the dense clay and the silt, although the field observations of naturally varved materials in northern Manitoba by Johnston *et al.* (1963) revealed no such simple pattern. Under natural conditions of repeated freeze–thaw, such as occur in the active layer, ice lenses possibly form at the same locations time after time. Consolidation of the soil layers between ice lenses (see 9.7.1) may reduce the hydraulic conductivity.

When a soil is initially subject to freezing, the rate of frost penetration is high and the interstitial water freezes rapidly. Although the *rate* of heaving may be high, the cumulative heave may be relatively small. This is because the rate of heat removal is so great that water cannot migrate at a rate sufficient to create large, continuous ice lenses, although some ice enrichment does occur in the form of tiny, closely-spaced hair-like lenses, which may not be visible to the eye (Figure 8.8).

As the freezing front penetrates further into the soil and the rate of freezing slows, larger visible lenses can occur. The progressively slower rate of freezing allows sufficient time for water to migrate to the freezing front where a (large) ice lens can grow. As the lens continues to grow, however, the soil below will experience a falling moisture content if water flow is

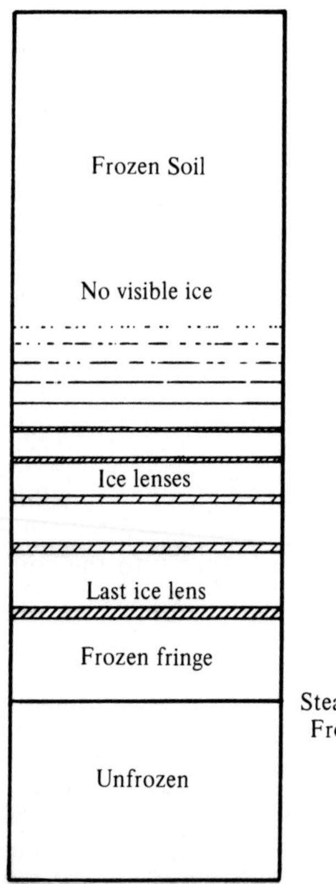

Figure 8.8 Schematic rhythmic ice lens formation (from Konrad & Morgenstern, 1980).

insufficient to replenish it. This eventually results in a thermal imbalance, soil cooling and an advance of the freezing front into the soil. A point is then reached where a new ice lens may form.

With time, as the freezing front penetrates further into the soil, the temperature gradient decreases and heat flow declines. However, while the rate of heaving decreases with depth, the slower rate of freezing allows for greater total heave within a soil layer, with the development of thicker and thicker ice lenses (Figure 8.8). Such *rhythmic ice banding* is seen frequently in nature, with Taber (1929, 1930) being amongst the first to describe it in detail.

Eventually, a point may be reached where a balance exists between heat flow away from the ice lens and the latent heat released by the freezing of the water migrating to it. Of course, this cannot last indefinitely, since the frozen zone is gradually extended by the growth of the ice lens, and the temperature gradient (and hence heat flow) decreases. However, under some special circumstances, very thick segregated ice masses may be formed (e.g. Mackay, 1971).

Harlan (1973) assembled the coupled flow equations into a computer model, but did not test it against experimental data. Instead he simulated soil moisture redistribution under freezing conditions, although his results apparently differ from experimental profiles determined by Dirksen & Miller (1966), Hoekstra (1966), Jame & Norum (1976) and others since. Taylor & Luthin (1978) used a version of the Harlan model to analyse laboratory frost heave data, but needed to introduce a correction factor to the mass transport coefficient in order to get agreement in the predicted results (see also Guymon *et al.* 1980). This factor was explained in terms of the presence of ice in the soil, even though the hydrodynamic model ascribes *no* role to the ice phase. Ice accumulation occurs simply as a *passive* consequence of liquid transport phenomena and the soil is viewed as a totally compliant sink. In addition, the rheological implications of frost heave are also ignored. A simplified version of a hydrodynamic model was developed by Konrad & Morgenstern (1980).

The hydrodynamic model of liquid film transport has been criticised by Miller and co-workers; they contend that the ice phase is *mobile* and that mass transport is not restricted to the liquid phase. Further, since liquid transport interacts strongly with ice movement, the conditions for using a Darcy-type equation for water flow are not fulfilled and the driving force for film flow is not the water pressure gradient alone (Miller *et al.*, 1975). There are also implications for heat flow, since phase changes create local sources and sinks which are superimposed on the macroscopic flow of heat by conduction. In addition to these considerations, Bresler & Miller (1975) contend that, in unsaturated soil, capillary effects form the primary mechanism for inducing water into the freezing zone and that this further confounds assumptions in the hydrodynamic model.

8.5.3 The rigid ice model

The alternative explanation for ice lensing put forward by Miller (1978, 1980), also argues that it takes place within the (partially) frozen soil at some distance behind the freezing front. According to the rigid ice model, the ice in the frozen soil, including segregated ice, forms a continuous rigid crystalline phase, i.e. it can be visualised as a lattice (Figure 8.6). Under a temperature gradient, ice in the frozen fringe moves from warmer to colder regions of the soil by thermally-induced regelation. The soil particles in the frozen fringe remain stationary as long as the ice pressure cannot push them apart. When the ice pressure rises sufficiently, soil particles are no longer pressed against their stationary neighbours below and will be pushed apart by the moving ice (Miller, 1978). The widening gap between the soil particles is filled by ice formed from the flow of water across the fringe, and an ice lens develops. Thus the ice lens is simply a particle-free zone within the developing ice lattice.

In this model, the flux of pore ice (regelation) forms an intrinsic part of the mass transport in the frozen fringe. Regelation has been summarised by Miller *et al.* (1975) in the following terms (see also section 8.4.3). Soil particles are covered by adsorbed films of unfrozen water that forms a tortuous liquid phase capable of transporting water through the frozen soil (Figure 8.9). The remaining space is fully occupied by pore ice, which can move with uniform translational velocity (arrows). Since the ice is bounded everywhere by the film phase, movement can be accommodated by the continuous transformation of film water to ice at the upstream ice/water interfaces in soil pores (A, A′, in Figure 8.9) and from ice into water at the downstream ends (B in Figure 8.9). The motion involves appropriate movements of water in the film phase (from B to A′), and this is added to whatever flow would be taking place in the films if the ice were immobile. Thus the mass transport is biased toward liquid transport in pore necks and ice movement in the pore body. Miller *et al.* (1975) referred to this process as *series–parallel* transport and they argue (pp. 1032–1033) that ice movement can enhance the mass transport by a very large factor indeed.

They also recognise that the ice pressure gradient may be large enough to cause significant *creep* deformation, and that this may be important in limiting the heaving pressure (cf. section 7.6). Further, they envision (p.

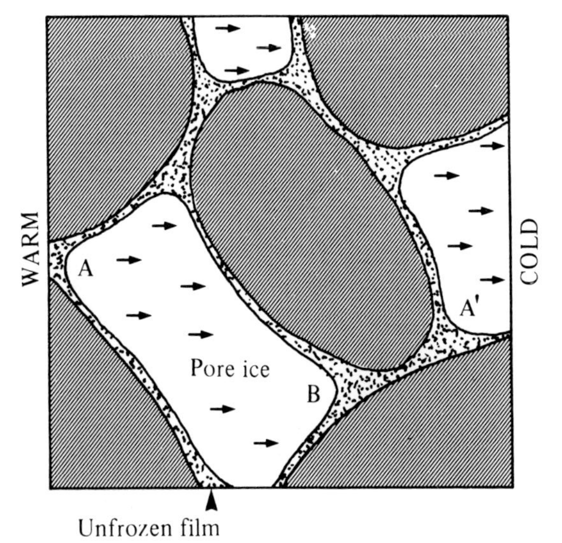

Figure 8.9 Series–parallel transport in a frozen non-colloidal soil (from Miller *et al.*, 1975).

1030) the possibility of circulatory transport in which the ice moves in one direction (toward higher temperature) and water in the other to produce little or no net transport.

Miller (1978) has explained the initiation of ice lenses in *cohesionless* (non-colloidal) soils in terms of the stress conditions in the frozen fringe. Prior to the lens forming, the load of the soil, P, is partially supported by the reaction stress of the soil matrix (the effective stress, σ') and partially by the reaction of the pore contents (the neutral or buoyant stress, σ_n):

$$P = \sigma' + \sigma_n \tag{8.15}$$

(Miller, 1978). In unfrozen soils, the effective stress equation is often written as $\sigma' = P - u$, where u is the pore water pressure (see section 9.2.2). When pore water and pore ice are continuous phases, Miller proposed the following relationship for σ_n:

$$\sigma_n = P_i - \chi(\Omega)\,(P_w - P_i) \tag{8.16}$$

where $\chi(\Omega)$ is a partition factor that varies with the unfrozen water content (temperature), although little is known about it.

Ignoring unnecessary detail, we may examine the ice and water pressure distribution in a general way by considering a simple example (following Miller, 1978). Let us presume that an ice lens is growing at some distance behind the freezing front and that the soil pores within the frozen fringe are filled with ice, as discussed previously. For simplicity, we may assume a linear gradient of temperature. If there is a constant flux of water, q_w, through the fringe in order to supply the growing lens, then P_w must fall at an increasing rate across the fringe (Figure 8.10), owing to the essentially exponential decrease in hydraulic conductivity with distance (temperature) (see Figure 7.10). But since P_w and P_i are related by the temperature (equation (7.9)), P_i must also vary through the fringe, as shown in Figure 8.10. From equation (8.16), σ_n increases through the fringe to some point where it is equal to P (the load of the soil). At that point, σ' falls to zero, soil particles are no longer held together and an ice lens forms.

As the lens continues to thicken, P_w falls as the soil below gradually becomes desiccated, and in accordance with the Clapeyron equation P_i must fall also. When P_i is less than P, the ice lens can no longer displace the overlying soil and it ceases to grow at that point. A new lens forms below, where $\sigma' = 0$, and the process may repeat itself in a rhythmic fashion, as represented in Figure 8.8.

Presumably, in *cohesive* soils P_i would need to be much higher than shown in Figure 8.10, since it must be sufficient to overcome the soil cohesion as well as the load; in fact, the presence of ice imparts cohesion to all soils (see section 9.2), so that it cannot be ignored. Gilpin (1982) refers

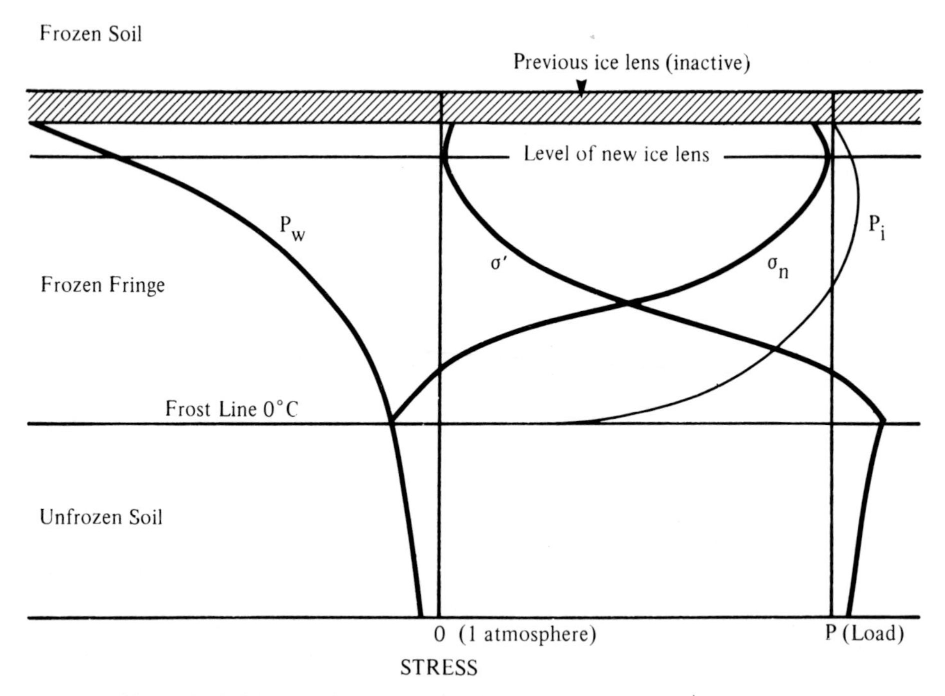

Figure 8.10 Stress and pressure distributions in freezing soil just at the initiation of a new ice lens (from Miller, 1978).

to a pressure in the ice necessary to separate soil particles, and observations by Williams & Wood (1985) indicate that the ice pressure in freezing soil can be considerable. (See also section 7.6).

While this model represents the most comprehensive attempt to explain ice segregation, there is yet no evidence of how well it may describe the process in reality. Attempts to use the rigid ice model for simulation ('prediction') have necessitated simplifications or excessive numerical demands (Gilpin, 1980; Holden, 1983; O'Neill & Miller, 1985). Recently, however, Black & Miller (1985) reported on a simplified numerical version of the model.

Although we have concentrated on models which are concerned with the physical nature of soil freezing at a 'microscopic' scale (and which relate to the theme of this book), it should be noted that heat and mass flow models based on continuum mechanics have been developed which take account of the thermal, hydraulic and mechanical aspects of frost heave (e.g. Nixon *et al.*, 1983; Blanchard & Fremond, 1985; Blanchard *et al.*, 1985).

Finally, observations discussed in section 8.6 indicate that frost heave may not originate solely at a single, 'primary' active ice lens but that vertical displacement of soil involves continuing internal deformation of the frozen soil itself. Thus a considerable thickness of soil may be significantly affected by moisture movement and ice accumulation during frost heaving. The continuing enlargement of a body of ice that is surrounded by frozen soil requires that the expansive work of frost heaving take place against the resistance offered by the semi-rigid soil. As a result, frost heave will depend on the interplay of thermodynamic conditions and soil rheological properties. Such considerations, however, are not presently represented in frost heave models.

8.5.4 Extreme forms of ice segregation

The excess ice produced by ice segregation is by no means confined to small ice lenses at shallow depths. The development of large heaving forces explains how segregated ice can form at depth and how pingos can grow, for example. Massive ground ice is widespread in North America and the Soviet Union and most of it is probably segregated ice, formed under somewhat special conditions of an 'unlimited' supply of water under pressure (cf. sections 7.4 and 6.2.1). Mackay (1971) had described such examples of massive segregated ice at depths exceeding 35 metres in the Mackenzie delta, with many ice bodies 10, 20, 30 m or more in thickness. In addition, substantial volumes of segregated ice are found in pingos (section 6.2.1), often at considerable depth. The presence of ice layers in permafrost at depths exceeding 30 m, presumably formed with simultaneous heaving (see Williams 1968), is further evidence of the magnitude of heaving pressures. The ice was apparently able to force an opening beneath tens of metres of overburden.

8.6 Seasonal hydrodynamics in permafrost

With the freeze-back of the active layer in the permafrost environment (and in seasonally frozen ground), moisture is redistributed towards the surface under the influence of a positive temperature gradient. This produces large increases in the moisture content of the upper layers of the ground, accompanied by frost heaving (e.g. see Parmuzina, 1978; Cheng, 1983; Smith, 1985). Interestingly, Parmuzina also reported that redistribution of moisture apparently continued at temperatures down to as low as −5 °C. Such redistribution and the accompanying frost heave could be important to a variety of natural processes and engineering problems.

Parmuzina (1978) determined that after the reversal of the temperature gradient in late winter to early spring, some redistribution of moisture occurred in the frozen active layer. This was evident in changes in cryogenic structure, with ice bands apparently increasing in thickness. Mackay *et al.* (1979b) observed that significant frost heave continued to occur (until May) following complete freeze-back of the active layer by December. Cheng (1983) reports observations from northeast China where up to one-sixth of the seasonal frost heave resulted from continuing ice segregation within frozen ground, a conclusion supported by the results of Smith (1985) from Inuvik, NWT.

8.6.1 Water migration beyond the fringe

In a fascinating experiment, E. Penner & Goodrich (1980) used X-ray photography to observe that icy layers could develop *within* frozen soil. Subsequently, Yoneyama *et al.* (1983) and Ohrai & Yamamoto (1985) also used the technique to document the apparent growth of ice lenses within frozen soil beyond the frozen fringe. In addition, Ohrai & Yamamoto determined values of permeability for a frozen silty clay loam down to −2.2 °C. Their values are similar to those inferred from field observations by Smith (1985), being about 10^{-13} to 10^{-14} m s^{-1}.

Ohrai & Yamamoto (1985) were able to observe that a number of ice lenses appeared to grow simultaneously within the frozen soil, and proposed that water was transferred across intervening lenses by regelation (Figure 8.11). They suggest that existing ice lenses would be the preferred locations for further ice accumulation within the frozen soil, since the

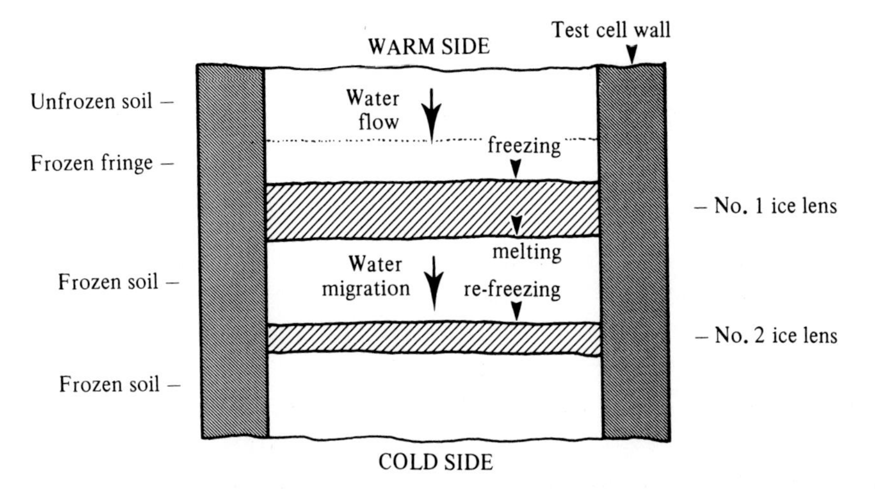

Figure 8.11 Schematic diagram of moisture transfer through frozen soil (from Ohrai & Yamamoto, 1985).

regelation transport of water across an ice lens will be limited by the lower permeability of the frozen soil on the cold side. This leads to a flux divergence there, with water accumulating and hence freezing on the warm side of the ice lens (recall the thermodynamic limitations of ice and water pressures discussed in 7.3).

The issue of frost-heave within frozen ground is important, not least because of the substantially greater heaving pressures that would result from ice growth at low temperatures and their significance to geocryologic processes and engineering. In addition, it raises important conceptual issues related to the understanding of frost-heaving. The enlargement of a body of ice that is surrounded by *frozen* soil requires that the confining material yields, providing the pressures are high enough. This will involve creep (deformation) of the frozen soil, the rate depending on the pressure the ice exerts. Conversely, the rate of creep will exert a control on the rate of accretion (expansion) of ice and thus on the frost heave. Consequently frost heave occurs at a rate which depends in a complex manner on the *thermodynamic* conditions of temperature and water and ice pressures as they are modified by the creep properties (*rheology*) of the soil (Figure 8.12). Interestingly, this view echoes a theme taken up in the following chapter, in which the mechanical properties are themselves shown to be affected by the thermodynamic conditions.

8.6.2 Development of aggradational ice

Cheng (1983) reported measurements made during the course of active layer thawing that revealed consistent increases in moisture content

Figure 8.12 Frozen soil showing ice lenses deformed by differential heave. The soil to the left had a higher silt content than the sandy material to the right. The deformation of the lenses illustrates the magnitude of the forces involved. Forces of similar magnitude are to be expected around ice lenses in uniform soils undergoing continuing heave. Approx. natural size.

occurring in the frozen soil *below* the thawing front. Parmuzina (1978) had observed that the moisture content in the lowest part of the active layer and in the upper horizon of permafrost increased continuously as the thawing front advanced into the ground. Ershov *et al.* (1979 – cited in Cheng, 1983) showed experimentally that ice lenses in the frozen ground below were enlarged during downward thawing. These observations are consistent with thermally induced moisture redistribution as outlined in previous sections. With the reversal of the temperature gradient from winter to summer, (unfrozen) water is drawn downwards and the total moisture content of the frozen ground is increased. As a result, frost heave may occur in the summer, as reported by Mackay (1983*b*).

Further evidence of the process of water movement within permafrost is provided by observations of tritium in groundwater (see Michel & Fritz, 1978; Chizhov *et al.*, 1983). The intake of tritium into groundwater is achieved through atmospheric precipitation; the tritium originates from nuclear testing in the atmosphere in the 1950s and 1960s. It reached a peak in 1963. Tritium has a half-life of about 12.5 years, therefore its absence in groundwater indicates isolation from access to atmospheric moisture for at least the past 50 years. Not surprisingly, Chizhov *et al.* (1983) found high tritium concentrations in wedge ice; more interestingly, however, they also found high concentrations within the permafrost itself. The latter results, and the similar observations reported by Michel & Fritz (1978), can be explained by the downward migration of contemporary atmospheric moisture from the active layer into the upper permafrost horizons under the 'summer' (negative) temperature gradient.

We should not think of frozen soil as presenting a sharp boundary, therefore, but rather as forming part of a continuous hydrologic system in which flows of water (and consequent changes in water content) depend on the direction of the temperature gradient. Further, this is of importance to the formation of the cryogenic structure and the ice content of the upper layers of permafrost, and one intriguing aspect of this has been described by Cheng (1983).

He argued that while water migrates upward in winter and downward in summer, the fluxes are *unequal.* This is explained by the fact that k_f may change by three or four orders of magnitude between 0° to -2°C. Therefore, we can expect a difference between the (relatively large) downward movement of water in summer (when ground temperatures are warm) and the (relatively small) upward flow in winter. The net result of the annual cycle, therefore, is an increment in moisture content in the uppermost permafrost layer, and the repetition of this process year-after-year

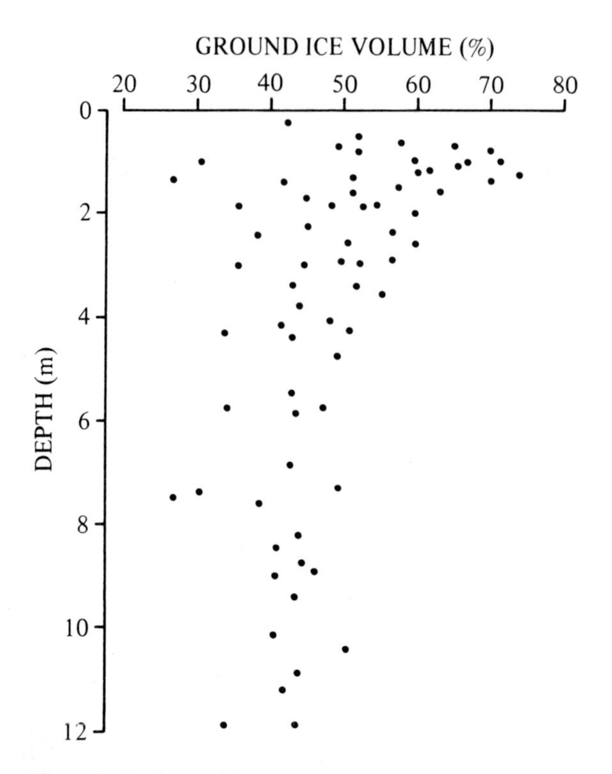

Figure 8.13 Ground ice content versus depth Richards Island NWT Canada (from Pollard & French, 1980).

gradually results in an ice-rich layer. This *aggradational* ice has been studied by Burn (1986); he found evidence in the Yukon Territory, Canada, that this process of ice accumulation has operated throughout post-glacial time. The fact that the top of permafrost is commonly observed to be ice-rich (Figure 8.13), indicates that permafrost is ubiquitously a long-term sink for atmospheric water, which may be important on a geological time-scale.

9

The mechanics of frozen ground

9.1 Introduction

As we begin this chapter, it is appropriate to consider first the physical features that make frozen ground a material of such challenging complexity. Without an appreciation of its special characteristics – including the peculiar properties of the ice – the mechanical properties of frozen ground will not be fully understood. This perspective finds much in common with the views of Tsytovich (1975), whose book is one of the major works on the subject.

Frozen soil is typically stronger than unfrozen soil or ice but it displays a time-dependent creep behaviour similar to ice. In addition to considerable time, temperature and strain rate dependent behaviour which is similar to ice, frozen soil displays frictional behaviour as in the unfrozen state. In contrast to unfrozen soil, however, its strength declines at high confining stresses after reaching a maximum (e.g. Chamberlain *et al.*, 1972).

In simple terms, the strength of frozen ground can be considered as consisting of the cohesion of the ice matrix and the frictional resistance of the soil particles. The viscoplastic strength and deformation characteristics may be attributed largely to the presence of the ice matrix and cementation bonds. The frictional resistance may be considered as a residual strength component (see Sayles, 1973), that is a continuing resistance found *after* such bonds have been overcome and deformation has commenced. However, the composite behaviour of the frozen soil may not be a simple sum of the structural components, and some complex interaction apparently exists between the ice and the soil components. For example, Ting (1981) found that the compressive strength of medium to dense frozen sand exceeds the sum of the strengths of the ice and the soil matrix.

Frozen soil is thus more than a simple amalgam of ice and soil; it is a complex multiphase system consisting of mineral particles, ice, water and air. In particular, the mechanical properties of frozen soil are undoubtedly influenced by the presence of a film of unfrozen water around the soil particles, since the ice exists adjacent to this adsorbed water layer and not directly in contact with soil particles (Figure 9.1). Unfrozen water is also present in bulk ice, on the external surfaces and at the grain boundaries (e.g. Jellinek, 1972). The behaviour of frozen soil retains the complexities of its main structural components – the ice and the soil matrix – but, in addition, is subject to complications arising from the unfrozen water interface between the ice and soil particles.

In fine-grained soils, in particular, with their high unfrozen water contents, the behaviour of the ice is only one aspect to be considered, and the thermodynamic conditions of the ice–water–soil system are equally important. Experimental observations indicate that any change in the temperature of the soil causes distinct and often major changes in mechanical properties. The effect of temperature is important not only because of its large influence on the deformational mechanisms in the ice but also as it determines the amount of unfrozen water in the soil. The quantity of unfrozen water changes with temperature and with stress conditions. As a result, the ice content and degree of cementation bonding also vary. Even for non-saline frozen sands which, in comparison to other soils, contain

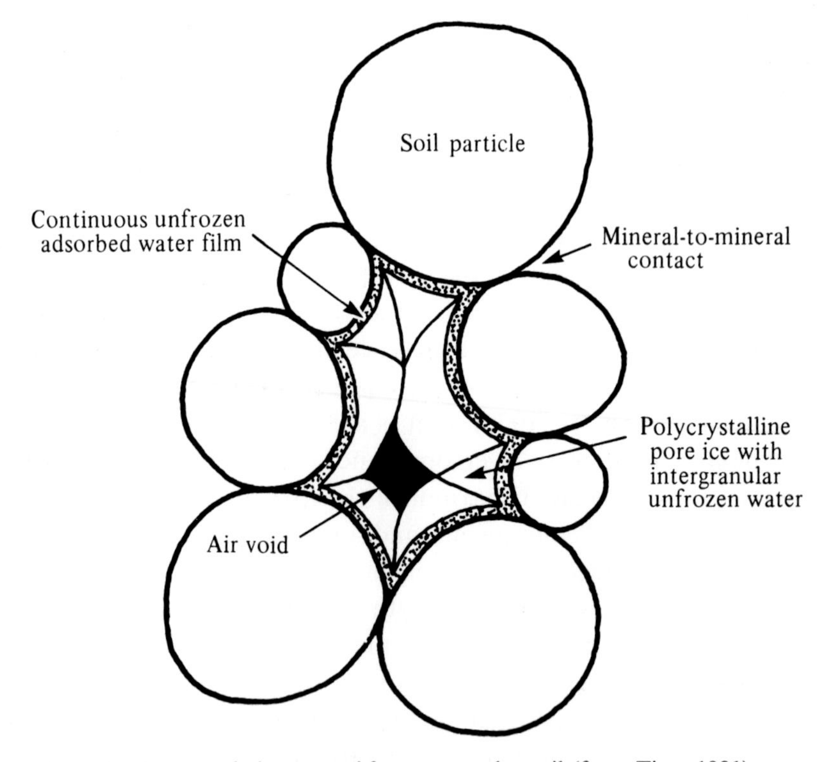

Figure 9.1 Structural elements of frozen granular soil (from Ting, 1981).

extremely small amounts of unfrozen water, a change in temperature results in a change in the strength properties. Because the unfrozen water is concentrated in extremely small capillaries and at contact points between particles, any change in the amount of this water affects the bonding between the mineral grains.

Even a relatively small transfer of water to ice or vice versa in a frozen soil involves, in addition to an exchange of latent heat, migration of water in thin films around particles, particle movement and other time-dependent processes. Thermal and hydraulic conductivities, and the viscosity of water are thus important properties. Such properties, of course, are also active in the transfer of ice, via the water, along stress gradients within the soil.

It follows that under natural conditions, where there are always variations in temperature and stress, the mechanical properties of frozen ground are not constant. In addition, the migration of unfrozen water along temperature-induced (or stress-induced) potential gradients must be considered, especially where such gradients persist over long periods of time. The distribution of moisture within frozen ground, and even within permafrost, will change over time, albeit slowly, and this will affect the mechanical properties and deformation behaviour.

While the precise nature of the frozen soil system is not understood adequately, a large amount of work has been carried out to quantify the strength and deformation characteristics. The results of numerous laboratory investigations show that the stress–strain–strength behaviour of frozen ground is highly time and temperature dependent, that is, the resistance to deformation under a *constant* stress (load) changes through time, as well as with changes in the temperature. Thus, while the term 'strength' is used in a general sense to refer to the mechanical resistance of soil to an applied stress, we cannot speak of the *strength* of frozen ground in simple terms.

Recent reviews of the strength and deformation properties of frozen soils by Andersland & Anderson (1978); Johnston (1981); Ladanyi (1981) and Ladanyi *et al.* (1981) relate to engineering considerations, but in this chapter more fundamental aspects will be considered, with particular regard to thermodynamics (dealt with in Chapter 7).

9.2 The frozen soil system

9.2.1 Properties of the ice

The fundamental feature of the freezing of ground is, of course, the formation of ice in soil pores (at the appropriate negative temperature) and as lenses. The area of contact between ice and soil is important, and this depends on the distribution as well as the amount of ice. The development of ice lenses, among other things, results in the formation of a frozen material which has a new and quite different structure than present before freezing. Indeed, the development of various forms of ice inclusions can lead to an extremely heterogeneous soil structure (Figure 9.2). The freezing of a soil gives rise to the development of a variety of small scale structural defects, such as cavities and microscopic cracks; under the action of an external load, these features are the foci of local irreversible shears and failures.

While the mechanical behaviour of ice is similar to most crystalline materials, it is significant that the ice in frozen soil is very close to its melting point under most natural conditions. In practice, the main scientific and engineering interest concerns the behaviour of ice at these high homologous temperatures, where pressure melting and regelation at stress concentrations, crystal reorientation and grain growth, greatly influence its rheological behaviour.

Ice also exhibits a pronounced time-dependent mechanical behaviour. It behaves elastically in small deformations under rapid loading, but under sustained loading, which induces shear stresses and strains, it behaves inelastically, experiencing continuing irreversible strains (Mellor, 1983). Ice will creep continuously under very low shear stresses.

This behaviour is explained by the fact that various deformational mechanisms occur in ice and almost all exhibit some degree of time-dependence related to the diffusion of heat and moisture. These mechanisms include migration of defects (microcreep), microcracking and grain boundary effects such as sliding and pressure melting (Ting, 1981, pp. 31–32). In addition, the relative importance of these mechanisms depends

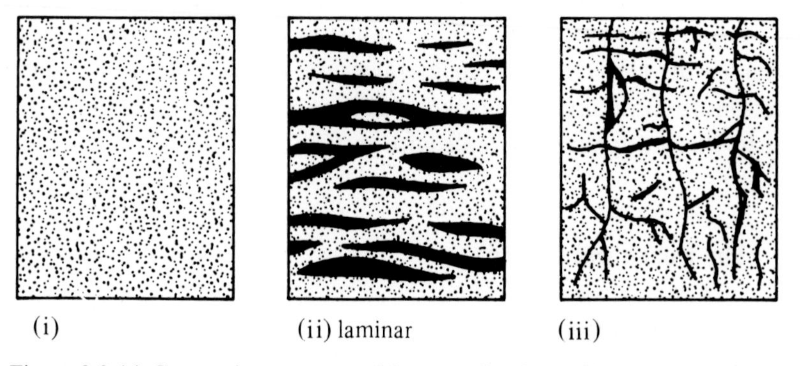

Figure 9.2 (*a*) Cryogenic structures of frozen soils: (i) massive; (ii) laminar; (iii) reticulate (from Tsytovich, 1975).

on the temperature, the applied stress (or the strain rate) and the ice structure. At sufficiently high levels of stress or strain rate, brittle failure is induced, whereas at lower levels ductile yielding occurs. The relationship between strain rate and stress is nonlinear, which means that ice is visco-plastic.

Much of what is known about the rheology and strength of ice has been determined from laboratory experiments. However, according to Mellor (1983, p. 41), the current state of knowledge is not adequate to predict accurately the entire deformation–time behaviour of ice as a function of stress level and temperature; more and better experimental data are required.

Monocrystalline ice exhibits anisotropy that becomes especially conspicuous in direct shear. The crystal structure of ordinary ice is planar with the water molecules forming wrinkled sheets consisting of hexagonal rings. The plane of the sheets is termed the basal plane, with the axis perpendicular to this called the C-axis. The shearing resistance is much lower in the basal plane (easy glide), while slip in other directions can only be induced at much greater stresses (hard glide). Gold (1962, 1966), in fact, determined that when ice was stressed in nonbasal plane directions, it often fractured before slip occurred.

In nature, ice typically exists in polycrystalline form and because of this the deformational mechanisms are different from monocyrstalline ice. In polycrystalline ice, the processes in the monocrystal are combined with grain boundary processes which tend to weaken the resistance to shear. Reorganisation of the crystal structure in individual grains due to imposed stress – by the melting of ice in high stress zones, liquid transport and resulting recrystallisation in lower-stressed regions – leads to crystals more favourably oriented to easy glide. Microcracking serves to relieve stress concentrations, but also results in a weakened structure, with possible new points of stress concentration.

The structure of ice in frozen soils is not well known, although thin-section data in Gow (1975) indicate that the pore ice is polycrystalline and presumably randomly oriented, while carefully prepared laboratory samples of segregated ice featured a strong prismatic (C-axis) alignment of crystals in the direction of heat flow (see section 2.2.8). Such alignment occurs in unidirectional freezing of bulk water but in nature segregated ice lenses may have a disorderly C-axis alignment. Within frozen ground, the properties of the polycrystalline pore ice may appear isotropic, although under prevailing temperature and stress gradients, recrystallisation and alignment of the ice may occur gradually.

Apart from their anisotropy and time-dependence, the internal bonds of ice are also sensitive to changes in temperature, becoming stronger as the temperature is lowered (Figure 9.3). Studies by Mellor & Testa (1969), for example, show that the creep rate of ice is especially temperature-dependent above −10 °C (and even more so above −2 °C). Barnes *et al.* (1971)

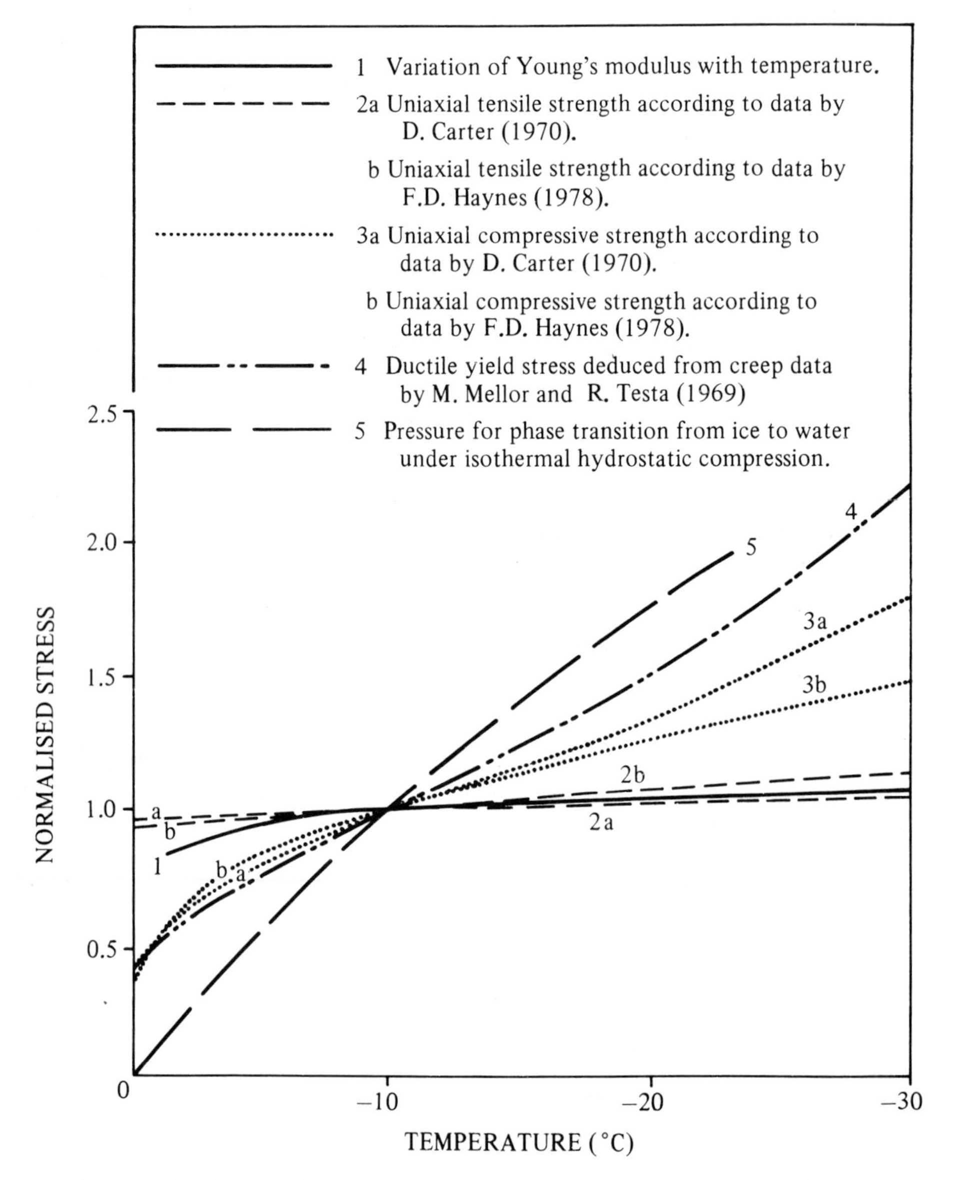

Figure 9.3 Influence of temperature on strength properties of ice (from Mellor 1983). Normalised stress refers to the yield stress (strength) relative to that at −10 °C, for each curve. A normalised stress value of, for example, 2, means the stress was twice that at −10 °C on the curve.

postulate that grain boundary mechanisms dominate above $-10\,°C$, while pressure melting and recrystallisation become important above $-3\,°C$. Thus the importance of temperature, already significant to the unfrozen water content of the soil, is made more apparent.

Unlike the behaviour of ice under compressive loading, the tensile strength of ice is almost independent of temperature and strain rate when the strain rate is above $10^{-5}\,s^{-1}$. Essentially, failure is brittle. The tensile strength of ice is lower than the strength under compressive loading.

9.2.2 The strength of soils

The nature and mechanical behaviour of *unfrozen* soil has been widely treated, see for example, Lambe & Whitman (1979) and Mitchell (1976). In studies of soil mechanics, *strength* is normally considered as the maximum resistance of a material to applied stress. When the stresses, such as in an earth slope or under a building foundation, exceed a certain value, a landslide or other soil displacement occurs. This is often referred to as '*failure*'. However, as regards the behaviour of materials, the meaning of the term 'failure' is arbitrary; in other words, a material may be said to have 'failed' when it ceases to meet certain arbitrary requirements. For example, Mellor (1983) points out that whereas a typical engineering material is deemed clearly to have failed when it fractures, materials can fail without fracturing; the definition of failure might be based on a transition from elastic deformation to plastic yielding, or on irreversible acceleration of strain rate.

The mechanical properties of soils, as for any solid, are chiefly governed by the shear resistance of the particles with respect to one another. According to the Mohr–Coulomb failure criterion, the strength of dry, particulate material depends on two basic quantities:

$$\tau = C + \sigma \tan \phi \qquad (9.1)$$

where

τ = shear stress at failure

C = cohesion

σ = normal stress at failure

ϕ = friction angle

The frictional component of shear strength increases as the normal stress (confining stress) increases (see section 5.2). The cohesion may be due to physical cementation (as in rocks) and/or the existence of interparticle attractive forces. In soils with low specific surface areas, such as sands and silts, the frictional component of strength dominates and cohesion is negligible. In clays, however, surface forces (cohesion) strongly influence mechanical behaviour.

When water is present in the soil, the normal stress is expressed in terms of an effective stress:

$$\sigma' = \sigma - u \tag{9.2}$$

where u is the pore water pressure (see also section 5.2). While positive values of u imply weakening, soil strength is developed by the presence of *negative* pore water pressure, giving a positive effective stress even when the total normal stress is zero. Particle surface forces are associated with the adsorbed water adjacent to the particle surfaces, typically several molecular layers thick, and the double layer water associated with the negatively charged mineral surfaces, up to several hundred Ångströms thick. Such forces are responsible for negative pore water pressures at low water contents, while capillary forces are important at relatively high water contents. In frozen soils also, the effective stress is important, although difficult to evaluate because of the presence of both ice and unfrozen water in the pores (see section 9.7.1).

When water freezes in soils, the cohesion increases many times due to cementation (bonding) between ice crystals and soil particles, even though a film of unfrozen water intervenes. The frictional component of strength is attributed primarily to the soil phase, with some contribution from fractured ice crystals.

Goughnour & Andersland (1968) studied the strength properties of ice–sand mixtures. Compared to pure ice, the presence of soil particles impedes the movement of dislocations within ice crystals, thereby supressing ice creep (see also Phukan, 1985). They noted an increase in peak strength with increasing sand content, with a particularly sharp increase above about 40% sand volume, corresponding to the change in the structure from sandy ice to a frozen soil with the particles in contact (see also Hooke *et al.*, 1972).

9.3 Deformation of frozen ground

The shear strength of *unfrozen* soils is normally determined with an apparatus such as the shear box or triaxial cell by applying force which is increased until the specimen fails. Usually this happens suddenly, and a plane or planes of shear develop in the sample. If a sample of frozen soil is tested in this way, it initially appears to be hard and rigid. However, this belies the extent to which the response of frozen soil to applied stress varies.

When a frozen soil is deformed (strained) very rapidly, it may break along clearly defined surfaces. However, this occurs infrequently in nature, and usually only if a load is applied abruptly. In nature, stresses do not usually build up suddenly and relaxation (see section 6.4) generally prevents the soil from failing by brittle fracture. Instead, it deforms in a *ductile* manner, that is, there is an absence of well-defined shear or rupture surfaces, and the deformation is slow and continuous through the medium. Whereas the soil appears to be very strong under rapid loading, its strength declines considerably under prolonged loading.

The action of a constant load on frozen ground results in recrystallisation of the ice inclusions, the development of microscopic cracks, and, at a certain stress level, growth of these cracks to macroscopic dimensions. These processes lower the shear strength of the soil and promote *creep* deformations. The extent to which such deformations develop depend on the magnitude and duration of the load, the temperature, soil composition, and ice content and distribution.

9.3.1. *Characteristics of creep*

The characteristic nature of creep behaviour is most easily appreciated by considering a sample of ice-rich frozen soil to which is applied a constant compressive stress, which may be simply due to an applied weight. The general form of the creep curve, plotted according to long-established conventions, is shown in Figure 9.4; the designation of primary, secondary and tertiary phases of creep is traditional although somewhat arbitrary.

The immediate response to the application of the stress is a strain, a

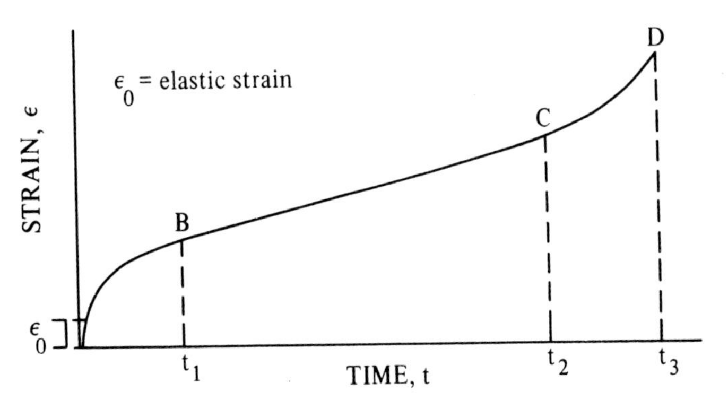

Figure 9.4 The classical creep curve.

shortening of the sample. If the stress is not too great and is quickly removed, the original length may be immediately restored. Such a deformation is an *elastic* one. If the stress is large enough and is applied for a considerable time, deformation continues, but at a decreasing rate (deceleration). Conventionally, all this is called primary creep. After a period of perhaps hours, the deformation may assume (B, on Figure 9.4) a fairly steady rate for a (considerable) length of time. If conditions (especially temperature) do not change, then after a certain period (which could be days, weeks or longer, depending on the applied stress) the strain rate will again increase (acceleration, C, on Figure 9.4), leading inevitably to failure. The 'constant' rate of strain is referred to as secondary creep, and the accelerating part as tertiary creep.

Ting (1981) pointed out that no single deformational mechanism controls the creep rate of ice (or frozen soil) over the entire interval from initial loading to final rupture, but instead, the strain rate at any time in a creep test is an aggregate which is influenced by various mechanisms. Consequently, the existence of a true steady state (secondary) stage of creep – corresponding to a single dominant mechanism – is unlikely. He also suggested that the secondary 'stage' should be viewed simply as the point of inflection on the creep curve (see also section 9.4.1).

According to Andersland, Sayles & Ladanyi (1978) dense soils with a low ice content do not show the typical creep behaviour of Figure 9.4, but rather tend toward brittle behaviour with cracking and rupture after only small strains. This is particularly the case for coarse-grained materials where the amount of unfrozen water is small.

9.3.2 The origins of creep

The creep behaviour of frozen soil follows from a variety of time-dependent processes. Among the constituents of the soil, the ice is the first to enter creep, since this requires the smallest shear stress. In ice, the spread of crystal dislocations, processes of re-crystallisation leading to the reorientation of crystals aligned to promote flow, and intercrystalline liquid migrations are involved.

When the ice is within a porous medium such as soil, there is present a far larger proportion of liquid water than in pure ice (especially if the soil is finely particulate). Since ice and water coexist in equilibrium for a given temperature and pressure, any load applied to the soil causes a disturbance to this equilibrium condition, resulting in partial melting of the ice phase. Under a stress gradient, unfrozen water will migrate to regions of lower stress and freeze there. Roggensack & Morgenstern (1978) documented changes in moisture distribution in frozen soils under an applied shear stress. As moisture is redistributed, and the ice itself flows, structural bonds are broken, weakening the soil and promoting creep deformation.

Tsytovich (1975, p. 109) describes some of the details of the creep process, as determined from microscopic and crystal-optical studies. In the initial (decelerating) stage, the dominant process is one of closing of micro-cracks and the 'healing' of structural defects by moisture migration from points of high stress to these zones of lower stress. This causes some rearrangment of the soil particles. Subsequently, some 'balance' may be established between the healing of existing structural defects and the generation of new ones, mainly in the form of microcracks. This stage may be characterised by sustained viscous flow at a practically constant rate of deformation (strain).

After some period of time, which may be very long under low stress conditions, when the strains reach a certain value and sufficient restructuring of the soil has occurred, the behaviour may pass point C on Figure 9.4 into the stage of accelerating strain. New microscopic cracks develop at a steadily increasing rate and their transition to macrocracks results in progressive weakening of the soil (i.e. failure). In addition, the ice inclusions have, by now, become reoriented with the basal planes parallel to the direction of shear, causing a substantial decrease in their shear strength and, consequently, the strength of the frozen soil as a whole. This stage eventually terminates in rupture or plastic loss of stability of the soil.

9.4 Strength characteristics of frozen soils

The essential characteristics of the strength properties of frozen ground can be demonstrated by examining the outcome of a series of shear tests. The principles of a shear test are shown diagrammatically in Figure 9.5. To begin with, let us assume that the normal stress and the temperature are both held constant. The application of a shear stress, τ, to the specimen results in a shear strain, which will vary characteristically with time as shown in Figure 9.6 (see the creep curve for τ_3, for example). We see that after some period of more-or-less steady strain – the middle portion on the curve – the soil loses stability and goes into the stage of accelerating creep leading to failure. Thus, after a certain elapsed time, t, the applied shear stress is equivalent to the failure stress, τ_f. If the test is repeated using larger (τ_1, τ_2) and smaller (τ_4, τ_5) values of applied shear stress, we get a family of creep curves (Figure 9.6), which show that the time to failure is progressively greater the smaller the shearing load. In other words, the resistance to shearing decreases with the duration of the applied stress (loading).

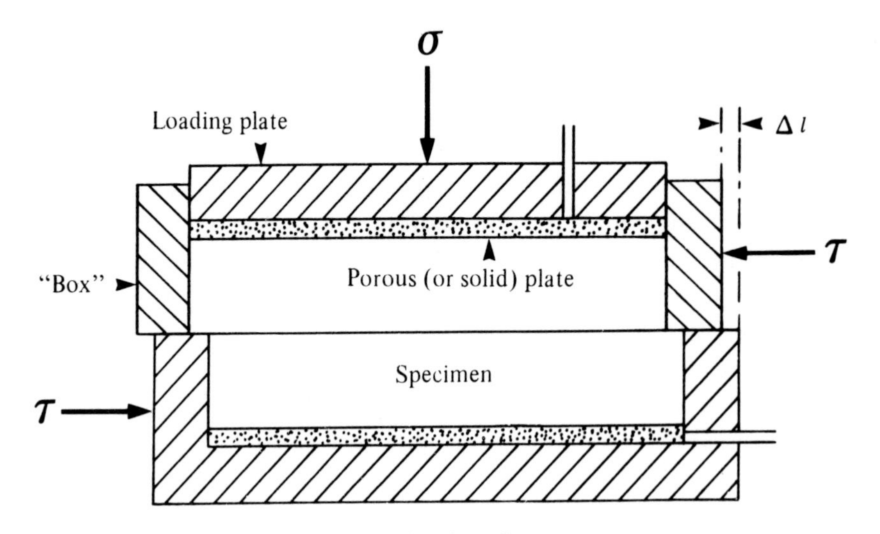

Figure 9.5 Principles of a shear test with shear box.

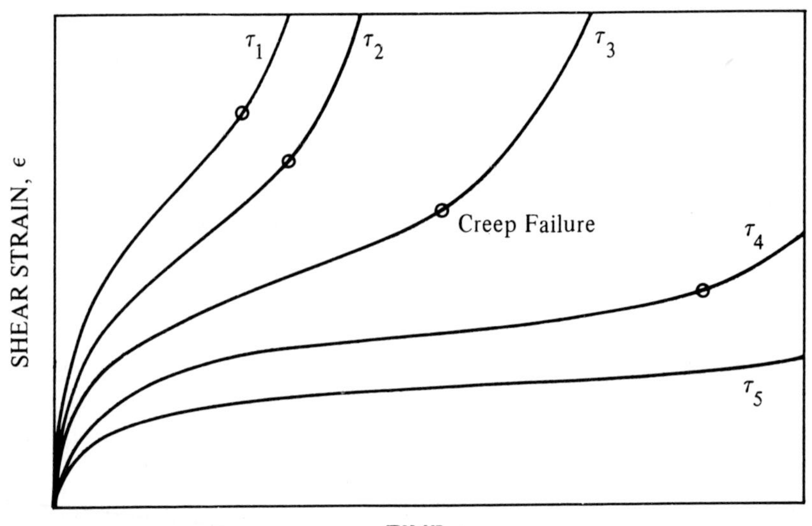

Figure 9.6 Creep curves for various applied shear stresses.

Information of the type shown in Figure 9.6 can be presented in the form of Figure 9.7*a*, where strength (resistance) is shown to be dependent on time. The largest value is often referred to as the *short-term* strength, and the lowest as the *continuous*, *long-term* or *residual* strength. When we regard frozen soil as rigid we think only of high stresses applied for a very short time. Adfreeze strength (the strength of the bond of frozen ground to wood, steel or other materials) shows the same general pattern as shown in Figure 9.7*a* for shear strength (Tsytovich 1975, p. 161).

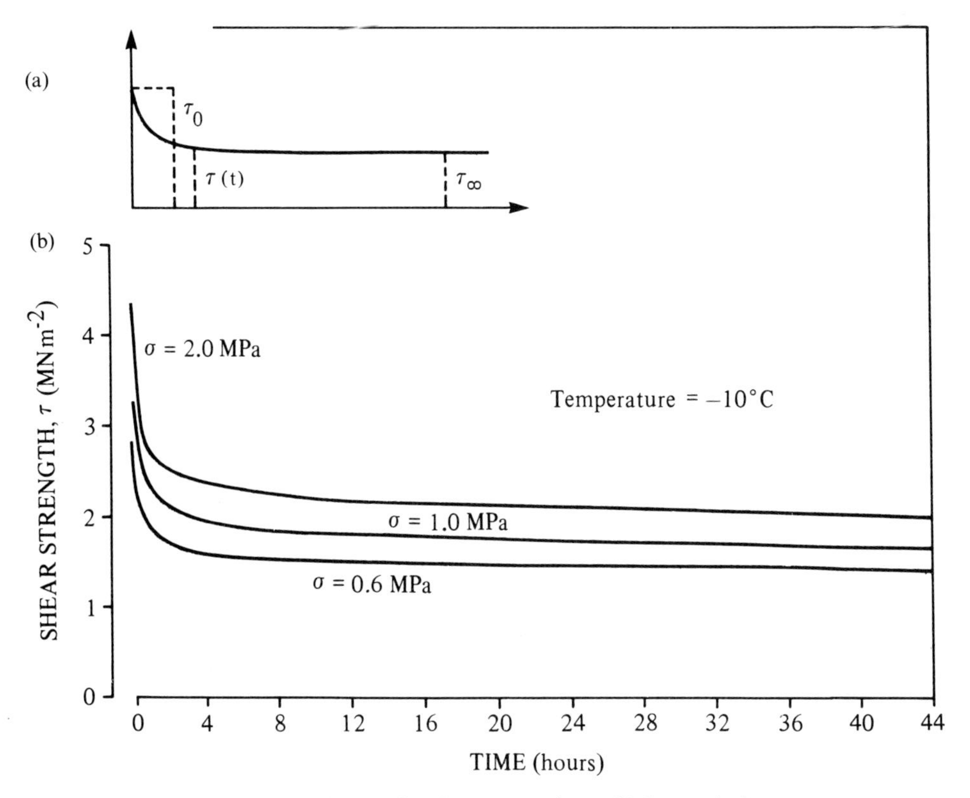

Figure 9.7 (a) Dependence of resistance on time. (b) Strength-time characteristics of frozen sand loam (−10 °C) under various normal stresses (from Vialov, 1965a).

If we were to repeat the set of shear tests for different values of normal stress, σ_2, σ_3, etc, we would obtain a pattern of results as shown in Figure 9.7(*b*). As implied by equation (9.1), the shear strength increases with σ. With the information contained in Figure 9.7(*b*), it is possible to construct a family of curves showing the dependence of τ on σ for particular values of (elapsed) time (Figure 9.8). This Figure shows that for any value of normal stress, the shear stress sufficient to cause failure is lower the longer the duration of loading. In the examples shown, the lines are approximately linear, and from equation (9.1) we may infer that the slope is equal to $\tan\phi$ and the intercept to the cohesion, C. Any variation in the slopes of the curves describe the time variation of $\tan\phi$ and the intersection with the τ-axis represents the variation of C.

The various curves in Figure 9.8 are essentially parallel, indicating that the frictional component of strength does not vary significantly with the duration of stress. In contrast, the cohesion (intercept) *does change* significantly with time, indicating that the characteristic decrease in shearing

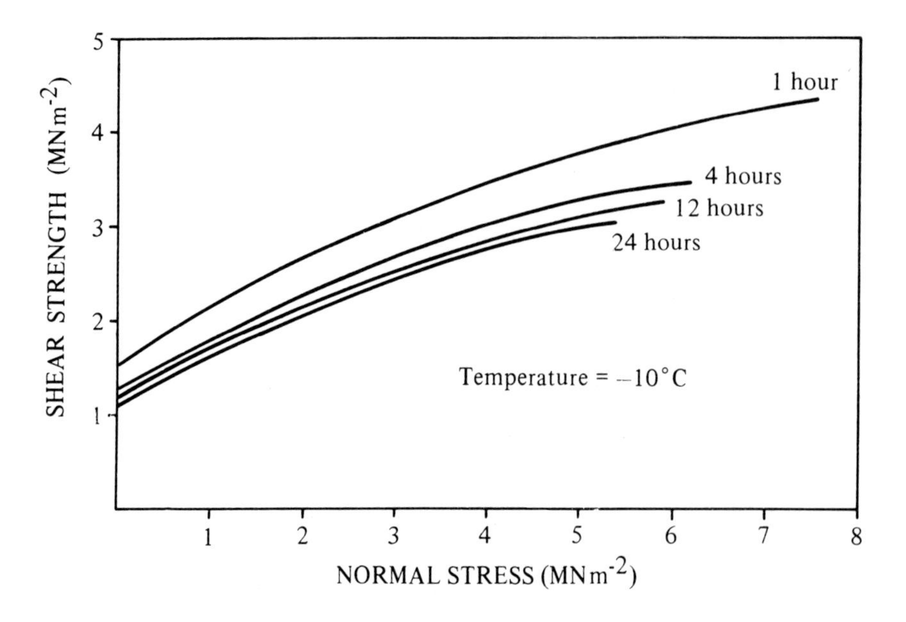

Figure 9.8 Dependence of shear resistance on normal stress, for various durations of applied shear stress (temperature − 10 °C). After Vialov, 1965a.

resistance under the long-term application of a constant load occurs at the expense of the cohesion of the ice. Thus, in contrast to unfrozen soils, the cohesion of frozen soils depends on the duration of applied stress.

Early investigations of frozen soil strength usually involved measuring the effects of loads applied to produce failure in some short, practical period of time, the length of which was not thought important. In nature, and in most geotechnical applications, stresses persist over essentially unlimited time. Thus, in due course, frozen materials may fail and demonstrate a strength far less than that measured in short-term tests. In fact, Sayles (1973) has suggested that for materials which are 'purely' frictional in the unfrozen state, the ultimate creep strength in the frozen state would be roughly equal to that measured in (triaxial) tests on the freely drained *unfrozen* material. This is because the cohesive strength of the pore ice tends to zero in the long term, and the only remaining source of strength is the friction between soil particles. Similarly, one might argue that for frozen materials which are essentially cohesive in the unfrozen state, the ultimate strength derives from the unfrozen cohesion.

If the entire test procedure described above is repeated at various temperatures, T, then we get another family of curves for τ versus σ (for a particular t) as shown in Figure 9.9. This example shows that the cohesion (of a frozen sandy loam) increases as the temperature decreases, but that,

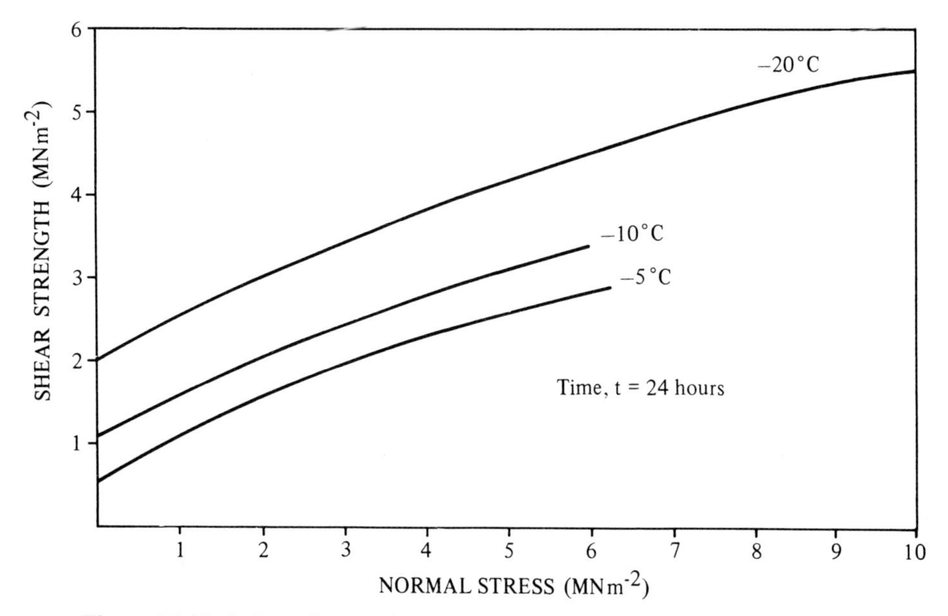

Figure 9.9 Variation of strength with normal stress, for various temperatures (time to failure in each case: 24 hours. Same soil as Figure 9.8 – after Vialov, 1965a).

as in the case of the time variable, the angle of internal friction can be regarded as constant for most practical purposes.

From all of this we can conclude that *variations* in the shear strength of frozen soils *with time and temperature* result primarily from variations in the ice cohesion, and that the frictional resistance may be regarded as essentially constant. In practical terms, therefore, variations in the shear strength of frozen ground may be investigated in terms of their cohesion, and equation (9.1) can be written as follows:

$$\tau(t, T) = C(t, T) + \sigma\tan\phi \qquad (9.3)$$

9.4.1 Rates of strain and processes of deformation

Because the deformation behaviour of frozen soil is highly dependent on the duration and magnitude of the applied stress (as well as upon temperature, soil type and ice content), laboratory testing often uses the procedure of applying a constant rate of strain instead of a constant stress. This is achieved with a piston ('ram') moving at a constant speed (Figure 9.10). The stress being applied at any time (that necessary to maintain the rate of strain) may be shown by a proving ring.

The results of such tests with the application of different, but constant, rates of strain and plotted as stress versus strain curves provide insights to

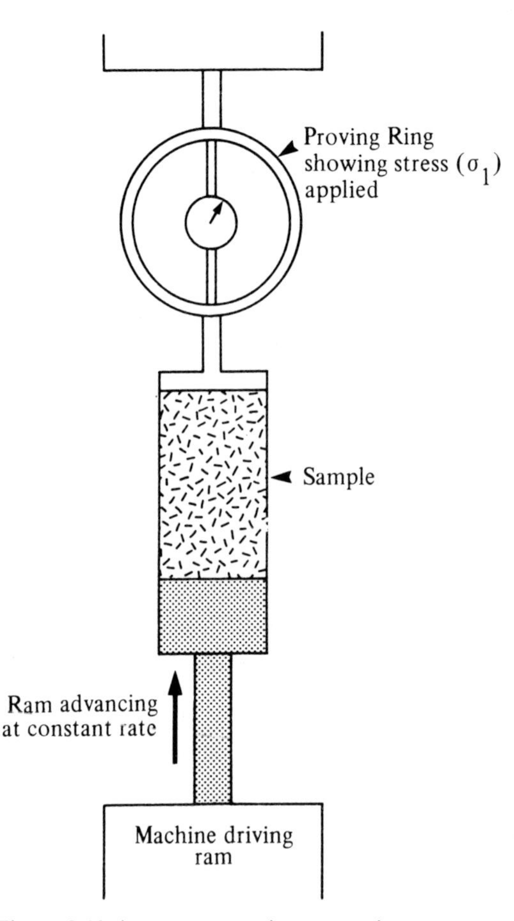

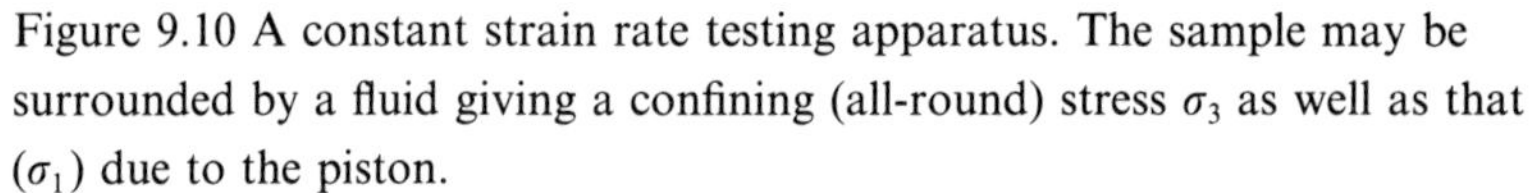
Figure 9.10 A constant strain rate testing apparatus. The sample may be surrounded by a fluid giving a confining (all-round) stress σ_3 as well as that (σ_1) due to the piston.

the deformation processes. As deformation progresses the stress will vary characteristically as shown in Figure 9.11(*a*). The highest stress value is greater for greater strain rates. For sufficiently high strain rates, the highest stress value represents the maximum short term strength. While it would be expected that failure would occur earlier for rapid strain rates, it is also apparent that failure occurs at lower strains under these conditions (Figure 9.11(*b*)). There is less deformation prior to failure, and for the frozen sands shown in the figure, brittle failure occurred at strain rates above about $10^{-4}\ s^{-1}$. Tests on samples of ice, specially prepared to resemble that in ice segregations in soil, have also exhibited brittle failure at strain rates above about $10^{-4}\ s^{-1}$ (Ohrai *et al.* 1983).

Brittle failure is associated with abrupt rupture, but the drop-off in applied stress (strength) from the peak value is followed by a continuing

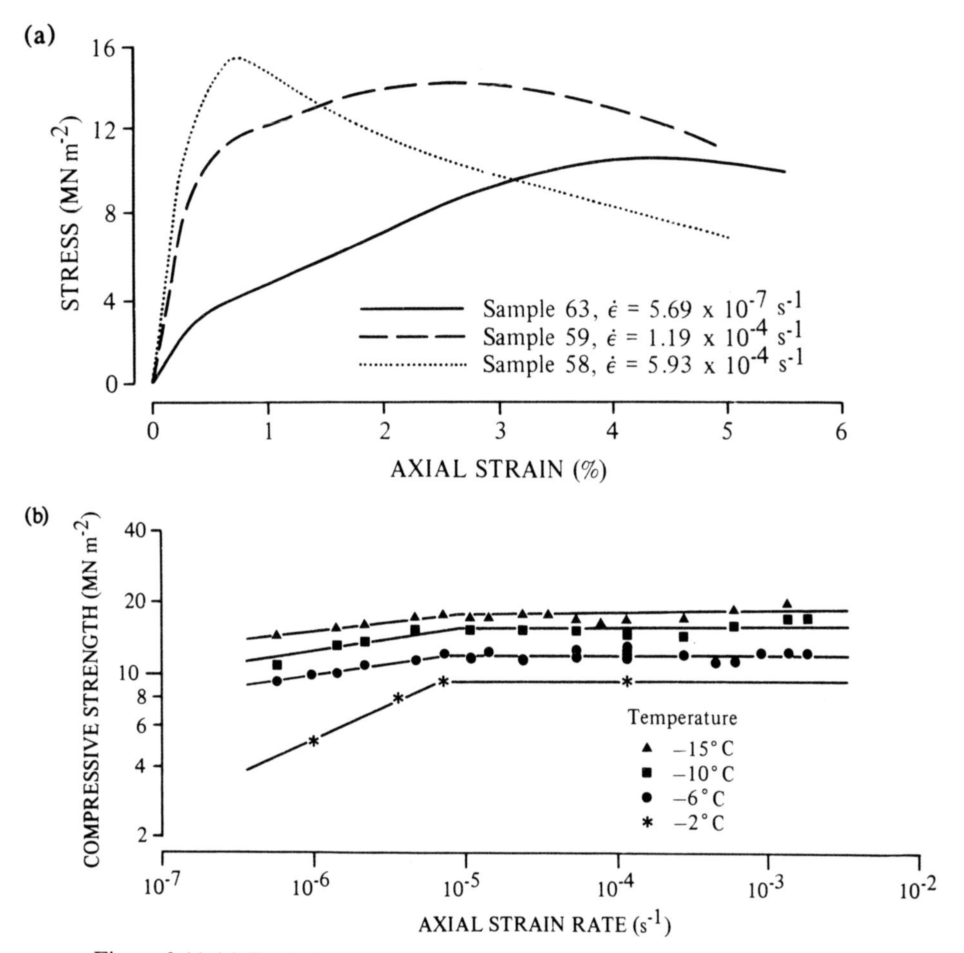

Figure 9.11 (*a*) Typical stress–strain curves for frozen sand (*b*) Compressive strength versus strain rate (from Bragg & Andersland, 1980).

residual strength (Figure 9.11(*a*)). Thus although the strongest bonds have been overcome, movement along fracture surfaces still encounters a continuing resistance. A similar pattern of behaviour is often seen in clays when unfrozen.

Sayles (1973), from tests carried out on frozen sand, identified the residual strength with the frictional resistance between the soil grains. He noted that at confining stresses above about 2.7×10^6 N m^{-2}, two peak stresses were observed (Figure 9.12). The first peak occurs at strains below 10^{-2}, close to the failure strain for columnar-grained ice. The second peak, about 10^{-1} strain, corresponds to the mobilisation of frictional strength between soil grains and/or ice crystals as the strain progresses. Thus the short term strength is essentially a consequence of the ice matrix and the residual strength derives from internal friction as noted earlier. Roggensack

& Morgenstern (1978) also concluded that at low strain rates or long times to failure, an increasing proportion of the shear strength can be attributed to friction. Since the strength due to friction increases with confining stress (equation (9.1)), and, for practical purposes, the cohesion of the ice does not, then the second peak may be the smaller at low confining stresses, but larger at high confining stresses (Figure 9.12).

Mellor & Cole (1982, 1983), amongst others, have questioned the classical view of creep outlined in section 9.3.1, drawing attention to the correspondence which should exist between creep curves, where a *constant stress* is applied, and stress/strain curves, where the stress is progressively changed in order to maintain a *constant rate of strain*. Stress/strain curves (e.g. Figure 9.11*a*) typically show a distinct point of maximum stress, this highest stress corresponding to the point of failure; at this point, the ratio between stress and strain rate ($\tau/\dot{\varepsilon}$) attains its maximum value. This result implies that there should be a distinct *point* of minimum strain rate on the creep curve, the minimum value of ($\dot{\varepsilon}/\tau$) on the creep curve corresponding to the maximum value of ($\tau/\dot{\varepsilon}$) on the stress/strain curve. The results of Zhu & Carbee (1987) show that there may be a direct correspondence between the results from the two types of tests, under some conditions.

The point of minimum strain-rate on a creep curve can be viewed as the transition from elastic (decelerating) to plastic (accelerating) deformation, i.e. it is a point of inflection which can be used to define 'failure' (see also Fish 1983). To some degree, therefore, the appearance of a 'linear' (secon-

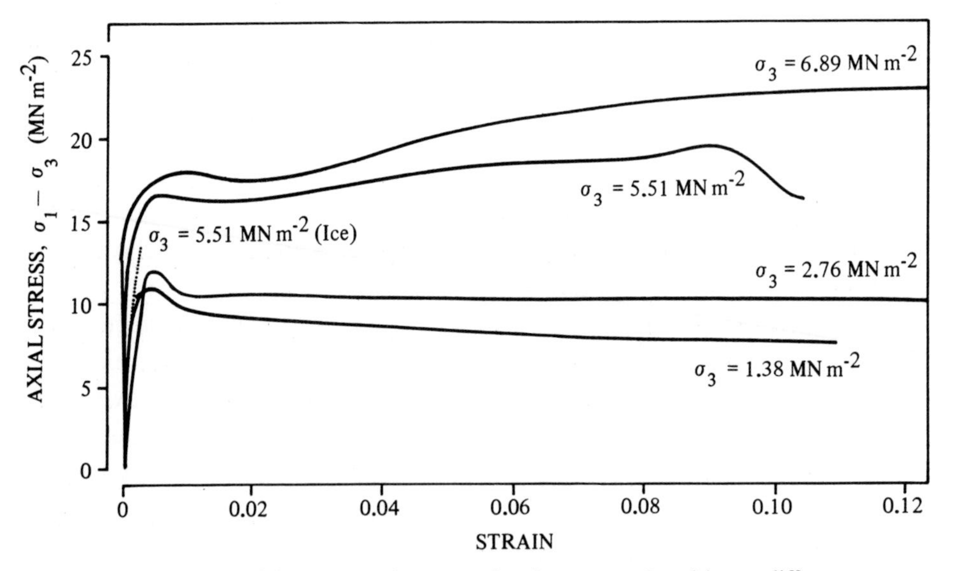

Figure 9.12 Axial stress–strain curves for Ottawa sand and ice at different confining stresses, σ_3, and at large strains (from Sayles, 1973).

dary) stage in the classical creep curve may be an artifact of the way the data are plotted.

Finally, constant strain-rate tests are valuable in interpreting field situations, not least because the rate of deformation, that is, of soil movement, is more easily measured than are the stresses existing in the ground. They are also useful in geotechnical engineering, where designs can be based on maximum allowable stresses giving a defined strain for a given period – for example, the design life of a structure. However, it is difficult to apply rates of strain in tests that are as slow as those usually occurring in the field, and the processes of deformation observed may not be those that are most important under field conditions.

9.5 Temperature dependence of creep rate and strength

As with ice, temperature is one of the main factors influencing the mechanical properties of frozen ground, deriving its influence from three main effects. First, it determines the amount and effects of unfrozen water present in the soil. Secondly, as a result of the lowering of temperature, the strength of the ice bonds, as well as those of the unfrozen water, increases, endowing the soil with a greater resistance to deformation. Thirdly, at lower temperatures, a stress increment results in less thawing, the rate of water migration will be much slower, and this is important to the process of stress redistribution.

It follows from this that as the temperature is lowered, the time-dependent strength of frozen ground should increase, as illustrated in Figure 9.13. The increase in strength is manifested in lower rates of creep deformation and a prolongation of the so-called secondary stage of creep. The temperature-dependency of strength for frozen soil may also be portrayed in the manner shown in Figure 9.14, where data for instantaneous compressive strength for three soils are compared. In all cases, one notes a more-or-less steady increase in strength with decreasing temperature. The data for polycrystalline ice (from Haynes 1978) indicate that while the curve for the sand follows closely that of the ice, *the change in unfrozen water content is the dominant factor in the finer-grained soils*. Thus the differences between the soils are due primarily to their unfrozen water content characteristics. At very low temperatures, below about −50 °C, strength is probably little dependent on temperature.

Within the temperature range of significant phase transformation – from 0 °C to −2 °C, or so (commonly the range of interest) – the strength characteristics can change greatly for temperature changes of less than a degree and the rate of creep increases sharply as the temperature approach-

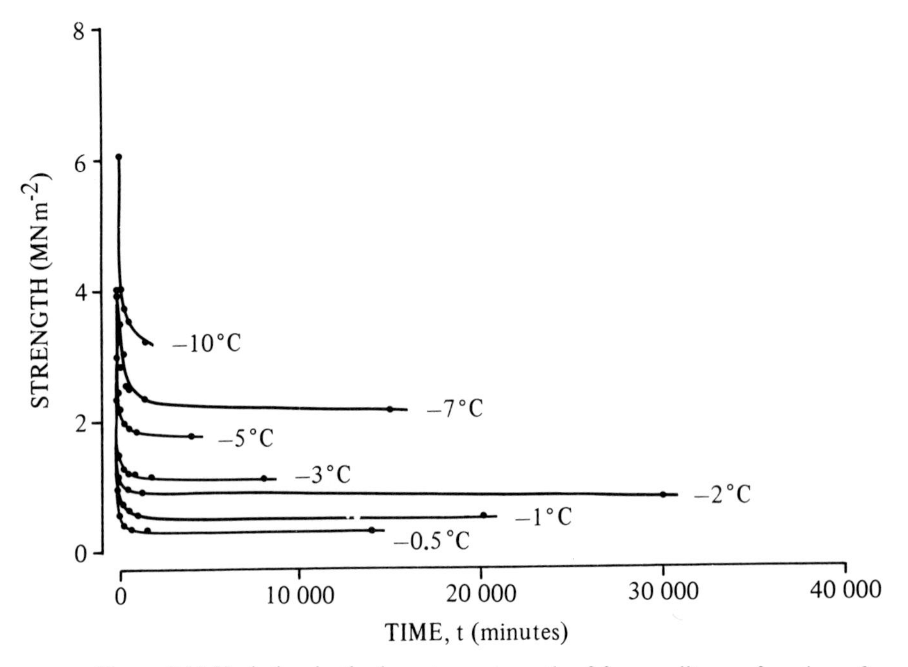

Figure 9.13 Variation in the long-term strength of frozen silt as a function of temperature (from Zhu and Carbee, 1983).

es 0 °C (Ladanyi *et al.*, 1981). This behaviour is principally related to the rapid change in the unfrozen water content, at the higher temperature (cf. Figure 1.4). This may well be more important than the dependence on temperature of the behaviour of the ice. The creep resistance of frozen fine-grained soils is lower at these temperatures than that of ice, due to the unfrozen water content.

Finally, from a comparison of Figures 9.7 and 9.13 we see that the influence of temperature, through its thermodynamic implications, is as fundamental as that of stress. In attempting precise assessment of the strength and deformation properties of frozen ground, therefore, temperatures must always be stated.

9.6 Effect of soil composition

Frozen ground is rarely, if ever, a simple, uniform material, but a complex amalgam of mineral grains of varying size and lithology, ice and water. In natural situations, the composition can vary spatially even over very small intervals.

In endeavouring to establish the manner in which the composition determines deformation behaviour and strength, direct and indirect effects must be distinguished. Soils with large and abundant ice inclusions behave

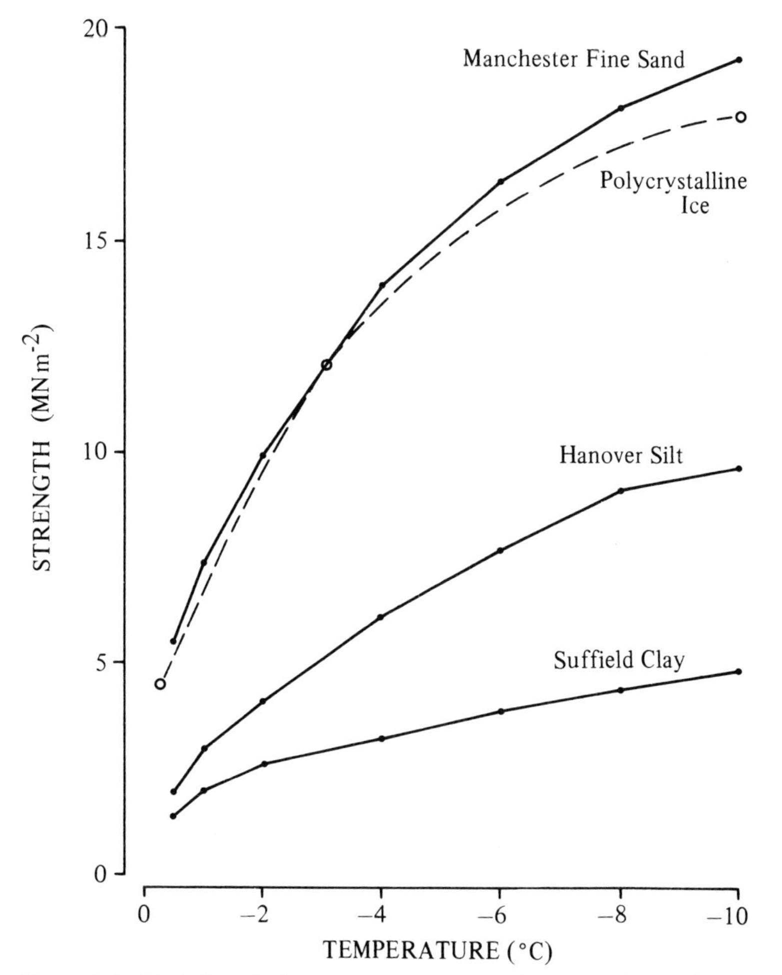

Figure 9.14 Variations in instantaneous compressive strength as a function of temperature (compiled from Sayles, 1968; Sayles & Haines, 1974 and Haynes, 1978).

in a characteristic manner. Silt-rich soils often have excess ice – but not always. Thus it is essential in field interpretations to assess the ice contents and their significance as well as the granular composition (and, of course, the temperature).

Tsytovich (1975, p. 11) differentiates between hard-frozen and plastic-frozen ground. The former is firmly cemented by pore ice and at a temperature low enough that there is little unfrozen water present. In sands, this limit is about −0.5 °C or so, whereas for clays it may be much lower. Plastic-frozen soils exhibit creep properties by virtue of their high unfrozen water contents; they are characterised by relatively high compressibility in the frozen state. The greatest engineering difficulties are encountered when frozen ground is ice-rich and is in the plastic-frozen and high temperature states (Tsytovich, 1975, p. 11).

9.6.1 Size of soil particles

The importance of the grain size composition is clearly seen when the relationship between stress and strain is examined. The stress necessary to maintain a fixed strain rate on ice-rich clay soil characteristically rises only slowly as strain increases (Figure 9.15). A frozen sand, on the other

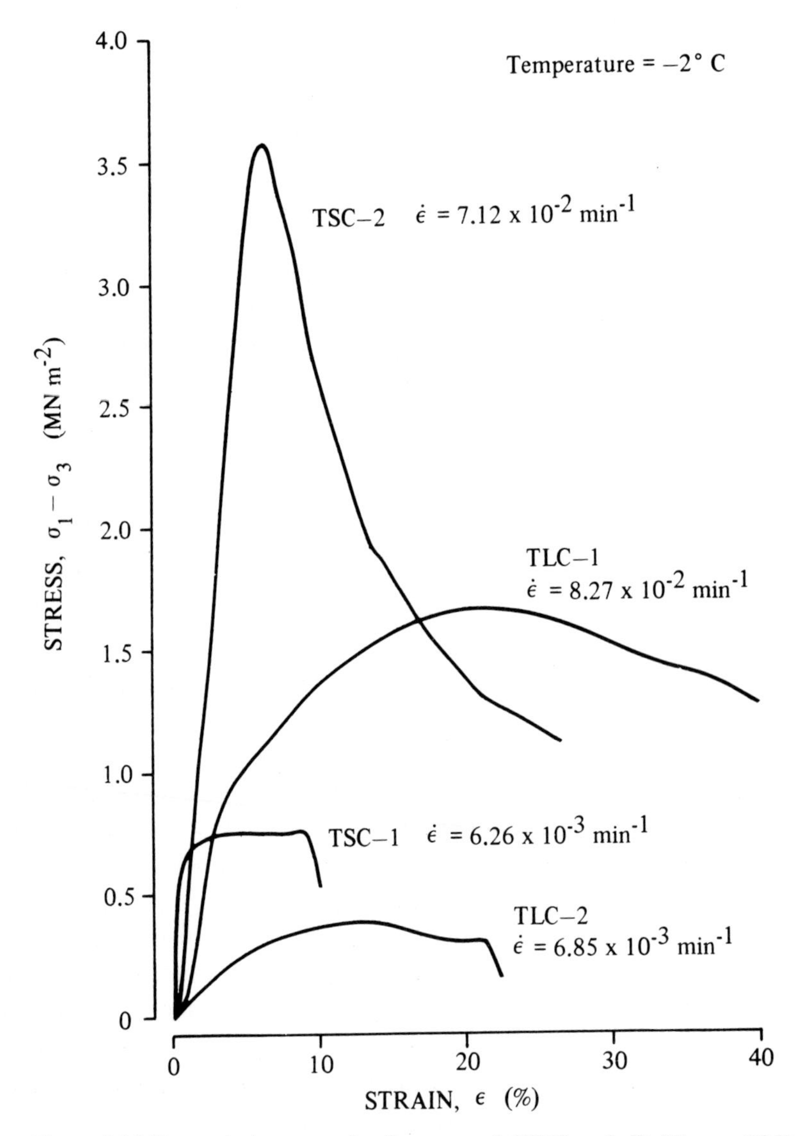

Figure 9.15 Stress-strain curves for frozen sand (TSC) and silt ($T = -2$°C) (source: Ladanyi & Lauzon, 1986).

hand, tested at a similar strain rate, shows a rapid rise of resistance. After only a small strain, however, collapse occurs in the sand with rapidly falling resistance (that is, a rapid fall in the applied stress). The behaviour of the fine-grained soil is a manifestation of creep, which is greater in soils of higher ice content and higher unfrozen water content, and of course at temperatures near to 0 °C. Because the unfrozen water content depends on temperature, it was considered under that heading. However, it should be remembered that the unfrozen water content–temperature relationship depends greatly on soil lithology as well (Figure 1.4).

9.6.2 Ice content

Small samples taken from ground in which the mineral soil is localised rather than uniformly dispersed (e.g. where there are massive icy beds) may well be essentially pure ice. The tendency of ice to deform slowly even under very low stresses implies that very ice-rich soil will exhibit creep (as does ice itself) even on slopes of very low angle (see Figure 5.6). More uniformly dispersed mineral matter in low concentration may cause only small modifications to the ice behaviour. Sand particles are harder than ice and will have a retarding effect on crystal dislocation and other deformation processes in ice. Experiments reported by Hooke *et al.* (1972) showed that with a 40% volume fraction of sand, the creep rate was less by an order of magnitude than that of pure ice at the same temperature.

The influence of ice content was first reported on by N.A. Tsytovich in 1937 (in Vialov, 1965*b*). He showed that the strength of frozen soil increased with an increase in the ice content, up to a certain level, which corresponds to the complete filling of the pores with ice. With further increases in ice content, the strength decreases. The latter may be explained as follows. When the ice content is below that sufficient to form an integrated network, any external load is borne by the framework of the mineral particles. As the volume of ice increases, however, it forms a skeletal framework of its own which carries a greater proportion of the load relative to the mineral matrix. Since the long-term cohesion of ice is small compared to soil, the result is a drop in the strength of the frozen soil.

Goughnour & Andersland (1968) found that strength increased sharply when there was sufficient sand (about 40% or more by volume) for friction between grains to develop as well as dilatancy (the expansive tendency as grains ride over each other). Dilatancy is in turn resisted by the cohesion of the enveloping ice. Again, the rate of deformation is important because with slow deformation this cohesive resistance would be (very) small.

Alkire & Andersland (1973), investigating sand and ice mixtures at

$-12\,°C$, showed that the frictional component became important only when mineral grains were in direct contact. They suggest this occurs in ice-rich materials after the strength of the ice is exceeded – leaving the frictional residual strength mentioned earlier, identified by Sayles (1973). The break of slope at approximately 1% strain (see Figure 9.11(*a*)) occurs at stresses corresponding to the strength of pure ice. They also suggest the cohesive component (at least for frozen dense sands) is very small when confining stresses exceed about $7\,MN\,m^{-2}$ (see section 9.7).

Whether the ice is present in discrete bodies or is confined to pores is also fundamental and should be taken into consideration when terms such as 'high ice content' are used. Ice is present in frozen soils either in the form of ice cement or as interlayers – lenses, veins, inclusions, etc. Tsytovich (1975, p. 48–49) recognises three basic cryogenic structures: massive (fused), laminar (layered) and cellular (reticulate) (see Figure 9.2). The massive cryotexture forms only on rapid freezing, and is characterised by a practically uniform distribution of ice in the frozen soil. Such soils exhibit the strongest resistance to external forces. Frozen soils with laminar and reticulate structures exhibit lower strength in the frozen state and settle most on thawing. Roggensack & Morgenstern (1978) subjected natural permafrost samples to direct shear tests, and found that failure occurred along ice laminations.

9.6.3 Pore water salinity

Ogata *et al.* (1982) investigated the effects of dissolved salt and found that even small concentrations produced a sharp decrease in the compressive strength (Figure 9.16). (The compressive strength was defined as the maximum stress reached in a constant strain rate test of 0.016% compressive strain s^{-1}.) They demonstrate a strong dependence of strength on the degree of ice saturation (the extent to which the voids are filled with ice) and conclude that the effect of the salt follows from the greater amount of unfrozen water. The salt lowers the freezing 'points', displacing the unfrozen water content curves towards lower temperatures (see section 7.1.2). Furthermore, creep rates were maintained with smaller applied stresses when the salt was present. The eutectic point for salt solution (the temperature at which further concentration of salt does not further lower the freezing point) is $-21\,°C$, and at temperatures below this it appears that the strength is actually increased by the salt.

9.7 Effect of normal and confining stresses on strength and deformation

The dependence of the frictional strength on normal stress was described in section 9.2.2, and can usefully be discussed further. Tsytovich

(1975) reports observations of the increase in strength of frozen soils under increasing normal load and therefore due to friction. (Table 9.1: see also, Figure 9.8). Normal stresses result from confining stresses, as well as uniaxial stresses, and their effect on the strength of frozen soil has been studied by Chamberlain *et al.* (1972), Alkire & Andersland (1973), Sayles (1973) and Ladanyi (1985), among others. The tests by Alkire & Andersland were carried out by applying a constant rate of strain to samples which

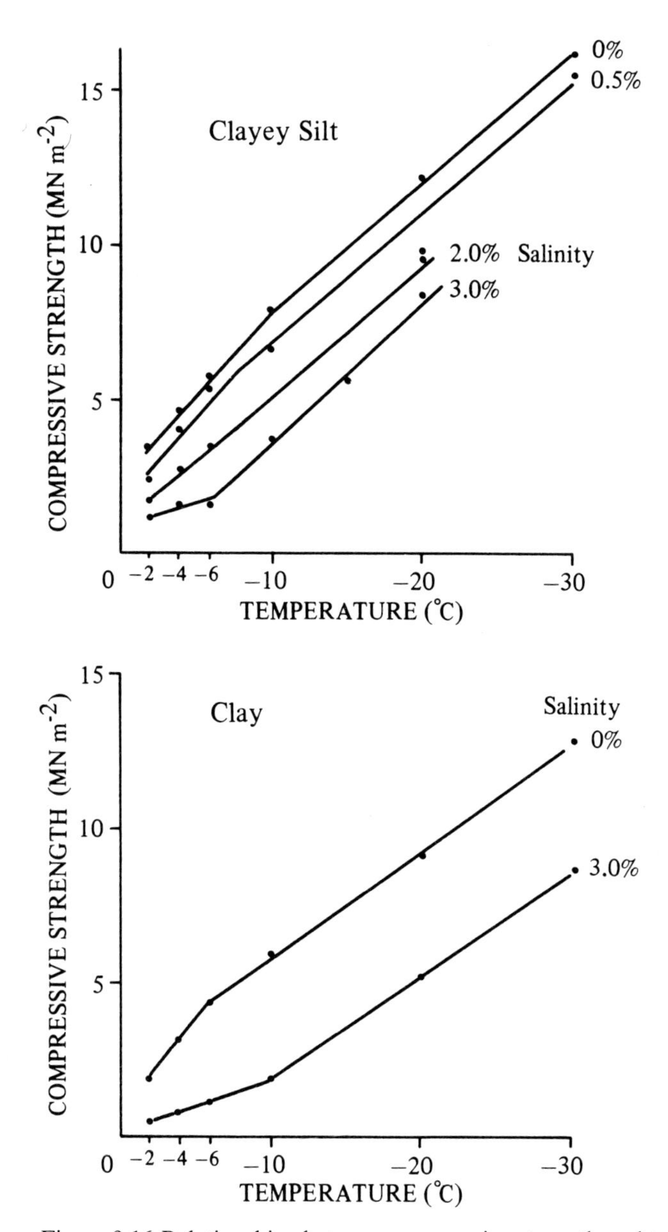

Figure 9.16 Relationships between compressive strength and temperature at various salinities (from Ogata *et al.*, 1982).

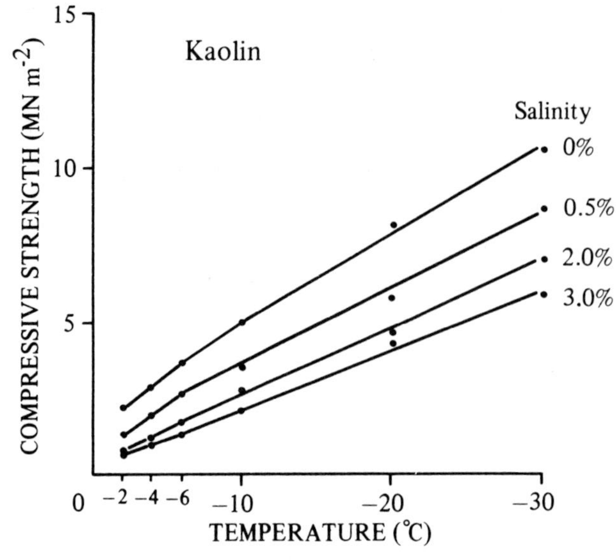

Figure 9.16 (Cont.)

were under various confining (all-round) stresses, σ_3, (Figure 9.17). The highest so-called *deviator stress* ($\sigma_1 - \sigma_3$) applied represents the short-term strength. The rate of increase of the deviator stress – that is, of soil resistance – prior to the peak (especially after the break in the slope of the

Table 9.1. Friction and cohesion in % of total shear strength for soils in the frozen and unfrozen but solid states (cover clay) (after Tsytovich, 1975)

Bulk density ρ, g cm^{-3}	Moisture content w, %	Normal load σ, kg cm^{-2}	Shear stress τ, kg cm^{-2}	Cohesion C, %	Friction $\sigma \tan \phi$, %
At $T = -1$°C					
—	—	1	5.5	94.7	5.3
—	—	3	6.0	86.7	13.3
1.84	26.5	4	6.2	83.9	16.1
1.88	34.8	8	7.3	71.2	28.8
1.85	29.1	12	8.3	62.6	37.4
At $T = -2$°C					
—	—	1	7.6	94.7	5.3
—	—	3	8.4	85.7	14.3
1.86	32.1	4	8.9	81.0	19.0
1.84	32.3	8	10.5	68.5	31.5
—	—	12	12.2	59.8	40.2
At $T = +20$°C					
2.04	23.2	1	0.84	91.7	8.3
2.07	22.8	3	0.99	77.7	22.3
2.11	22.7	5	1.10	70.0	30.0

Note: The unfrozen specimens were first compacted by a load of 5 kg cm^{-2}. 1 kg cm^{-2} ≏ 100 kN m^{-2}.

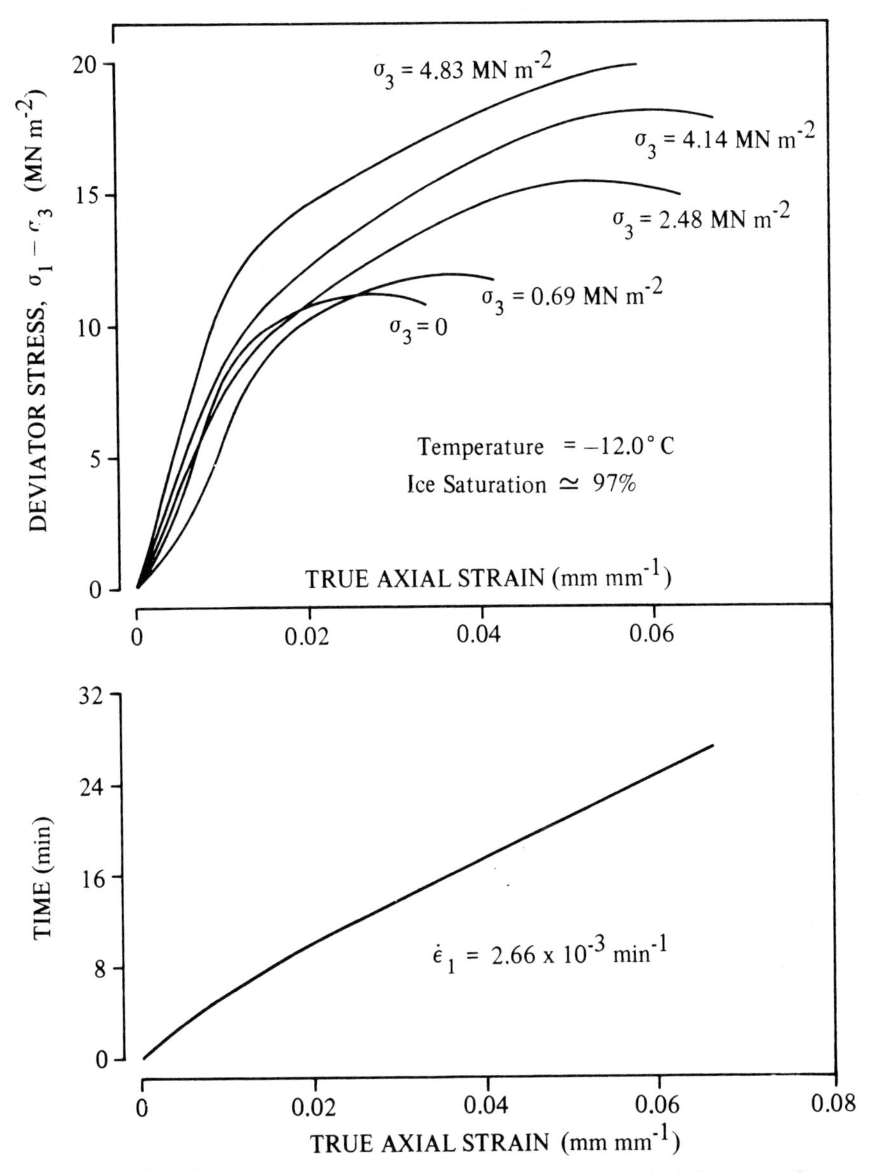

Figure 9.17 Influence of confining pressure on the stress–strain behaviour of a frozen sand–ice mixture (from Alkire & Andersland, 1973).

curves) depends on the manner in which mineral particles gradually interact, push together or ride over one another. The closeness of the particle packing affects frictional resistance, and this was confirmed by tests with different initial void ratios.

The strength and creep of isotropic polycrystalline ice was examined by Jones (1982), who found that at high strain rates, confining stresses of up to 34 MN m^{-2} (at -11 °C) increased the strength by preventing cracking. Deformation associated with cracking involves dilation so that it is logical it be resisted by confining stress. However, at higher confining stresses the strength was somewhat lower. This is probably explained by localised pressure melting; melting of the ice as a whole is expected when the confining stress exceeds about $120\,\mathrm{MN\,m^{-2}}$ at $-11\,°\mathrm{C}$ (cf. equation (7.5)). At low rates of strain, there was relatively little increase in strength with confining stress because the deformation is plastic and the strength, by implication, is cohesive.

Normally, when a continuing confining stress is applied to frozen soil, there is a small increase in water content at the expense of the ice. At temperatures of $-1\,°\mathrm{C}$ or lower the effect of confining stresses of 0.5 to $2\,\mathrm{MN\,m^{-2}}$ is to increase the unfrozen water content by at most a few percent (Figure 7.3). Such small increases apparently do not usually lead to a weakening of the soil since, in general, the effect of confining stress at these levels is to strengthen the frozen ground by increasing friction, as discussed above. In fact, the small amount of thawing may bring more mineral particles into contact, and thus be partly responsible for the increase of frictional strength. However, there may be some increase in creep deformation under these conditions (see section 9.8.1). Figure 9.18 shows that there is a loss of strength when confining stresses exceed $28\,\mathrm{MN\,m^{-2}}$ for the clay-rich soil and this is easily understood from the broader range of pressure and temperature over which unfrozen water is present in significant quantity in such materials (cf. Figure 1.4). However, the water content *per se* is probably less important than its energy status (represented by the suction).

Chamberlain *et al.* (1972) showed the strength of frozen sand at $-10\,°\mathrm{C}$ fell when a confining stress of $62\,\mathrm{MN\,m^{-2}}$ or more was applied (Figure 9.18). This is a very large value and would not frequently be met in the region of the earth's surface, although it might occur where ice or saturated soil was trapped in a small space. The lowest strength occurred at about $110\,\mathrm{MN\,m^{-2}}$ for both the sand and the clay-rich soil, and this approximates to the value predicted by equation (7.5) (when allowance is made for temperature-dependent modifications to certain of the terms) as the pressure causing total melting at $-10\,°\mathrm{C}$. When the confining stress was greater still, the strength increased, presumably because of friction between mineral particles in the now ice-free soil.

Finally, work done by Smith & Cheatham (1975) on frozen sand shows that as the confining stress is increased, the temperature-dependency is reduced considerably. Apparently, at higher stresses the interactive behaviour of the sand and the ice is more important than the behaviour of the ice *per se*. Again, this indicates the importance of friction in soil systems

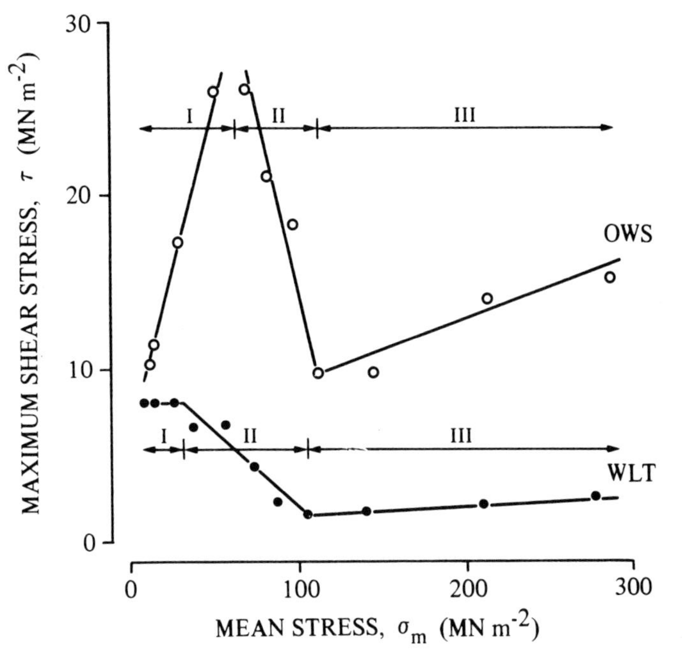

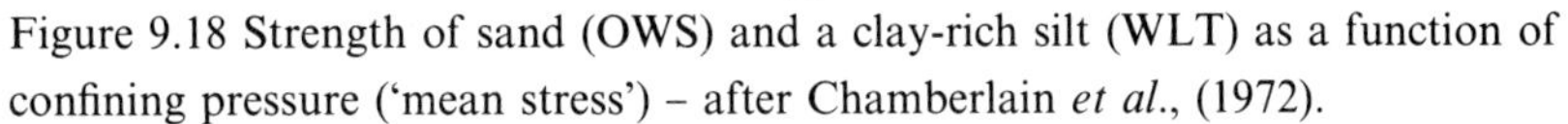

Figure 9.18 Strength of sand (OWS) and a clay-rich silt (WLT) as a function of confining pressure ('mean stress') – after Chamberlain *et al.*, (1972).

compared with the cohesive nature of pure ice. Further study is required, however.

9.7.1 Internal, thermodynamically controlled stresses

Much deformation of frozen ground (including frost heave in the most general sense) does not occur in response to externally applied stresses but as a result of internal thermodynamic processes. In some situations a small external stress will modify the magnitude and/or direction of such deformation (strain), and may thus appear to have a disproportionately large effect. In any case, the mechanical behaviour of frozen ground cannot be properly considered without attention to thermodynamic conditions.

The fundamental characteristic of the water in frozen soil is its suction, the magnitude of which is essentially defined by the temperature (see section 7.2.2). This is so because the suction, which can be regarded as a state of pressure reduction (the negative pore pressures considered in 9.2.2), or of tensile stress, affects the temperature at which the water freezes. The suction has its origin in capillarity or adsorption and thus occurs, of course, in unfrozen soils as well, under conditions of decreasing water content. In such soils, the development of suction results in an effective stress, σ', in accordance with equation (9.2). The suction in frozen soils similarly gives rise to effective stresses (Williams, 1963; Chamberlain & Gow, 1979) and these may be very large. Evidence for this is seen in the structure of newly thawed, compressible soils, especially clays. Such soils show a large *pre-consolidation* effect (a term used in engineering for compaction due to some previous effective stress) in spite of their loose nature. They often have a flaky or shaley structure, the discontinuities between the soil pieces being the site of former ice lenses. During freezing, the water for the ice layers comes largely from the soil pores. The accompanying compaction (consolidation – see section 7.7.2) of the soil layers is associated with the suction and the accompanying effective stress developed during the freezing.

If the soil is frozen and the temperature is near to 0 °C, the soil layers themselves (flakes, orthogonal pieces, etc) may be totally free of ice. Under such a condition the value of the suction developed within the soil will be given by $(P_i - P_w)$, according to equation (7.9) and Figure 7.7, and the effective stress developed on this account will have the same value (Williams, 1963). The soil pieces are consolidated accordingly.

At only a degree or so below 0 °C, however, ice will occur throughout the pores quite generally, as it is then thermodynamically stable even in such small spaces (equation (7.12)). Under these conditions the prediction of the effective stress is more difficult. Indeed, as has been indicated in this and earlier chapters, the distribution of stresses, considered on the scale of pore sizes, is complex and still uncertain. With the ice occupying the pores, further consolidation would be unlikely, and while a variety of microstructural changes may take place, there will not be any significant additional preconsolidation to be observed after thaw. The spreading of ice throughout the pore system has an analogy in the intrusion of air into a drying (unfrozen) soil that occurs at the shrinkage limit.

The effective stresses existing in frozen soil are probably always large, even though at lower temperatures they may not be as great as the value $(P_i - P_w)$ in equations (7.8) and (7.9). They are likely to be large relative to externally applied stresses. For soil at a temperature of − 1 °C, for example, the thermodynamically determined effective stress arising within the sample could be as high as $1.2\,\mathrm{MN\,m^{-2}}$ (following equation (7.9)). Obviously, the increment of effective stress due to the thermodynamic relations within the soil can have a very significant effect on the strength of the frozen soil. Indeed, part of the increase in strength of frozen soils (where frictional strength is involved) as the temperature is lowered must be due to this effect.

When there is an externally applied stress as well, the suction $(P_i - P_w)$

and the increment of effective stress associated with it retain approximately the same value (appropriate to the temperature – equation (7.9)). If such increments were added to the values for externally applied stress σ given in Table 9.1, the much greater effective normal stresses would, in fact, lead to very different conclusions as to the relative proportions of cohesive to frictional strength. This point serves to illustrate the fundamental complexity of the mechanical behaviour of frozen soil: explanations must recognise the thermodynamic constraints arising from the temperature-dependence of the ice and water contents and the associated pressures and stresses, as well as the purely mechanical and hydraulic conditions (Ershov, 1986).

9.8 Field situations

The gravity-induced shear stresses acting down slopes (see section 5.2) are usually far smaller than the typical short-term resistance of frozen ground. Even vertical faces of frozen ground can be stable, at least for a limited time, so high is the short-term strength, as is also the case for glaciers and ice shelves. However, the shear stresses in a slope are such that deformation due to creep may be substantial. In the analysis of natural conditions, as well as of structures on frozen ground, we are concerned especially with slow deformation (creep) and what we have referred to as the long-term strength (section 9.4).

Some authors suggest that the lowest stress that produces a continuing, secondary creep represents the ultimate long-term strength of the soil, since at some time this would be expected to move into accelerating creep, the state of failure. This may be a convenient approach for geotechnical design but this concept of strength is complicated under natural conditions by changes of temperature in causing, accelerating or retarding creep. Such changes occur continuously in the near-surface layers of the ground. It must be remembered, too, that in many field situations the deformation results in a redisposition of material by which the shear stresses are relieved, temporarily at least.

There has been rather little consideration of the role of persistent creep of frozen ground in natural slopes, although such deformation may be widespread (see section 5.3.2). McRoberts (1975) compared the rates of strain in response to a range of shear stresses for various soils and for pure ice (Figure 9.19). If a uniform ('ideal') natural slope, of angle β, is then considered, the value of τ, the shear stress, at any depth Z, is given by $\rho g Z \sin\beta \cos\beta$ (see section 5.2). The information in Figure 9.19 can then be used to predict creep, in the form of downslope velocity for each depth in the profile, or, as in Figure 5.6, the cumulative effect as the surface velocity. If

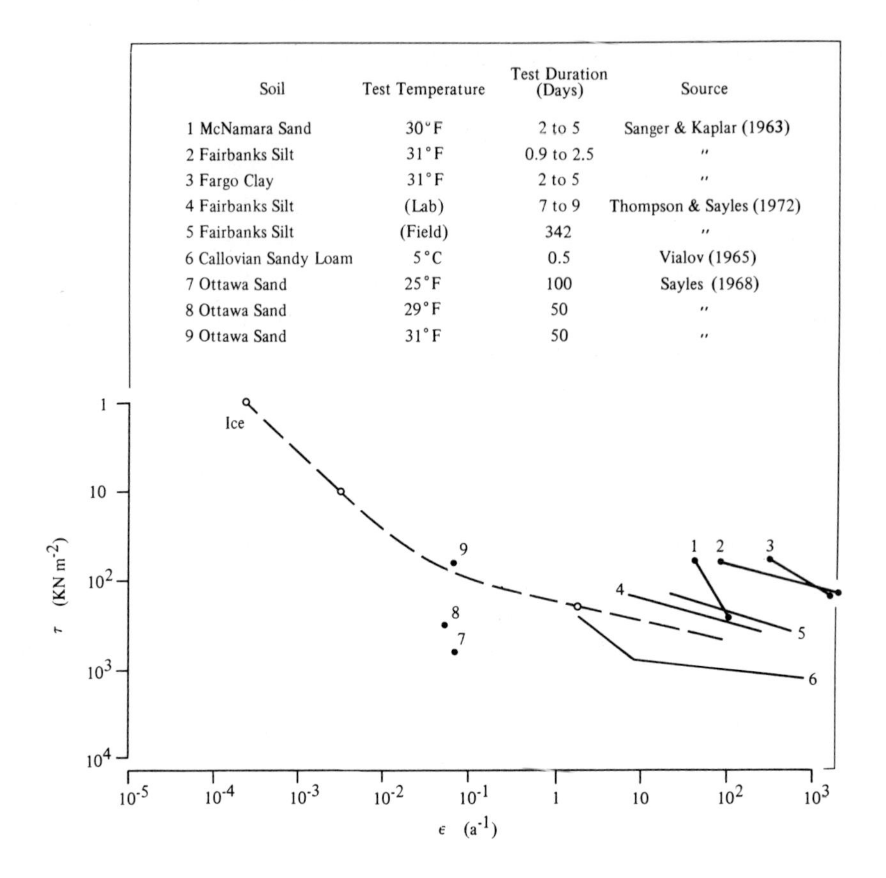

Figure 9.19 Rates of strain (creep) of various soils and of ice, for various applied shear stresses, τ. (after McRoberts, 1975).

the soil material is very ice rich, with the soil being more or less discrete inclusions in the ice, then the creep rates may be approximated using the creep properties for pure ice (as McRoberts did). The creep rates of ice-rich soil may exceed those of ice while soil without much ice may have much lower creep rates. McRoberts used empirical values for creep at temperatures within a few degrees of 0 °C, although without specific regard for the dependence on temperature. Perhaps even more significant would be the effect of *temperature changes* and the associated changes of temperature gradient within the frozen soil. There seem to have been few, if any, studies of the creep of frozen ground under conditions of changing temperature even though this is the normal condition of the near-surface layers under natural conditions.

9.8.1 Role of temperature and pressure variations

TEMPERATURE VARIATIONS

Almost everywhere in the cold regions the surface of the ground is exposed to freezing and thawing annually. Commonly, of course, there is an active layer decimetres or metres in thickness. The changing temperatures and annual temperature cycle characteristic of this layer also extend into underlying permafrost. The uppermost part of the permafrost (that which, along with the base of the permafrost, experiences the warmest temperatures) is exposed to a constantly changing and annually reversing, temperature gradient (Chapter 4). The upper part of the active layer, even when frozen, is also exposed to shorter term, weather-induced changes. All this greatly complicates the assessment of the creep properties, not only because of their dependence on temperature *per se*, but because of the direct effects of temperature gradients.

Moisture movements in frozen ground result from the potential gradients associated with temperature gradients. In slopes these movements are modified (even if only slightly) by gravitational (elevation) potential and thus can be expected to have a downslope component. The translocation of water and ice associated with the growth or diminution of ice lenses may well produce some net downslope movement of the mineral soil. If the values reported, for example, by Mackay (1983*b*) are typical (see also section 8.6.1), the cumulative seasonal movements of moisture and modification of distribution would probably be sufficient to give a significant annual component of downslope creep. It would seem that seasonal, or weather-induced temperature changes may promote periodic creep, even though stresses are below those that would appear necessary to cause significant creep if the temperature were to remain constant.

Extending the perspective, one notes that the variations in arrangement and proximity of mineral particles, of the thickness of adsorbed water layers, and of the disposition of ice in pores and lenses are all likely to be important mechanisms in the creep induced in this way. Deceleration of creep constitutes the development of a more stable arrangement as a consequence of limited displacements. Conversely, changes in the proportions of ice and water and the attendant migrations of water associated with even a degree or two of warming and recooling near 0 °C, would appear more than sufficient to destroy a quasi-stable state and cause renewed creep.

PRESSURE VARIATIONS

Deformation of materials occurs as a response to application of stresses (or pressures) and thus of stress gradients, and it may seem strange to consider the effect of pressure gradients as though they were an independent variable like temperature. However, we have already seen how the normal stresses, whether of external or internal thermodynamic origin, play an important role in controlling resistance to deformation. The thermodynamic conditions in frozen soil are such that we can indeed profitably consider the effects of pressure (or stress) *gradients* on the mechanisms that themselves constitute the process of creep. This effect of the gradients is analogous to that of temperature gradients and has similar implications for the creep process.

Indeed, it is because the effect relates to the general thermodynamics of the phase relations (equation (7.9) for example) that we use 'pressure' in this context, rather than 'stress' with its implication of a direction of applied forces. If the pressure on a sample of frozen, saturated soil is raised, the immediate effect is to melt a small amount of the ice. The curve of water content as a function of temperature (Figure 1.4) is displaced towards lower temperatures by $-0.0074\,°C$ per $100\,kN\,m^{-2}$ (this is assuming, the applied pressure is uniformly transmitted to the ice phase and that, simultaneously, the pressure of the water, P_w, rises correspondingly to maintain thermodynamic equilibrium). The suction ($P_i - P_w$) will be reduced by only a small amount, with a small increase in the amount of unfrozen water.

If a body of frozen ground under pressure is continuous with ground (at the same temperature) not carrying such pressure – that is, if there is a gradient in pressure – it follows that there will be a gradient of water potential. The unfrozen water will tend to flow from the region of high pressure to that of lower pressure. Its arrival there increases the unfrozen water content and thus reduces the suction. This means the equilibrium is disturbed – and an equivalent amount of water will freeze to restore equilibrium by increasing the local ice content. The process is continuous because some ice melts to replace the water leaving the region of elevated pressure.

Movements of ice (in addition to its re-formation) may be involved in such pressure-induced moisture migration just as in temperature-induced migration (sections 8.4, 8.5). It appears that the processes will continue at least until ice bodies larger than pore size have disappeared from the stressed region, although this may take a long time. The loss of ice from a region of high pressure and the corresponding accumulation in a region of lower pressure was demonstrated in the classic experiments of Vialov (1965b). The slow intrusion of a punch some centimetres in diameter resulted in the reforming of ice lenses, beyond the region stressed by the intrusion of the punch.

Both the temperature-gradient and pressure-gradient driven transfers occur at a rate controlled by the *hydraulic conducitivity* of the frozen material. The rate is also limited by the rate of heat transfer required for the exchange of latent heat in the melting and refreezing of the ice, and this is controlled by the *thermal conductivity*. These quantities are therefore of importance for the rate of creep and the creep of frozen soils under natural conditions may be as much dependent on inputs of thermal energy as on the gravitationally induced stresses.

In concluding this chapter, the following quotation from Tsytovich (1975, p. 135) is still appropriate:

'. . . experimental data on the quantitative values of the flow characteristics of frozen soils . . . are now totally inadequate, especially for frozen soil at temperatures near 0 °C, i.e. for the conditions that are most likely to develop undesirable processes. This calls for further research on the deformation and strength properties of frozen ground as functions of time with determination of their parameters at various temperatures and especially for . . . temperatures near 0 °C. . .'

10

Geocryology past and future

10.1 Geocryology and geotechnique

In this book we have been concerned with the scientific understanding of the cold regions. In doing so we build upon a long tradition that dates back to the early explorers, who were struck by the distinct terrain of these remote regions and the obvious contrasts with the temperate and inhabited regions from which they came. We have noted the diversity of terrain associated with cold climates, and that such conditions are not limited to the polar regions. Freezing of the ground also occurs to an important extent in regions that have been quite heavily populated for centuries. Parts of the Soviet Union, even Moscow itself, and much of southern Canada and the northern United States are characterised by very cold winters. In the cold regions more generally, however, including much of the Arctic, temperatures remain low in the summer as well, limiting plant growth, and native, nomadic populations have held sway until this century. Only with the arrival of European settlers intent on introducing urban centres and military, industrial and technological activities of various kinds, have these typical cold regions of the North been subject to more intense and practical investigation.

Technological innovation has also led to frozen ground having an importance quite unrelated to natural climatic conditions: artificial freezing to provide soil strength in civil engineering procedures such as tunnel and mine shaft construction is assuming an importance that makes studies of the behaviour of frozen ground as an engineering material imperative. It seems that the first utilization of artificial freezing for such purposes occurred in 1876 when the still-extant Brunkeberg tunnel in Stockholm was constructed and there was a fear that the weight of overlying buildings and the loose nature of the morainic soils would result in collapse of the tunnel roof (The Engineer, 1886). The technique is widely used for subway (underground railway) construction in Japan, for example, and very recently was used in the construction of a tunnel below the Seine in Paris.

Scientific investigation of the remote and apparently useless cold regions, characterised by treeless tundra, permafrost and many periglacial features were of a largely descriptive and exploratory nature well into this century. It was the Second World War that gave impetus to the first studies that might properly be called geotechnical – investigations made necessary by the problems of vehicular traffic, of the construction of permanent buildings serviced to acceptable standards, of storage tanks and other structures associated with oil and gas or mining, and of airports, dams and communication facilities.

Such investigations, inspired by engineering needs, differed greatly from the earlier descriptive, natural history approach. The mechanical properties of the terrain, such as its bearing capacity and stability, were of paramount importance. Of course, this depended on the temperatures of the ground, and whether it was freezing, frozen or thawing. Thus the climatic conditions, and the state of the ground, its lithology and ice content, became the general subject of investigation. The particular and the curious, the patterned ground, ice wedges, pingos and solifluction that had previously been the main subjects of scientific reporting, now were frequently overlooked and sometimes quite unknown to those responsible for geotechnical investigations. Nevertheless, we should recognise that such natural features are commonly the product of those same forces responsible for the unusual geotechnical behaviour of the ground and thus can reveal the origin of its unique nature.

At present, therefore, our knowledge of the surface of the ground and its behaviour in cold regions stems from two main sources. Firstly, there is the long-standing and comprehensive description of the variety and extent of natural features unique to the cold regions, and secondly, the already quite substantial experience gained in only a few decades, of the behaviour of the ground from the point of view of man's activities – both the experience gained from construction and that from the unwanted effects of man's activities in modifying the natural environment.

Within this century, too, there has been development of the scientifically based analysis of the effects of cold climates, of freezing and thawing of the ground, which is the subject of this book. The relatively short period over which such knowledge has been developed, when construction practices were usually dictated by expediency, means that we are not dealing with a very advanced or sophisticated branch of science. In spite of the evident complexity of some of the theoretical discussion (for example that concerning heat and moisture transfers in frozen soil, or that concerning the rheological properties of the material), important discoveries are still being made through the application of elementary scientific principles. Certain well-described phenomena still remain poorly understood in respect to the nature of the processes involved. In this category are several forms of patterned ground, in spite of the enormous descriptive literature on the subject generally. Similarly the mechanisms of deformation and the origin of stresses, which have proved a very serious problem for buried structures and especially pipelines (Williams, 1986) are only now being revealed.

There are also broader considerations, concerning the cold regions and man's activities, to which a book such as ours directs attention. It is evident that there are a number of issues which can, indeed in some cases certainly will, be of very great importance in the quite near future. These issues are not limited to eventual industrial development of the cold regions, but concern the role of the cold regions in a global context. It is appropriate to consider briefly those aspects of research in geocryology which appear to be the most significant with respect to both the cold regions themselves and to their global importance.

10.2 Energy exchange and climate instability

The distribution of permafrost, its eventual thawing or increase in extent, as well as the depth of seasonal freezing and thawing of the ground, all depend on exchanges of heat energy. Because the flow of energy from the earth's interior at any place is effectively constant with time, it is the atmospheric climate which controls the energy exchanges through the ground surface and in the ground itself, at least for the near-surface layers with which we are concerned. The nature of the surface of the ground affects the energy exchange greatly, so that changes in the surface, whether man-made or natural, result in temperature changes in the ground. The complexity and variability of the energy exchange at the surface defines the microclimate, the strictly local conditions which distinguish one small piece of ground from neighbouring areas.

The understanding of the local conditions is important since so much geotechnical design for cold regions is dependent upon them. The greater complexity of those geotechnical structures being built or envisaged for the cold regions, large dams, improved highways and various structures associated with the oil and gas industry, requires a fuller knowledge and greater predictive ability concerning the effects of changing the surface of the ground.

Comparatively little attention has been given to the effects of change in the atmospheric climate, even though scientists have known for many years that the atmospheric climate is not constant. But it is only recently, with the realisation that human activities can accelerate change, that thc implications are being recognised and attempts made at rational prediction of future trends.

Atmospheric climate change differs in its relative uniformity over wide areas, compared with the always local, if ubiquitous, variations in microclimate. Differences in microclimate are responsible for much of the variety of terrain in the cold regions. Often features associated with cooling, the aggradation of permafrost, the growth of a pingo, for example, may occur quite close to terrain features associated with warming – mudflows, and other effects of thermal erosion, for example. Changes of atmospheric climate, by contrast, result in regional trends of either increased thawing, or increased freezing, but not both together. The effects will be widespread and contiguous and not restricted to small areas of active disturbance.

The relatively short time of significant geotechnical activity in the cold regions, coupled with the usually small areas under consideration, has meant that such widespread effects of climatic changes have largely gone unrecognised. Nevertheless, Soviet and North American scientists believe that the extent of thermokarst topography especially in the Soviet Union, reflects the climatic warming that is proven to have occurred over recent decades and centuries.

Because such large areas of the earth's surface have average temperatures within a degree or so of 0 °C, it appears that a general warming, or a general cooling of the ground, even if this were to be of the order of only 1 °C or so, would have particularly serious effects for large or extended structures such as highways, and pipelines. The latter, especially, require careful thermal design to avoid stresses due to frost-heave or thaw settlement. Such design must therefore take into account the possible magnitudes of naturally or man-induced modifications of ground thermal conditions.

There is growing public concern over the 'greenhouse' effect, the increase in world air temperatures due to the accumulation of CO_2 and other gases from burning of industrial and domestic fuels and other changes possibly being induced in the atmosphere by mankind (Davies, 1985). The direct and indirect effects on populous areas have been widely discussed and it is recognised there are certain areas of the world that are particularly liable to dramatic changes as a result of even small climatic changes. It has been rarely pointed out that the cold regions, especially those where permafrost occurs in a discontinuous fashion, are such an area. There is an additional point: there is a general agreement that changes of climate tend to be greatest towards the poles and, while opinions differ, there are estimates of one to several degrees of warming within the next half century or so. The extent of the cold regions and the fact that much lies within the polar areas, leads to the conclusion that this is a topic of importance not only with respect to environmental conditions and industrial development in the cold regions but probably for the world as a whole.

It is important therefore to have as well-based estimates as possible of the magnitude, direction and rate of change of climate, and this is as important as having an understanding of the effects of microclimatic change. The changes of atmospheric climate might well be of such magnitude and extent as to have effects extending beyond the cold regions themselves. Hydrological regimes would be modified, especially runoff characteristics. The changing nature of the ground surface occurring over such extensive areas might well have a powerful feed-back effect further modifying climatic patterns. The same applies to the changes below the surface of the ground and in the permafrost itself. Clearly, because of the enormous variety of natural surface (microclimatic) conditions, prediction will be difficult.

In the immensely complex thermal system of which the earth's surface is part, where temperatures vary through tens of degrees in a day, we should not overlook how small a change of temperature 1 °C is. We might rather expect such changes, or greater ones, as being inherently more probable than more stable conditions. The evidence is in the historical record, in shifts in agricultural production, famines, and, for example, the visual and written evidence of glaciers, which in Norway advanced over farm land in the eighteenth century.

It is important to note that climatic instability is not simply a matter of temperature change (although it is commonly referred to in such terms). All other elements of climate, precipitation, winds, cloudiness etc. are likely to be modified as well.

10.3 Thermodynamic and mechanical properties of frozen ground

We have examined in detail the thermodynamics of freezing soils and the manner in which the porous structure of soils exerts controls on the freezing process. It is the confinement of the ice in the soil pores and the extensive mineral surfaces which are responsible for the potentials which find expression in stresses and migrations of water or ice.

This approach builds on the work of physical chemists and other scientists concerned with phase relations. It helps explain the characteristics of frost heave and provides a basis of geotechnical procedures to overcome the adverse effects of soil displacements and heaving pressures. But geotechnical problems often require a knowledge of the strength and deformation properties of the frozen soil. A central theme in this book is the degree to which the thermodynamic behaviour and mechanical behaviour of frozen soils are integrated (at least, for those temperatures of most common interest). It appears that neither can be studied to full effect without the other. The processes of deformation involve phase change and the temperature dependence of mechanical properties depends on the equilibrium water/ice contents at each temperature. While the properties of the constituents individually are important – those of the water, ice and soil minerals – it is their *interaction* which gives rise to the unique characteristics of frozen soil. Thus, while there are a variety of theoretical and experimental approaches to the investigation and characterisation of freezing soil for geotechnical purposes, only those that recognise this dual nature of the medium will be fully useful. As yet there are few published papers which do so, but this seems to be the most important thrust needed for research into the material properties.

A related phenomenon is that of the propensity of frozen soils to creep. Although the dependence of strength, or the resistance to deformation, is now generally understood as being time-dependent, much remains to be done to clarify the mechanical properties and their dependence on materials and environmental conditions. The problems of the time required for a test, and of scaling the laboratory results against the field situation, often go unrecognised. Most laboratory measurements of frost-heave are still carried out at freezing rates which are far above those that actually occur in field conditions.

The increasing complexity of construction in cold regions, as well as the use of frozen ground as a construction material (using techniques of artificial freezing), require a knowledge of mechanical properties of frozen ground comparable to that which exists for unfrozen earth materials. The relatively limited attention given to this matter, coupled with the complexity introduced by the thermodynamic behaviour of frozen ground, means that this is one of the most pressing research problems within geotechnology.

10.4 Submarine and other extreme conditions for permafrost

It became known only some twenty years ago that there was widespread permafrost below some of the far Northern seas; these seas have also proved to be the location of substantial submarine oil and gas reserves. The geotechnical problems of submarine structures for oil and gas rigs and pipelines are well-known from experience in the North Sea and other off-shore areas. But there has been little experience with facilities for oil and gas extraction from the regions with submarine permafrost, and only a limited amount of exploration.

It is clear, however, that the properties and geotechnical behaviour of frozen ground containing salt in significant quantity are modified compared with normal terrestrial permafrost. The salt has to be considered in the thermodynamics of the phase change and thus for its effects on the relations of the ice and water.

The fundamental nature of the soil water system can still be approached along similar lines with respect to its thermodynamic behaviour but the same cannot be said for the submarine permafrost environment more generally. Compared with past experience, on land, it is an exotic environment which has become, in the course of little more than a decade one of the frontier regions for the oil and gas industry.

The absence of an interface between atmosphere and ground, of the complex microclimate associated with vegetation and ground surface, is a simplifying element, but the conditions associated with submergence under often great depths of water and the additional complications of sea ice which may reach to the sea bed are challenging prospects. Renewal of interest in Arctic off-shore oil and gas is stimulating research but with the exception of Vigdorchik's (1980) book, information is generally available only through relatively few scientific papers (often dealing with a narrow aspect, such as salt movement in sediments (Osterkamp, 1987)) or through reports prepared, sometimes hastily, for governmental enquiries into the effects of Arctic submarine pipeline or drilling activities.

The thermodynamic system represented by a freezing soil, is also modified by the presence of air, as well as ice in the soil pores. The study of unsaturated soils, more generally, became a major branch of soil physics when soil scientists realised the special nature and significance of such soils in warm or temperate lands. Relatively dry, frozen soils (in which the presence of air modifies the behaviour) occur extensively in parts of the high Arctic, and especially in Antarctica. It is, perhaps, not surprising, that, with our limited understanding of the wet (saturated) frozen soils, little attention has been given to soils in dry, cold conditions.

At the height of the United States' space programme some consideration was given to the extent and occurrence of permafrost on extraterrestial bodies, and especially Mars. These, too, represent exotic – and dry – environments where similar basic scientific principles apply yet where conditions are such as to stretch the imagination and challenge the originality of the scientist. The study of the effects of extremely low temperatures on soil or rock, perhaps in combination with extremely low humidity may have an immediate importance as well. These conditions occur with storage of gases at low temperatures. Gas hydrates, also discovered in the polar regions only in the last twenty years, are the product of another extreme:

that of the high pressures associated with great depth in the ground, coupled with low temperatures. They represent a particular danger to those searching for oil and gas and yet may prove also to be a valuable source of gas.

It is not necessary to contemplate space travel in order to find challenging problems in the scientific study of soils and rocks at freezing temperatures. The vast areas of the earth which are affected by freezing and thawing, the cold regions so-defined, are, even today, a largely untrampled field in which the flowers of scientific discovery have just begun to appear.

References

ALKIRE, B.D. & ANDERSLAND, O.B. (1973) The effect of confining pressure on the mechanical properties of sand–ice materials. *Journal of Glaciology*, **12**, 66, 469–81.

ANDERSLAND, O.B. & ANDERSON, D.M. (EDS.) (1978) *Geotechnical engineering for cold regions*. New York: McGraw–Hill, 566 pp.

ANDERSLAND, O.B., SAYLES, F.H. & LADANYI, B. (1978) Mechanical properties of frozen ground. In *Geotechnical engineering for cold regions*, ed. O.B. Andersland and D.M. Anderson, New York: McGraw–Hill, 566 pp.

ANDERSON, D.M. & TICE, A.R. (1972) Predicting unfrozen water contents in frozen soils from surface area measurements. *Highway Research Record*, **393**, 12–18.

ANNERSTEN, L.J. (1964) Investigations of permafrost in the vicinity of Knob Lake, 1961–1962. *McGill Sub-Arctic Research Papers*, **161**, 51–137.

ARE, F.E. (1984) Soviet studies of the subsea cryolithozone. *Proceedings of the Fourth International Permafrost Conference*, Fairbanks, Alaska. Final vol., p. 87. Washington D.C.: Nat. Acad. Sci – Nat. Acad. Press.

ARNBORG, L., WALKER, H.J. & PEIPPO, J. (1967). Suspended load in the Colville River, Alaska. *Geografisker Annaler, Ser. A*, **49**, 131–44.

BALLANTYNE, C.K. & MATTHEWS, J.A. (1982) The development of sorted circles on recently deglaciated terrain, Jotunheimen, Norway. *Arctic and Alpine Research*, **14**, 4, 341–54.

BALLANTYNE, C.K. & MATTHEWS, J.A. (1983) Dessication cracking and sorted polygon development, Jotunheimen, Norway. *Arctic and Alpine Research*, **15**, 3, 339–49.

BANIN, A. & ANDERSON, D.M. (1974) Effects of salt concentration changes during freezing on the unfrozen water content of porous materials. *Water Resources Research*, **10**, 1, 124–6.

BARNES, P., TABOR, D. & WALKER, J. (1971). The friction and creep of polycrystalline ice. *Proceedings of the Royal Society of London*, **A324**, 127–55.

BESKOW, G. (1935). *Tjälbildningen och tjällyftningen med särskild hänsyn til vägar och järnvägar*. Stockholm, 242 pp. Statens väginstitut, Stockholm, Meddelande 48. (Also published as Sveriges geologiska undersökning. *Avh. och Uppsats. Ser. C*, **375**, and translated into English: Tech. Inst. Northwestern Univ. Evanston, Ill. November 1947.)

BILY, C. & DICK, J.W.L. (1974). Naturally occurring gas hydrates in the Mackenzie Delta, N.W.T. *Can. Soc. Petr. Geol. Bull.*, **22**, 340–52.

BIRCH, F. (1948). The effects of Pleistocene climatic variations upon geothermal gradients. *American Journal of Science*, **238**, 529–58.

BIRD, J.B. (1967). *The physiography of Arctic Canada*, Johns Hopkins Press. 336 pp.

BJERRUM, L. & JÖRSTAD, F. (1966). Stabilitet av fjellskråninger i Norge (with English summary). *Nor. Geot. Inst. Publication 67*, 60–78.

BLACK, P.B. & MILLER, R.D. (1985). A continuum approach to modelling of frost heaving. In *Freezing and thawing of soil-water systems*, ed. D.M. Anderson and P.J. Williams, pp. 36–45. American Society of Civil Engineers, Technical Council on Cold Regions Monograph.

BLACK, R.F. (1973). Growth of patterned ground in Victoria Land, Antarctica. *Second International Conference on Permafrost, Yakutsk, USSR, North American Contribution.* pp. 193–203. Washington D.C.: Nat. Acad. Sci.

BLACK, R.F. (1976). Periglacial features indicative of permafrost: Ice and soil wedges. *Quat. Res.*, **6**, 3–26.

BLACK, R.F. (1983). Three superposed systems of ice wedges at McLeod Point, Northern Alaska, may span most of the Wisconsinian stage and Holocene. *Proceedings of the Fourth International Conference on Permafrost.* pp. 68–73. Washington D.C.: Nat. Acad. Sci. – Nat. Acad. Press.

BLANCHARD, D. & FRÉMOND, M. (1985). Mécanique des sols et milieux poreux-gonflement des sols gelés. *C.R. Acad. Sc. Paris*, t. **300**, Serie II, no. 14, 637–42.

BLANCHARD, D., FRÉMOND, M. & WILLIAMS, P.J. (1985). *Comportement des sols et des structures en Zone Arctique (Behaviour of soils and structures in the Arctic).* Int. Assoc. Bridges and Structural Engr., Vancouver Congress, 5 pp.

BOUYOUCOS, G.J. & MCCOOL, M.M. (1916). *Further studies of the freezing point lowering of soils.* Michigan Agricultural College Exper. Station, Technical Bulletin No. 31. 51 pp.

BRAGG, R.A. & ANDERSLAND, O.B. (1980). Strain rate, temperature and sample size effects on compression and tensile properties of frozen sand. *Proceedings (Preprints) of the Second International Symposium on Ground Freezing, Trondheim, Norway*, 34–7.

BRANDON, L.V. (1965). *Groundwater hydrology and water supply in the District of Mackenzie, Yukon Territory and adjoining parts of British Columbia.* Geological Survey of Canada Paper 64-39. 102 pp.

BRANDON, L.V. (1963). Evidences of groundwater flow in permafrost regions. *Proceedings, Permafrost International Conference*, pp. 176–7. Washington, D.C.: N.A.S. Publication 1287.

BRESLER, E. & MILLER, R.D. (1975). Estimation of pore blockage induced by freezing of unsaturated soil. *Proceedings, Conference on soil-water problems in cold regions, Calgary, Alberta, Canada*, May 6–7, 1975, pp. 161–75.

BREWER, M.C. (1958a). Some results of geothermal investigations of permafrost in northern Alaska. *American Geophysical Union, Transactions*, **39**, 19–26.

BREWER, M.C. (1958b). The thermal regime of an arctic lake. *American Geophysical Union, Transactions*, **39**, 278–94.

BROWN, J. & GRAVE, N.A. (1978). Physical and thermal disturbance and protection of permafrost. *Third International Conference on Permafrost, Edmonton, Canada*, vol. 2, pp. 51–91. Ottawa: N.R.C.C.

BROWN, R.J.E. (1963). Influence of vegetation on permafrost. *Proceedings, Permafrost International Conference*, pp. 20–5. Washington D.C.: NAS/NRC Publication 1287.

BROWN, R.J.E. (1965). Some observations on the influence of climatic and terrain features on permafrost at Norman Wells, N.W.T. *Canadian Journal of Earth Sciences*, **2**, 15–31.

BROWN, R.J.E. (1967). Permafrost in Canada. *Geol. Surv. Canada*, Map 1246 A.

BROWN, R.J.E. (1973). Influence of climatic and terrain factors on ground temperatures at three locations in the permafrost region of Canada. *Proceedings of the Second International Conference on Permafrost, Yakutsk, USSR, North American Contribution*, pp. 27–34. Washington D.C.: Nat. Acad. Sci.

BROWN, R.J.E. (1978a). Permafrost map of Canada. Plate 32, in *Hydrological Atlas of Canada*, Ottawa: Dept. of Fisheries and Environment.

BROWN, R.J.E. (1978b). Influence of climate and terrain on ground temperature in the continuous permafrost zone of northern Manitoba and Keewatin District, Canada. *Proceedings of the Third International Conference on Permafrost, Edmonton, Alberta*, vol. 1, pp. 15–21. Ottawa: N.R.C.C.

BROWN, R.J.E. & PÉWÉ, T.L. (1973). Distribution of permafrost in North America and its relationship to the environment: a review, 1963–1973. *Proceedings of the Second International Conference on Permafrost, Yakutsk, USSR, North American Contribution*, pp. 71–100. Washington D.C.: Nat. Acad. Sci.

BROWN, W.G. (1963). *Graphical determination of temperature under heated or cooled areas on the ground surface.* National Research Council of Canada, Division of Building Research, Technical Paper no. 163, 39 pp. + Figures.

BROWN, W.G. (1964). Difficulties associated with predicting depth of freeze or thaw. *Canadian Geotechnical Journal*, **1**, 4, 215–26.

BROWN, W.G., JOHNSTON, G.H. & BROWN, R.J.E. (1964). Comparison of observed and calculated ground temperatures with permafrost distribution under a northern lake. *Canadian Geotechnical Journal*, **1**, 147–54.

BRUTSAERT, W.H. (1982). *Evaporation into the atmosphere: Theory, history and applications*, Holland: Reidel. 308 pp.

BUNTING, B.T. (1983). High Arctic soils through the microscope: Prospect and retrospect. *Annals Am. Assoc. Geographers*, **73**, 4, 609–16.

BURN, C.R. (1982). Investigations of thermokarst development and climatic change in central Yukon Territory. Unpublished M.A. thesis, Carleton University, Ottawa, Ontario, 142 pp.

BURN, C.R. (1986). On the origin of aggradational ice in permafrost. Ph.D. Thesis. Carleton University, Ottawa, Ontario, 222 pp.

BURN, C.R. & MICHEL, F. (1988) Evidence for recent temperature-induced water migration into permafrost from the tritium content of ground ice near Mayo, Yukon Territory, Canada. *Canadian Journal of Earth Sciences* (in press).

BURN, C.R. & SMITH, C.A.S. (1988). Observations of the 'thermal offset' in near-surface mean annual ground temperatures at several sites near Mayo, Yukon Territory, Canada. *Arctic*, **41**, 2, 99–104.

BURN, C.R., MICHEL, F.A. & SMITH, M.W. (1986). Stratigraphic, isotopic and mineralogical evidence for an early Holocene thaw unconformity at Mayo, Yukon Territory. *Canadian Journal of Earth Sciences*, **23**, 6, 794–803.

BURT, T.P. & WILLIAMS, P.J. (1976). Hydraulic conductivity in frozen soils. *Earth Surface Processes*, **1**, 3, 349–60.

BYERS, R.B. (1965). *Cloud Physics*. Univ. Chicago Press. 191 pp.

Canada Soil Survey Committee, Sub-Committee on Soil Classification. (1978). *The Canadian System of soil classification.* Can. Dep. Agric. Publ. 1646. Ottawa, Ontario: Supply and Services Canada. 164 pp.

CAREY, K.L. (1973). *Icings developed from surface water and ground water.* United States Army Cold Regions Research and Engineering Laboratory (CRREL). Hanover, New Hampshire, Monogr. 111-D3; 67 pp.

CARSLAW, H.S. & JAEGER, J.C. (1959). *Conduction of heat in solids.* 2nd. edn. Oxford: Clarendon Press.

CARSON, C.E. (1968). Radiocarbon dating of lacustrine strands in Arctic Alaska. *Artic*, **21**, 12–26.

CARSON, M.A. & KIRKBY, M.J. (1972). *Hillslope Form and Process.* Cambridge University Press. 475 pp.

CARSON, J.E. & MOSES, H. (1963). The annual and diurnal heat exchange cycles in upper layers of soil. *J. App. Meteorol.*, **2**, 397–406.

CARTER, D. (1970). Brittle fracture of snow ice. *Proceedings, IAHR Symposium, Reykjavik, Iceland*, **5**, 2.

CARY, J.W. & MAYLAND, H.F. (1972). Salt and water movement in unsaturated frozen soil. *Soil Science Society of America, Proceedings*, **36**, 4, 549–55.

CERMAK, V. (1971). Underground temperature and inferred climatic temperature of the past millenium. *Palaeogeography, Palaeoclimatology, Palaeoecology*, **10**, 1–19.

CHAMBERLAIN, E.J. (1980). Overconsolidation effects of ground freezing. *Second International Symposium on Ground Freezing, Trondheim, Norway*. pp. 325–37.

CHAMBERLAIN, E.J. (1981). *Frost susceptibility of soil and review of index tests*. United States Army Cold Regions Research and Engineering Laboratory. Hanover, New Hampshire, Monogr. 81–82; 110 pp.

CHAMBERLAIN, E.J., GASKIN, P.N., ESCH, D. & BERG, R.L. (1985). Survey of methods for classifying frost susceptibility. In *Frost action and its control*, ed. Berg & Wright, pp. 104–41. Tech. Conc. Cold Reg. Engg. Monogr., Amer. Soc. of Civil Engineers.

CHAMBERLAIN, E., GROVES, C. & PERHAM, R. (1972). The mechanical behaviour of frozen earth materials under high pressure triaxial conditions, *Geotechnique*, **22**, 3, 469–83.

CHAMBERLAIN, E.J. & GOW, A.J. (1979). Effect of freezing and thawing on the permeability and structure of soils. *Engineering Geology*, **13**, 73–92.

CHENG, G. (1983). The mechanism of repeated segregation for the formation of thick layered ground ice. *Cold Regions Science and Technology*, **8**, 1, 57–66.

CHIZHOV, A.B., CHIZHOVA, N.I., MORKOVINA, I.K. & ROMANOV, V.V. (1983). Tritium in permafrost and ground ice. *Proceedings of the Fourth International Conference on Permafrost*, Fairbanks, Alaska. pp. 147–51. Washington, D.C.: Nat. Acad. Sci. - Nat. Acad. Press.

CHURCH, M. (1972). Baffin Island sandur: A study in arctic fluvial processes. *Geol. Surv. Can., Bull.*, **216**.

CHURCH, M. (1974). Hydrology and permafrost with reference to northern North America. In *Permafrost Hydrology*, pp. 7–20. Proceedings of a workshop seminar, Canadian National Committee for the International Hydrological Decade, Ottawa, Ontario.

COGLEY, J.G. & MCCANN, S.B. (1976). An exceptional storm and its effects in the Canadian High Arctic. *Arctic and Alpine Research*, **8**, 1, 105–10.

CORTE, A.E. (1962). *The frost behaviour of soils: laboratory and field data for a new concept – II, Horizontal sorting*. U.S. Army Cold Regions Research and Engineering Laboratory, Research Report, Hanover, New Hampshire, 85 (2) 20 pp.

CORTE, A.E. (1976). Rock glaciers. *Biul. Perygl.*, **26**, 175–97.

COUTARD, J.P. & MUCHER, H.J. (1985). Deformation of laminated silt loam due to repeated freezing and thawing cycles. *Earth Surface Processes and Landforms*, **10**, 309–19.

CRAIG, R.F. (1974). *Soil Mechanics*. Van Nostrand Reinhold. 275 pp.

DAVIDSON, D.A. (1978). *Science for physical geographers*. Arnold. 187 pp.

DAVIDSON, D.W., EL-DEFRAWY, M.K., FUGLEM, M.D. & JUDGE, A.S. (1978). Natural gas hydrates in Northern Canada. *Proc. 3rd Int. Conf. Permafrost.*, vol. 1, pp. 938–43.

DAVIES, J.A. (1985). Carbon dioxide and climate: a review. *The Canadian Geographer*, **29**, 1, 74–85.

DAVISON, C. (1889). On the creep of the soilcap through the action of frost. *Geol. Mag.*, **6**, 255–61.

DAY, R.R., BOLT, G.H. & ANDERSON, D.M. (1967). *Nature of soil water: irrigation of Agricultural Lands.* Agronom. Series, Monogr. 11, American Society of Agronomy, 1, 180 pp.

DEFAY, R. & PRIGOGINE, I. (1951). *Tension superficielle et absorption, Traité de thermodynamique conformement aux méthodes de Gibbs et de Donder*, Editions Desoer, vol. 3, Liège, 295 pp.

DINGMAN, S.L. (1973). Effects of permafrost on streamflow characteristics in the discontinuous permafrost zone of central Alaska. *Proceedings of the Second International Conference on Permafrost*, Yakutsk, USSR. vol. 1, pp. 447–53. Washington, D.C.: Nat. Acad. Sci.

DINGMAN, S.L. & KOUTZ, F.R. (1974). Relations among vegetation, permafrost and potential insolation in central Alaska. *Arctic and Alpine Research*, **6**, 1, 37–47.

DIRKSEN, C. (1964). Water movement and frost heaving in unsaturated soil without an external source of water. Ph.D. thesis, Cornell Univ.

DIRKSEN, C. & MILLER, R.D. (1966). Closed system freezing of unsaturated soil. *Soil Science Society of America, Proceedings*, **30**, 2, 168–73.

DOSTOVALOV, B.N. & KUDRIAVTSEV, V.A. (1967). *General geocryology* (in Russian). Moscow: Izdatel'stvo Moskovskogo Universiteta. 403 pp.

DYKE, L.D. (1981). Bedrock heave in the central Canadian Arctic. In *Current Research, Part A*, pp. 157–67. Geol. Surv. Can. Paper 81-1A.

EDLEFSEN, N.E. & ANDERSON, A.B.C. (1943). *Thermodynamics of soil moisture.* Hilgardia. 298 pp.

The Engineer, (1886). *Tunnel for foot passengers in Stockholm.* (article in issue of 9 April).

ERSHOV, E.D. (1986) *Fiziko-Khimia i Mekhanika Merzlykh Porod.* Moscow, Izd-vo Moskovskogo Universiteta, 333 pp.

EVERETT, D.H. (1961). The thermodynamics of frost damage to porous solids. *Trans. Farad. Soc.*, **57**, 9, 1541–51.

FAROUKI, O.T. (1981). *Thermal properties of soils.* U.S. Army Cold Regions Research and Engineering Laboratory, Hanover, New Hampshire Monograph 81-1, 136 pp.

FAROUKI, O.T. (1982). *Evaluation of methods for calculating soil thermal conductivity.* U.S. Army Cold Regions Research and Engineering Laboratory, Hanover, New Hampshire, Report 82-08, 90 pp.

FISH, A.M. (1983). *Thermodynamic model of creep at constant stresses and constant strain rates.* United States Army Cold Regions Research and Engineering Laboratory, Hanover, New Hampshire, Report 83-33, 18 pp.

FITZGIBBON, J.E. (1981). Thawing of seasonally frozen ground in organic terrain in central Saskatchewan. *Canadian Journal of Earth Sciences*, **18**, 1492–6.

FRANKS, F. (1980). *Biophysics of water.* Wiley.

FRÉMOND, M. & VISINTIN, A. (1985). Mécanique des milieux continus (Problèmes mathématiques de la mécanique). *C.R. Acad. Sc. Paris*, t. **301**, Série II, no. 18, 1265–8.

FRENCH, H.M. (1974). Active thermokarst processes, Eastern Banks Island, Western Canadian Arctic. *Canadian Journal of Earth Sciences*, **II**, 6, 785–94.

FRENCH, H.M. (1976). *The Periglacial Environment.* Longman. 309 pp.

FRENCH, H.M. (ED.) (1986). Climate change impacts in the Canadian Arctic.

Proceedings of a Canadian Climate Program Workshop, March 3–5, 1986, Geneva Park, Ontario, 171 pp.

GEIGER, R. (1965). *Climate near the ground*. revd. edn. Harvard University Press. 611 pp.

GELL, W.A. (1978). Ice-wedge ice, Mackenzie Delta – Tuktoyaktuk Peninsular Area, N.W.T., Canada. *Journal of Glaciology*, **20**, 84, 555–61.

GIDDINGS, J.C. & LaCHAPELLE, E. (1962). The formation rate of depth hoar. *Journal of Geophysical Research*, **67**, 6, 2377–83.

GILPIN, R.R. (1980). A model for the prediction of ice lensing and frost heave in soils. *Water Resources Research*, **16**, 5, 918–30.

GILPIN, R.R. (1982). A frost heave interface condition for use in numerical modelling. *Proceedings of the Fourth Canadian Permafrost Conference, Calgary, Alberta*, pp. 459–65, Ottawa: N.R.C.C.

GOLD, L.W. (1962). Deformation mechanisms in ice. In *Ice and snow*, ed. W.D. Kingery, pp. 8–27. Cambridge, Mass.: M.I.T. Press.

GOLD, L.W. (1963). Influence of snow cover on the average annual ground temperature at Ottawa, Canada. *International Association Scientific Hydrology, Publ.*, **61**, 82–91.

GOLD, L.W. (1966). Dependence of crack formation on crystallographic orientation of ice. *Canadian Journal of Physics*, **44**, 2757–64.

GOLD, L.W. & LACHENBRUCH, A.H. (1973). Thermal conditions in permafrost: a review. *Proceedings of the Second Internatioanl Conference on Permafrost, Yakutsk, USSR, North American Contribution*, pp. 3–25. Washington, D.C.: Nat. Acad. Sci.

GOLD, L.W. & WILLIAMS, G.P. (1963). An unusual ice formation in the Ottawa River. *Journal of Glaciology*, **4**, 35, 569–73.

GOODRICH, L.E. (1978). Some results of a numerical study of ground thermal regimes. *Proceedings of the Third International Conference on Permafrost, Edmonton, Canada*, vol. 1, pp. 29–34. Ottawa: N.R.C.C.

GOODRICH, L.E. (1982a). *An introductory review of numerical methods for ground thermal calculations*. National Research Council of Canada, Division of Building Research, Paper 1061. 32 pp.

GOODRICH, L.E. (1982b). The influence of snow cover on the ground thermal regime. *Canadian Geotechnical Journal*, **19**, 421–32.

GOODRICH, L.E. & GOLD, L.W. (1981). Ground thermal analysis. In *Permafrost: Engineering design and construction*, ed. G.H. Johnston, pp. 149–72. John Wiley and Sons.

GOODWIN, C.W., BROWN, J. & OUTCALT, S.I. (1984). Potential responses of permafrost to climatic warming. In *The potential effects of carbon dioxide-induced climatic changes in Alaska*, ed. J.H. McBeath, pp. 92–105.

GOUGHNOUR, R.R. & ANDERSLAND, O.B. (1968). Mechanical properties of a sand–ice system. *American Society of Civil Engineers, Soil Mechanics and Foundations Division Journal*, **94**, SM4, 923–50.

GOW, A.J. (1975). *Application of thin section techniques to studies of the internal structure of frozen silts*. United States Army Cold Regions Research and Engineering Laboratory, Hanover, New Hampshire, Technical note.

GRAVE, N.A. (1983). Cryogenic processes associated with development in the permafrost zone. *Proceedings of the Fourth International Conference on Permafrost, Fairbanks, Alaska*, final vol. pp. 188–93. Washington D.C.: Nat. Acad. Sci.– Nat. Acad. Press.

GRAY, H.J. & ISAACS, D.M. (1975). *A New Dictionary of Physics*. Longman, 619 pp.

GRAY, J.T. & BROWN, R.J.E. (1982). The influence of terrain factors on the distribution of permafrost-bodies in the Chic-Choc mountains, Gaspesie, Quebec. *Proceedings of the Fourth Canadian Permafrost Conference, Calgary, Alberta*, pp. 23–35. Ottawa, N.R.C.C.

GUYMON, G.L., HROMADKA, T.V. II & BERG, R.L. (1980). A one-dimensional frost heave model based upon simultaneous heat and water flux. *Cold Regions Science and Technology*, **3**, 253–62.

HANSEN, J.E., LACIS, A., RIND, D., RUSSELL, G., STONE, P., FUNG, I., RUEDY, R. & LERNER, J. (1984). Climate sensitivity: Analysis of feedback mechanisms. In *Climate Processes and Climate Sensitivity*, ed. J. Hansen and T. Takahashi, pp. 130–63. Maurice Ewing Series, 5, Washington, D.C.: Amer. Geophys. Union.

HARLAN, R.L. (1973). Analysis of coupled heat–fluid transport in partially frozen soil. *Water Resources Research*, **9**, 5, 1314–23.

HARLAN, R.L. (1974). Dynamics of water movement in permafrost: A review. In *Permafrost Hydrology*, pp. 69–77. Proceedings of a workshop seminar, Canadian National Committee for the International Hydrological Decade, Ottawa, Ontario.

HARRIS, C. (1981). *Periglacial mass-wasting: A review of research.* BGRG Res. Monogr. 4, Geo Abstracts, Norwich, 204 pp.

HARRIS, C. (1983). Vesicles in thin sections of periglacial soils from north and south Norway. *Proceedings of the Fourth International Conference on Permafrost*, Fairbanks, Alaska. pp. 445–9. Washington, D.C.: Nat. Acad. Sci. – Nat. Acad. Press.

HARRIS, S.A. (1986). *The Permafrost Environment.* Croom Helm. 276 pp.

HARRIS, S.A. & BROWN, R.J.E. (1982). Permafrost distribution along the Rocky Mountains in Alberta. *Proceedings of the Fourth Canadian Permafrost Conference, Calgary, Alberta*, pp. 59–67. Ottawa: N.R.C.C.

HARRISON, W.D. (1972). Temperature of a temperate glacier. *Journal Glaciol.*, **11**, 610, 15–29.

HARVEY, R.C. (1982). *The climate of arctic Canada in a 2 × CO_2 world.* Atmospheric Environment Service Report 82–5. Ottawa, Ontario: Environment Canada. 21 pp.

HAYNES, F.D. (1978). *Effect of temperature on the strength of snow-ice.* United States Army Cold Regions Research and Engineering Laboratory, Hanover, New Hampshire, Report 78-27, 28 pp.

HEGINBOTTOM, J.A. (1973). Some effects of surface disturbance on the permafrost active layer at Inuvik, N.W.T., Canada. *Second International Conference on Permafrost, Yakutsk, USSR, North American Contribution*, pp. 649–57. Washington, D.C.: Nat. Acad. Sci.

HENOCH, W.E.S. (1960). Fluvio-morphological features of the Peel and Lower Mackenzie Rivers. *Geog. Bull.*, **15**, 31–45.

HESSTVEDT, E. (1964). The interfacial energy ice/water. *Norw. Geotech. Institute Publication*, **56**.

HIGASHI, A. & CORTE, A.E. (1971). Solifluction: A model experiment. *Science*, **171**, 480–2.

HILLEL, D. (1971). *Soil and water, physical principles and processes.* New York: Academic Press, 288 pp.

HOBBS, P.V. (1974). *Ice physics.* Clarendon Press: Oxford. 837 pp.

HOEKSTRA, P. (1966). Moisture movement in soils under temperature gradients with the cold-side temperature below freezing. *Water Resources Research*, **2**, 241–50.

HOEKSTRA, P. (1969). Water movement and freezing pressures. *Soil Science Society of America Proceedings*, **33**, 512–8.

HOLDEN, J.T. (1983). Approximate solutions for Miller's theory of secondary heave. *Proceedings of the Fourth International Conference on Permafrost, Fairbanks, Alaska*, pp. 498–503. Washington, D.C.: Nat. Acad. Sci. - Nat. Acad. Press.

HOOKE, R. LE B., DOLLIN, B.B. & KAMPER, M.T. (1972). Creep of ice containing dispersed fine sand. *Journ. Glaciol.*, **63**, 327–36.

HOPKINS, D.M., KARLSTROM, T.N.V. & OTHERS. (1955). Permafrost and groundwater in Alaska. *U.S. Geological Survey, Professional Paper*, **264**, 113–46.

HORIGUCHI, K. & MILLER, R.D. (1980). Experimental studies with frozen soil in an 'ice sandwich' permeameter. *Cold Regions Science and Technology*, **3**, 177–83.

HORIGUCHI, K. & MILLER, R.D. (1983). Hydraulic conductivity functions of frozen materials. *Proceedings of the Fourth International Conference on Permafrost, Fairbanks, Alaska*, pp. 504–8. Washington, D.C.: Nat. Acad. Sci. - Nat. Acad. Press.

HORTON, R., WIERENGA, P.J. & NIELSEN, D.R. (1983). Evaluation of methods for determining the apparent thermal diffusivity of soil near the surface. *Soil Science Society of America Journal*, **47**, 25–32.

HUTCHINSON, J.N. (1974). Periglacial solifluxion: an approximate mechanism for clayey soils. *Geotechnique*, **24**, 438–43.

HWANG, C.T. & YIP, F.C. (1977). *Advances in frost heave prediction and mitigation methods for pipeline application.* Am. Soc. Mech. Eng., Paper #77-WA/HT-19, 11 pp.

INGERSOLL, L.R., ZOBEL, O.J. & INGERSOLL, A.C. (1954). *Heat conduction: with engineering, geological and other applications.* University of Wisconsin Press. 325 pp.

Institut Merzlotovedenia, im. V.A. Obrucheva, (1953–57). *Materialy po laboratornym issledovaniiam merzlykh gruntov*. Izd. Akad. Nauk SSSR, Moskva, Sb. 1, 2, 3.

International Society for Soil Science (1963). Soil physics terminology, Commission 1. (Soil Physics). *Bull. Int. Soc. Soil Sci.*, **23**, 7–10, also, **48**, (1975), 16–22.

ISAACS, R.M. & CODE, J.A. (1972). Problems in engineering geology related to pipeline construction. In *Proc. Can Northern Pipeline Res. Conf.* ed. Legget and MacFarlane. Tech. Memo. Nat. Res. Council Canada. 331 pp.

IVES, J.D. (1973). Permafrost and its relationship to other environmental parameters in a midlatitude, high-altitude setting, Front Range, Colorado Rocky Mountains. *Proceedings of the Second International Conference on Permafrost, Yakutsk, USSR, North American Contribution*, pp. 121–5. Washington, D.C.: Nat. Acad. Sci.

JACKSON, R.D. (1965). Water vapor diffusion in relatively dry soil: IV. Temperature and pressure effects on sorption diffusion coefficients. *Soil Science Society of America Proceedings*, **29**, 144–8.

JAHN, A. (1975). *Problems of the Periglacial Zone.* 223 pp. (translated form Polish).

JAHN, A. (1985). Experimental observations of periglacial processes in the Arctic. In *Field and Theory. Lectures in Geocryology*. Univ. Brit. Col. Press. 213 pp.

JAME, Y. & NORUM, D.I. (1976). Heat and mass transfer in freezing unsaturated soil in a closed system. *Proc., Second Conf. on Soil Water Problems in Cold Regions*. Edmonton, Canada. 46–62.

JELLINEK, H.H.G. (1972). The ice interface. In *Water and Aqueous Solutions*, ed. R.A. Horne, pp. 65–107. John Wiley and Sons.

JOHANSEN, O. (1972, 1973). Beregningsmetode for varmeledningsevne av fuktige og frosne jordarter, Del 1, Teoretisk grunnlag, *Frost i Jord*. pp. 17–25, Del 2, Frost i Jord 10, pp. 13–28 (includes English summaries: A method for calculation of thermal conductivity of soils, Parts 1 and 2).

JOHANSEN, O. (1977). *Vermeledningsevne av jordarter. Thermal conductivity of soils*. Trondheim-NTH: Institutt f. kjoleteknikk. (also as U.S. Cold Regions Research and Engineering Laboratory, Draft Translation, TL 637, 291 pp.)

JOHNSON, P.G. (1983). Rock glaciers. A case for a change in nomenclature. *Geogr. Ann.*, **65A**, 1–2, 27–34.

JOHNSTON, G.H. (ED.) (1981). *Permafrost: Engineering design and construction*. New York: John Wiley and Sons. 540 pp.

JOHNSTON, G.H. & BROWN, R.J.E. (1964). Some observations on permafrost distribution at a lake in the Mackenzie Delta, N.W.T., Canada. *Arctic*, **17**, 3, 163–75.

JOHNSTON, G.H., BROWN, R.J.E. & PICKERSGILL, D.N. (1963). *Permafrost investigations at Thompson, Manitoba: Terrain studies*. Division of Building Research, National Research Council of Canada, Technical Paper 158 (NRC 7568), 105 pp.

JONES, S.J. (1982). The confined compressive strength of polycrystalline ice. *Journal of Glaciology*, **28**, 98, 171–7.

JUDGE, A. (1973). The prediction of permafrost thicknesses. *Canadian Geotechnical Journal*, **10**, 1, 1–11.

JUDGE, A. (1982). Natural gas hydrates in Canada. *Proceedings of the Fourth Canadian Permafrost Conference, Calgary, Alberta*. pp. 320–8. Ottawa: N.R.C.C.

JUMIKIS, A.R. (1977). *Thermal geotechnics*. New Brunswick, New Jersey: Rutgers University Press. 375 pp.

KANE, D.L., CARLSON, R.F. & BOWERS, C.E. (1973). Groundwater pore pressures adjacent to subarctic streams. *Proceedings of the Second International Conference on Permafrost, Yakutsk, USSR*. vol. 1, pp. 453–8. Washington, D.C.: Nat. Acad. Sci.

KANE, D.L. & SLAUGHTER, C.W. (1973). Recharge of a central Alaska lake by subpermafrost groundwater. *Second International Conference on Permafrost. Yakutsk, USSR, North American Contribution*, 458–62.

KANE, D.L. & STEIN, J. (1983a). Field evidence of groundwater recharge in interior Alaska. *Proceedings of the Fourth International Conference on Permafrost, Fairbanks, Alaska*. pp. 572–7. Washington, D.C.: Nat. Acad. Sci. – Nat. Acad. Press.

KANE, D.L. & STEIN, J. (1983b). Water movement into seasonally frozen soils. *Water Resources Research.*, **19**, 6, 1547–57.

KAY, B.D., FUKUDA, M., IZUTA, H. & SHEPPARD, M.I. (1981). The importance of water migration in the measurement of the conductivity of unsaturated frozen soils. *Cold Regions Science and Technology*, **5**, 95–106.

KERSTEN, M.S. (1949). *Thermal properties of soils*. University of Minnesota, Institute of Technology, Engineering Experiment Station, Bulletin No. 28.

KING, L. (1984). Permafrost in Skandinavien. Untersuchungsergebnisse aus

Lappland, Jotunheimen und Dovre/Rondane. Dept. of Geography, University of Heidelberg, *Heidelberger Geogr. Arb.*, **76**, 174 pp.

KLYUKIN, N.K. (1963). Questions related to ameliorating the climate by influencing the snow cover. *Problems of the North*, **7**, 67–90.

KONISCHEV, V.N. (1982). Characteristics of cryogenic weakening in the permafrost zone of the European U.S.S.R. *Arctic and Alpine Research*, **14**, 3, 261–5.

KONRAD, J-M. & MORGENSTERN, N.R. (1980). A mechanistic theory of ice lens formation in fine-grained soils. *Canadian Geotechnical Journal*, **17**, 4, 473–86.

KOOPMANS, R.W.R. & MILLER, R.D. (1966). Soil freezing and soil water characteristics curves. *Soil Science Society of America Proceedings*,**30**, 680–5.

KUDRIAVTSEV, V.A. (1965). *Temperature, thickness and discontinuity of permafrost.* National Research Council of Canada, TT-1187 (technical translation).

KUDRIAVTSEV, V.A. (1978). *Obchchee Merzlotovedeniie – Geokryologiia.* Izd. Moskovskogo Universiteta. 463 pp. (*'General Permafrost Science – Geocryology.'*).

KVENVOLDEN, K.A. (1982). Occurrence and origin of marine gas hydrates. *Proceedings of the Fourth Canadian Permafrost Conference, Calgary, Alberta.* pp. 305–11. Ottawa: N.R.C.C.

LACHENBRUCH, A.H. (1957a). *Three-dimensional heat conduction in permafrost beneath heated buildings.* U.S. Geological Survey Bulletin 1052-B, 19 pp.

LACHENBRUCH, A.H. (1957b). Thermal effects of the ocean on permafrost. *Geological Society of America, Bulletin*, **68**, 1515–30.

LACHENBRUCH, A.H. (1957c). A probe for measurement of the thermal conductivity of frozen soils in place. *American Geophysical Union, Transactions*, **38**, 5, 691–7.

LACHENBRUCH, A.H. (1959). *Periodic heat flow in a stratified medium with application to permafrost problems.* U.S. Geological Survey, Bulletin 1083-A, 36 pp.

LACHENBRUCH, A.H. (1963). Contraction theory of ice wedge polygons: A qualitative discussion. *Proceedings, Permafrost International Conference.pp. 63–71.* Washington, D.C.: Nat. Acad. Sci. – Nat. Res. Counc. publication 1287.

LACHENBRUCH, A.H. & MARSHALL, B.V. (1969). Heat flow in the Arctic. *Arctic*, **22**, 300–11.

LACHENBRUCH, A.H. & MARSHALL, B.V. (1986). Changing climate: Geothermal evidence from permafrost in the Alaskan Arctic. *Science*, **234**, 689–96.

LACHENBRUCH, A.H., BREWER, M.C., GREENE, G.W. & MARSHALL, B.V. (1962). Temperatures in permafrost. In Temperature: *Its measurement and Control in Science and Industry*, vol. 3, pp. 791–803, New York: Reinhold Publ. Co.

LACHENBRUCH, A.H., SASS, J.H., MARSHALL, B.V. & MOSES, T.H. JR. (1982). Permafrost, heat flow and the geothermal regime at Prudhoe Bay, Alaska. *Journal of Geophysical Research*, **87**, B11, 9301–16.

LADANYI, B. (1981). Mechanical behaviour of frozen soils. *Mechanics of structured media. Proceedings of an International Symposium*, Part B, pp. 205–45.

LADANYI, B. (1985). Stress transfer mechanism in frozen soils. *Can. Conf. App. Mech.*, vol. 1, pp. 11–23. London, Ontario.

LADANYI, B. *et al.* (1981). Engineering characteristics of frozen and thawing soils. In *Permafrost: Engineering design and construction*, ed. G.H. Johnston, New York: John Wiley and Sons.

LADANYI, B. & LAUZON, M. (1986). In *Investigations of frost heave as a cause of pipeline deformation.* Report for Earth Physics Branch, Energy, Mines and Resources, Canada (Report IR 50, Geotechnical Science Laboratories, Carleton University) 71 pp.

LAMBE, T.W. & WHITMAN, R.V. (1979). *Soil mechanics.* SI Version. New York: John Wiley and Sons, 553 pp.

LARSSON, S. (1982). Geomorphological effects on the slopes of Longyear Valley, Spitsbergen, after a heavy rainstorm in July, 1972. *Geogr. Ann.*, **64**, A, 105–25.

LAWSON, D.E. (1983). *Erosion of perennially frozen streambanks.* United States Army Cold Regions Research and Engineering Laboratory, Hanover, New Hampshire, Report 83-29, 22 pp.

LEWKOWICZ, A.G. (1983). Erosion by overland flow, Central Banks Island, Western Canadian Arctic. *Proceedings of the Fourth International Conference on Permafrost, Fairbanks, Alaska.* pp. 701–6. Washington, D.C.: Nat. Acad. Sci. – Nat. Acad. Press.

LEWKOWICZ, A.G. & FRENCH, H.M. (1982). The hydrology of small runoff plots in an area of continuous permafrost, Banks Island, N.W.T. *Proceedings of the Fourth Canadian Permafrost Conference, Calgary, Alberta.* pp. 151–62. Ottawa: N.R.C.C.

LIESTÖL, O. (1977). Pingoes, springs and permafrost in Spitsbergen. *Årb. Norsk Polarinstitutt 1975*, 7–29.

LINDSAY, J.D. & ODYNSKY, W. (1965). Permafrost in organic soils of northern Alberta. *Canadian Journal of Soil Science*, **45**, 265–9.

LINNELL, K.A. (1973). Long term effects of vegetation cover on permafrost stability in an area of discontinuous permafrost. *Proceedings of the Second International Conference on Permafrost, Yakutsk, USSR, North American Contribution*, pp. 688–93. Washington, D.C.: Nat. Acad. Sci.

LINNELL, K.A. & KAPLAR, C.W. (1959). *The factor of soil and material type in highway pavement design in frost areas: a symposium – Part 1, Basic Considerations.* Highway Res. Board Bulletin 225, 131 pp.

LINNELL, K.A. & TEDROW, J.C.F. (1981). *Soil and Permafrost Surveys in the Arctic.* Oxford: Clarendon Press. 279 pp.

LIST, R.J. (1968). *Smithsonian meteorological tables.* Washington, D.C.: Smithsonian Institution Press. 527 pp.

LOCH, J.P.G. (1980). Frost action in soils, state of the art. *Proceedings, Second International Symposium on Ground Freezing*, vol. 1, pp. 581–96. Trondheim: Norwegian Institute of Technology.

LOCH, J.P.G. & KAY, B.D. (1978). Water distribution in partially frozen saturated silt under several temperature gradients and overburden loads. *Soil Science Society of America Journal*, **42**, 3, 400–6.

LOCH, J.P.G. & MILLER, R.D. (1975). Tests of the concept of secondary frost heaving. *Soil Science Society of America Journal*, **39**, 6, 1036–41.

LOCKWOOD, J. (1979). *Causes of climate.* London: Edward Arnold. 260 pp.

LORRAIN & DEMEUR. (1985). Isotopic evidence for relic Pleistocene glacier ice on Victoria Island, Canadian Arctic Archipelago. *Arctic and Alpine Research*, **17**, 1, 89–98.

LOW, P., ANDERSON, D.M. & HOEKSTRA, P. (1968). Some thermodynamic relationships for soils at or below freezing point. Freezing point depression and heat capacity, *Water Resources Research*, **4**, 2, 379–94.

LUNARDINI, V. (1981). *Heat transfer in cold climates.* New York: Van Nostrand Reinhold. 731 pp.

LUTHIN, J.N. & GUYMON, G.L. (1974). Soil moisture–vegetation–temperature relationships in central Alaska. *Journal of Hydrology*, **23**, 233–46.

MACKAY, D.K. & LOKEN, O.H. (1974). Arctic hydrology. In: *Arctic and Alpine Environments*, ed. J.D. Ives and R.G. Barry, pp. 111–32. London: Methuen and Co.

MACKAY, J.R. (1963). *The Mackenzie Delta area, N.W.T.* Geographical Branch, Department of Mines and Technical Surveys, Canada, Memoir 8, 202 pp.

MACKAY, J.R. (1970). Disturbances to the tundra and forest tundra environment of the western Arctic. *Canadian Geotechnical Journal.*

MACKAY, J.R. (1971). The origin of massive icy beds in permafrost, western arctic coast, Canada. *Canadian Journal of Earth Sciences*, **84**, 397–422.

MACKAY, J.R. (1972). The world of underground ice. *Annals Am. Assoc. Geographers*, **62**, 1–22.

MACKAY, J.R. (1974a). Ice-wedge cracks, Garry Island, Northwest Territories. *Canadian Journal of Earth Sciences*, **11**, 10, 1366–83.

MACKAY, J.R. (1974b). Reticulate ice veins in permafrost, Northern Canada. *Canadian Geotechnical Journal*, **11**, 230–7.

MACKAY, J.R. (1975b). The closing of ice-wedge cracks in permafrost, Garry Island, Northwest Territories. *Canadian Journal of Earth Sciences* **12,** 9, 1668–74.

MACKAY, J.R. (1977a). Changes in the active layer from 1968 to 1976 as a result of the Inuvik fire. *Geological Survey of Canada*, Paper 77-1B, 273–5.

MACKAY, J.R. (1977b). Pulsating pingos, Tuktoyaktuk Peninsula, N.W.T., *Canadian Journal of Earth Sciences*, **14**, 209–22.

MACKAY, J.R. (1978a). Sub-pingo water lenses, Toktoyaktuk Peninsula, N.W.T., *Canadian Journal of Earth Sciences*, **8**,

MACKAY, J.R. (1978b). The use of snow fences to reduce ice-wedge cracking, Garry Island, Northwest Territories, In *Current Research, Part A; Geol. Surv. Can.* Paper 78-1A, pp. 523–4.

MACKAY, J.R. (1979). Pingos of the Tuktoyaktuk Peninsula area, Territories. *Geographie Physique et Quaternaire*, **33**, 1, 3–61.

MACKAY, J.R. (1980a). Deformation of ice-wedge polygons, Garry Island Northwest Territories. In *Current Research.* Paper 80-1A. Geol. Survey, Canada. 287–291.

MACKAY, J.R. (1980b). The origin of hummocks, western Arctic coast, Canada. *Canadian Journal of Earth Sciences*, **17**, 8, 996–1006.

MACKAY, J.R. (1981). Actice layer slope movement in a continuous permafrost environment, Garry Island, Northwest Territories, Canada. *Canadian Journal of Earth Sciences.*, **18**, 11, 1666–80.

MACKAY, J.R. (1983a). Oxygen isotope variations in permafrost, Tuktoyaktuk Peninsula area, Northwest Territories. *Geological Survey of Canada*, Paper 83-1B: 67–74.

MACKAY, J.R. (1983b). Downward water movement into frozen ground, western arctic coast, Canada. *Canadian Journal of Earth Sciences*, **201**, 120–134.

MACKAY, J.R. (1984a). The frost heave of stones in the active layer above permafrost with downward and upward freezing. *Arctic and Alpine Research*, **16**, 439–46.

MACKAY, J.R. (1984b). The direction of ice wedge cracking in permafrost: downward or upward? *Canadian Journal of Earth Sciences*, **21**, 5, 516–24.

MACKAY, J.R. (1985a). Pingo ice of the western Arctic coast, Canada. *Canadian Journal of Earth Sciences*, **22**, 10, 1452–64.

MACKAY, J.R. (1985b). Permafrost growth in recently drained lakes, western Arctic coast. *Geological Survey of Canada*, Paper 85-1B: 177–89.

MACKAY, J.R., KONISCHEV, V.N. & POPOV, A.I. (1979a). Geologic controls of the origin, characteristics, and distribution of ground ice. *3rd Int. conf. Permafrost*, vol. pp. 2, 1–18.

MACKAY, J.R. & LAVKULICH, L.M. (1974). Ionic and oxygen isotopic fractionation in permafrost growth. *Geol. Surv. Canada*, 74-1B, 255–6.

MACKAY, J.R., OSTRICK, J., LEWIS, C.P. & MACKAY, D.K. (1979b). Frost heave at ground temperatures below 0 °C, Inuvik, Northwest Territories. 79-1A, pp. 403–6.

MACKAY, J.R., RAMPTON, V.N. & FYLES, J.G. (1972). Relic pleistocene Permafrost Western Arctic Canada. *Science*, **176**, 404, 1321–3.

MAGEAU, D.W. & MORGENSTERN, N.R. (1980). Observations on moisture migration in frozen soils. *Canadian Geotechnical Journal*, **17**, 1, 54–60.

MAKOGON, YU. F. (1982). Perspectives for the development of gas-hydrate deposits. *Proceedings of the Fourth Canadian Permafrost Conference, Calgary, Alberta*. pp. 299–304. Ottawa, N.R.C.C.

MANNERFELT, C.M. (1945). Någran glacialmorfologiska formelement. *Geogr. Ann.*, **27**, 1–239.

MCBEATH, J.H. (ED.). (1984). *The potential effects of carbon dioxide-induced climate changes in Alaska: The proceedings of a conference*. School of Agricultural and Land Resources Management, University of Alaska-Fairbanks, Misc. Publ. 83-1, 208 pp.

MCCANN, S.B., HOWARTH, P.J. & COGLEY, J.G. (1972). Fluvial processes in a periglacial environment, Queen Elizabeth Islands, N.W.T., Canada. *Transactions of the Institute of British Geographers*, **55**, 69–82.

MCDONALD, B.C. & LEWIS, C.P. (1973). *Geomorphic and sedimentologic processes of rivers and coast, Yukon Coastal Plain*. Task Force on Northern Oil Development, Environmental–Social Committee, Report 73-39, Ottawa, Canada. 245 pp.

MCGAW, R., OUTCALT, S.I. & NG, E. (1978). Thermal properties and regime of wet tundra soils at Barrow, Alaska. *Proceedings of the Third International Conference on Permafrost, Edmonton, Canada* pp. 47–53. Ottawa: N.R.C.C.

MCROBERTS, E.C. (1975). Some aspects of a simple secondary creep model for deformations in permafrost slopes. *Canadian Geotechnical Journal*, **12**, 1, 98–105.

MCROBERTS, E.C. & MORGENSTERN, N.R. (1974). The stability of thawing slopes. *Canadian Geotechnical Journal*, **11**, 4, 447–69.

MELLOR, M. (1983). *Mechanical behaviour of sea ice*. Monogr. 83-1, U.S. Army Cold Regions Research and Engineering Laboratory, Hanover, N.H. 105 pp.

MELLOR, M. & COLE, D.M. (1982). Deformation and failure of ice under constant stress or constant strain-rate. *Cold Regions Science and Technology*, **5**, 3, 201–19.

MELLOR, M. & COLE, D.M. (1983). Stress/strain/time relations for ice under uniaxial compression. *Cold Regions Science and Technology*, **6**, 3, 207.

MELLOR, M. & TESTA, R. (1969). Effect of temperature on the creep of ice. *Journal of Glaciology*, **8**, 52, 131–45.

MICHEL, F.A. & FRITZ, P. (1978). Environmental isotopes in permafrost related waters along the Mackenzie Valley corridor. *Proceedings of the Third International Conference on Permafrost*, Edmonton, Canada. vol. 1, pp. 207–11, Ottawa: N.R.C.C.

MICHEL, F.A. & FRITZ, P. (1982). Significance of isotope variations in permafrost waters at Illisarvik, N.W.T. *Proceedings of the Fourth Canadian Permafrost Conference, Calgary, Alberta*. pp. 173–81. Ottawa: N.R.C.C.

MILLER, R.D. (1970). Ice sandwich: Functional semipermeable membrane. *Science*, **169**, 584–5.

MILLER, R.D. (1972). Freezing and thawing of saturated and unsaturated soils. *US Highway Research Board, Record*, **393**, 1–10.

MILLER, R.D. (1978). Frost heaving in non-colloidal soils. *Proceedings of the Third International Conference on Permafrost*, Edmonton, Canada. vol. 1, pp. 708–13, Ottawa: N.R.C.C.

MILLER, R.D. (1980). Freezing phenomena in soils. In *Applications of Soil Physics*. ed. D. Hillel, Academic Press.

MILLER, R.D., LOCH, J.P.G. & BRESLER, E. (1975). Transport of water and heat in a frozen permeameter. *Soil Science Society of America Proceedings*, **39**, 6, 1029–36.

MITCHELL, J.K. (1976). *Fundamentals of soil behaviour*. New York: John Wiley and Sons. 422 pp.

MOLOCHUSKIN, E.N. (1973). The effect of thermal abrasion on the temperature of the permafrost in the coastal zone of the Laptev Sea. *Proceedings of the Second International Conference on Permafrost, Yakutsk, USSR, USSR Contribution*, pp. 90–3. Washington, D.C.: Nat. Acad. Sci.

MONTEITH, J.L. (1973). *Principles of environmental physics*. London: Edward Arnold. 241 pp.

MORGENSTERN, N.R. (1985). Recent observations on the deformation of ice and ice-rich permafrost. In *Field and Theory. Lectures in Geocryology*. University of British Columbia Press. 213 pp.

MORGENSTERN, N.R. & NIXON, J.K. (1971). One-dimensional consolidation of thawing soils. *Canadian Geotechnical Journal*, **8**, 4, 558—65.

MÜLLER, F. (1963). Beobachtungen uber Pingos. Detailuntersuchungen in Östgrönland und in der Kanadischen Arktis. *Medd. om. Grönl.*, **153**, 3, 127 pp. (also translated into English: TT 1073, Nat. Res. Counc. Canada).

NAKANO, Y. & BROWN, J. (1972). Mathematical modeling and validation of the thermal regimes in tundra soils, Barrow, Alaska. *Arctic and Alpine Research*, **4**, 1, 19–38.

NAKANO, Y., TICE, A. & OLIPHANT, J. (1984). Transport of water in frozen soil: III. Experiments on the effects of ice content. *Advances in Water Resources*, **7**, 1, 28–34.

NAKANO, Y., TICE, A., OLIPHANT, J. & JENKINS, T. (1982). Transport of water in frozen soil: I: Experimental determination of soil-water diffusivity under isothermal conditions. *Advances in Water Resources*, **5**, 4, 221–6.

NAKANO, Y., TICE, A., OLIPHANT, J. & JENKINS, T. (1983a). Soil-water diffusivity of unsaturated soils at subzero temperatures. *Proceedings of the Fourth International Conference on Permafrost, Fairbanks, Alaska*. pp. 889–93. Washington, D.C.: Nat. Acad. Sci. – Nat. Acad. Press.

NAKANO, Y., TICE, A., OLIPHANT, J. & JENKINS, T. (1983b). Transport of water in frozen soil: II. Effects of ice on the transport of water under isothermal conditions. *Advances in Water Resources*, **6**, 1, 15–26.

NASH, L.K. (1970). *Elements of Chemical Thermodynamics*. 2nd edn, Addison-Wesley. 184 pp.

National Research Council Canada (1988). *Glossary of Permafrost and Related Ground-Ice Terms*. 156 pp. Ottawa. N.R.C.C.

NELSON, F. & OUTCALT, S.I. (1983). A frost index number for spatial prediction of ground frost zones. *Proceedings of the Fourth International Conference on Permafrost, Fairbanks, Alaska*, pp. 970–11. Washington, D.C.: Nat. Acad. Sci. – Nat. Acad. Press.

NELSON, F., OUTCALT, S.I., GOODWIN, C.W. & HINKEL, K.M. (1985). Diurnal thermal regime in a peat-covered palsa, Toolik Lake Alaska. *Arctic*, **38**, 4, 310–5.

NICHOLSON, F.H. (1978). Permafrost modification by changing the natural energy budget. *Proceedings of the Third International Conference on Permafrost*, Edmonton, Canada, vol. 1, pp. 61–7. Ottawa: N.R.C.C.

NICHOLSON, F.H. & GRANBERG, H.B. (1973). Permafrost and snow cover relationships near Schefferville. *Proceedings of the Second International Conference on Permafrost, Yakutsk, USSR, North American Contribution*, pp. 151–8. Washington, D.C.: Nat. Acad. Sci.

NIXON, J.F. & MORGENSTERN, N.R. (1973). The residual stress in thawing soils. *Can. Geotech. Jour.*, **10**, 571–80.

NIXON, J.F. & LADANYI, B. (1978). Thaw consolidation. In *Geotechnical engineering for cold regions*. ed. O.B. Andersland and D.M. Anderson, pp. 164–215. McGraw–Hill.

NIXON, J.F., MORGENSTERN, N.R & REESOR, S.N. (1983). Frost heave–pipeline interaction using continuum mechanics. *Canadian Geotechnical Journal*, **20**, 251–61.

NUMMEDAL, D. (1983). Permafrost on Mars: Distribution, formation and geological role. *Proceedings of the Fourth International Conference on Permafrost, Fairbanks, Alaska*. pp. 934–9. Washington, D.C.: Nat. Acad. Sci. – Nat. Acad. Press.

NYE, J.P. & FRANK, F.C. (1973). Hydrology of the intergranular veins in a temperate glacier, *International Association of Scientific Hydrology. Publication 95, Symposium on the Hydrology of Glaciers*, pp. 157–61.

OGATA, N., YASUDE, M. & KATAOKA, T. (1982). Salt concentration effects on strength of frozen soils. *Proceedings of the Third International Symposium on Ground Freezing*, Hanover, N.H. pp. 3–10.

OHRAI, T., TAKASHI, T., YAMAMOTO, H. & OKAMOTO, J. (1983). Uniaxial compressive strength of ice segregated from soil. *Proceedings of the Fourth International Conference on Permafrost, Fairbanks, Alaska*. pp. 945–50. Washington, D.C.: Nat. Acad. Sci. – Nat. Acad. Press.

OHRAI, T. & YAMAMOTO, H. (1985). Growth and migration of ice lenses in partially frozen soil. *Proceedings of the Fourth International Symposium on Ground Freezing, Sapporo, Japan*, vol. 1, pp. 79–84. Rotterdam: Balkema.

OKE, T.R. (1978). *Boundary Layer Climates*. London: Methuen and Co. 372 pp.

OLIPHANT, J.L., TICE, A.R. & NAKANO, Y. (1983). Water migration due to a temperature gradient in frozen soil. *Proceedings of the Fourth International Conference on Permafrost, Fairbanks, Alaska*. pp. 951–6. Washington, D.C.: Nat. Acad. Sci. – Nat. Acad. Press.

O'NEILL, K. & MILLER, R.D. (1985). Exploration of a rigid ice model of frost heave. *Water Resources Research*, **21**, 3, 281–96.

ONESTI, L.J. & WALTI, S.A. (1983). Hydrologic characteristics of small arctic-alpine watersheds, central Brooks Range, Alaska. *Proceedings of the Fourth International Conference on Permafrost, Fairbanks, Alaska.* pp. 957–61. Washington, D.C.: Nat. Acad. Sci. – Nat. Acad. Press.

OSTERKAMP, T.E. (1975). Structure and properties of ice lenses in frozen ground, *Proceedings of the Conference on Soil Water Problems in Cold Regions, Calgary, Alberta*, pp. 89–111.

OSTERKAMP, T.E. (1987). Freezing and thawing of soils and permafrost containing unfrozen water or brine. *Water Resources Research*, **23**, 12, 2279–85.

OSTERKAMP, T.E. & PAYNE, M.W. (1981). Estimates of permafrost thickness from well logs in northern Alaska. *Cold Regions Science and Technology*, **5**, 1, 13–27.

ØSTREM, G. (1963a). Ice-cored moraines in Scandinavia, *Geogr. Annal.*, **46**, 3, 282–337.

ØSTREM, G. (1963b). Comparative crystallographic studies on ice from ice-covered moraines, snowbanks and glaciers. *Geogr. Annal.* **45**, 210–40.

OUTCALT, S.I. (1972). The development and application of a simple digital surface climate simulator. *Journal of Applied Meteorology*, **II**, 629–36.

OUTHET, D.N. (1974). Progress report on bank erosion studies in the Mackenzie River Delta, N.W.T. *Task Force on Northern Oil Development, Environmental-Social Committee, Report #74-12*, Ottawa, Canada, 297—345.

PARMUZINA, O. YU. (1978). Cryogenic texture and some characteristics of ice formation in the active layer (in Russian). *Problemy kriolitologii*, **7**, 141–64. Translated in *Polar Geography and Geology*, July–September 1980, 131–52.

PAVLOV, A.V. (1973). Heat exchange in the active layer. *Proceedings of the Second International Conference on Permafrost, Yakutsk, USSR, USSR Contribution*, pp. 25–30. Washington, D.C.: Nat. Acad. Sci.

PENMAN, H.L. (1948). Natural evaporation from open water, bare soil and grass. *Proceedings of the Royal Society of London*, **A193**, 120–45.

PENNER, E. (1961). Ice-grain structure and crystal orientation in an ice lens from leda clay. *Geol. Soc. Amer. Bull.*, **72**, 1575–8.

PENNER, E. (1967). Heaving pressure in soils during unidirectional freezing. *Canadian Geotechnical Journal*, **4**, 398–408.

PENNER, E. (1970). Thermal conductivity of frozen soils. *Canadian Journal of Earth Sciences*, **7**, 982–7.

PENNER, E. (1986). Ice lensing in layered soils. *Canadian Geotechnical Journal*, **23**, 3, 334–40.

PENNER, E. & GOODRICH, L.E. (1980). Location of segregated ice in frost susceptible soil. *Proceedings, Second International Symposium on Ground Freezing*, vol. 1, pp. 626–39. Trondheim: Norwegian Institute of Technology.

PENNER, E. & WALTON, T. (1978). Effects of temperature and pressure on frost heaving, *Engineering Geology*, **13**, 1–4, 29–39.

PENNER, S.S. (1968). *Thermodynamics for Scientists and Engineers.* Addison–Wesley. 288 pp.

PERFECT, E. & WILLIAMS, P.J. (1980). Thermally induced water migration in frozen soils. *Cold Regions Science and Technology*, **3**, 101–9.

PÉWÉ, T.L. (1983). Alpine Permafrost in the contiguous United States. *Arctic and Alpine Research*, **15**, 2, 145–56.

PHUKAN, A. (1985) *Frozen Ground Engineering.* Prentice–Hall, 336 pp.

PISSART, A. (1973). Resultats d'experiences sur l'action du gel. *Biul. Perygl.*, **23**, 101–13.

PISSART, A. (1974). Determination experimentale des processus responsables des petits sols polygoneaux triés de haut montagne. *Abh. Akad. Wiss Gottingen, ser. 3*,. **29**, 86–101.

POLLARD, W.H. & FRENCH, H.M. (1980). A first approximation of the volume of ground ice, Richards Island, Pleistocene Mackenzie delta, Northwest Territories, Canada. *Canadian Geotechnical Journal*, **17**, 4, 509–16.

POLLARD, W.H. & FRENCH, H.M. (1984). The groundwater hydraulics of seasonal frost mounds. North Fork Pass, Yukon Territory. *Canadian Journal of Earth Sciences*, **21**, 10, 1073–81.

POLLARD, W.H. & FRENCH, H.M. (1985). The internal structure and ice crystallography of seasonal frost mounds. *Journal of Glaciology*, **31**, 108, 157–62.

PRICE, L.W. (1971). Vegetation, microtopography, and depth of active layer on different exposures in subarctic alpine tundra. *Ecology*, **52**, 4, 638–47.

RAMPTON, V.N. (1973). The influence of ground ice and thermokarst upon the geomorphology of the Mackenzie Beaufort region: In *Research in Polar and Alpine Geomorphology, Proceedings 3rd Guelph Symposium on Geomorphology, Guelph, Ontario*, pp. 43–60.

RAPP, A. (1960). Recent development of mountain slopes in Kärkevagge and surroundings, Northern Scandinavia. *Geogr. Annal.*, **XLII**, 71–200.

RAPP, A. (1985). Extreme rainfall and rapid snowmelt as causes of mass movements in high latitude mountains. In *Field and Theory. Lectures in Geocryology*. Univ. of Brit. Col. Press. 213 pp.

REIN, R.G., JR. & BURROUS, C.M. (1980). Laboratory measurements of subsurface displacements during thaw of low-angle slopes of a frost susceptible soil. *Arctic and Alpine Research*, **12**, 349–58.

RIEGER, S. (1983). *The Genesis and Classification of Cold Soils*. New York: Academic Press. 230 pp.

RISEBOROUGH, D.W. (1985). Modelling climatic influences on permafrost at a boreal forest site. Unpublished M.A. thesis, Carleton University, Ottawa, 172 pp.

RISEBOROUGH, D.W., SMITH, M.W. & HALLIWELL, D.H. (1983). Determination of the thermal properties of frozen soils. *Proceedings of the Fourth International Conference on Permafrost, Fairbanks, Alaska*, pp. 1072–7. Washington, D.C.: Nat. Acad. Sci. – Nat. Acad. Press.

RISSING, J.M. & THORN, C.E. (1985). Particle size and clay mineral distributions within sorted and non-sorted circles and the surrounding parent material Niwot Ridge, Front Range, Colorado, U.S.A. *Arct. & Alp. Res.*, **17**, 153–63.

ROGGENSACK, W.D. & MORGENSTERN, N.R. (1978). Direct shear tests on natural fine grained permafrost soils. *Proceedings of the Third International Conference on Permafrost, Edmonton, Alberta*, vol. 1, pp. 729–35. Ottawa: N.R.C.C.

ROMANOVSKIJ, N.N. (1973). Regularities in formation of frost fissures and development of frost fissure polygons. *Biul. Perygl.*, **23**, 237–77.

ROMKENS, M.J.M. & MILLER, R.D. (1973). Migration of mineral particles in ice with a temperature gradient. *Journal of Colloid and Interface Science*, **42**, 1, 103–11.

ROSENBERG, N.J., BLAD, B.L. & VERMA, S.B. (1983). *Microclimate: The biological environment* 2nd edn. Wiley. 495 pp.

ROUSE, W.R. (1976). Microclimatic changes accompanying burning in subarctic lichen woodland. *Arctic and Alpine Research*, **8**, 4, 357–76.

ROUSE, W.R. (1982). Microclimate of low Arctic tundra and forest at Churchill, Manitoba: *Proceedings of the Fourth Canadian Permafrost Conference, Calgary, Alberta*. pp. 68–80. Ottawa: N.R.C.C.

ROUSE, W.R. (1984). Microclimate of Arctic tree line. 2: Soil microclimate of tundra and forest. *Water Resources Research*, **20**, 1, 67–73.

RYDEN, B.E. (1981). Hydrology of northern tundra. In *Tundra Ecosystems*, ed. Bliss, Cragg, Heal and Moore, pp. 115–37. Cambridge Univ. Press.

SANGER, F.J. & KAPLAR, C.W. (1963). Plastic deformation of frozen soils. *Proceedings, Permafrost International Conference*. pp. 305–14. Washington, D.C.: N.A.S. - N.R.C. publication 1287.

SANTEFORD, H.S. (1978). Snow soil interactions in interior Alaska. In *Modelling of snow cover runoff*, ed. S.C. Colbeck and M. Ray, pp. 311–8. Hanover, NH.: U.S. Army Cold Regions Research and Engineering Laboratory.

SAYLES, F.H. (1968). *The creep of frozen sands*. United States Army Cold Regions Research and Engineering Laboratory, Hanover, New Hampshire. Technical Report no. 190.

SAYLES, F.H. (1973). Triaxial and creep tests on frozen Ottawa sand. *Proceedings of the Second International Conference on Permafrost, Yakutsk, USSR, North American Contribution*, pp. 384–91. Washington, D.C.: Nat. Acad. Sci.

SAYLES, F.H. & HAINES, D. (1974). *Creep of frozen silt and clay*. U.S. Army Cold Regions Research and Engineering Laboratory, Technical Report no. 252.

SCHOFIELD, R.K. (1935). The pF of the water in soil, *Third International Congress on Soil Science* vol. 2, pp. 37–48, vol. 3, pp. 182–6.

SCHOFIELD, R.K. & BOTELHO DA COSTA, J.V. (1938). The measurement of pF in soil by freezing point, *Agric. Science*, **28**, 645–53.

SELLERS, W.D. (1965). *Physical climatology*. Chicago: University of Chicago Press. 272 pp.

SELLMAN, P.V. & HOPKINS, D.M. (1984). Subsea permafrost distribution on the Alaskan shelf. *Proceedings of the Fourth International Conference on Permafrost, Fairbanks, Alaska*. final vol., pp. 75–82. Washington, D.C.: Nat. Acad. Sci. - Nat. Acad. Press.

SEPPÄLÄ, M. (1972). The term 'palsa'. *Zeitschr. Geomorph.*, **16**, 463.

SEPPÄLÄ, M. (1979). Recent palsa studies in Finland. In *Palaeohydology of the temperate zone*. Proc. Comm. Holoceine-Inqua. Acta Universit. Ouluensis, Ser. A, Sci. Rer. Natur. 82, Geologica 3, 81–7.

SEPPÄLÄ, M. (1982). An experimental study of the formation of palsas. *Proceedings of the Fourth Canadian Permafrost Confernece, Calgary, Alberta*. pp. 36–42. Ottawa: N.R.C.C.

SEPPÄLÄ, M. (1986). The origin of palsas. *Geogr. Ann.*, **68A**, 3, 141–7.

SEPPÄLÄ, M. & KOUTANIEMI, L. (1985). Formation of a string and pool topography as expressed by morphology, stratigraphy and current processes on a mire in Kuusamo, Finland. *Boreas*, **14**, 287–309.

SHERMAN, R.G. (1973). A groundwater supply for an oil camp near Prudhoe Bay, arctic Alaska, *Proceedings of the Second International Conference on Permafrost, Yakutsk, USSR, North American Contribution*, pp. 469–72. Washington, D.C.: Nat. Acad. Sci.

SHILTS, W.W. (1978). Nature and genesis of mudboils, central Keewatin, Canada. *Canadian Journal of Earth Sciences*, **15**, 7, 1053–68.

SHUMSKII, P.A. (1964). *Principles of Structural Glaciology* (trans. by D. Kraus, of Osnovy strukturnogo ledovedeniia, Moscow 1955). New York: Dover, 497 pp.

SKAVEN HAUG, S. (1959). Protection against frost heaving on Norwegian railways. *Geotechnique*, **9**, 3, 87–106.

SLAUGHTER, C.W., HILGERT, J.W. & CULP, E.H. (1983). Summer stream flow and sediment yield from discontinuous-permafrost headwaters catchments. *Proceedings of the Fourth International Conference on Permafrost, Fairbanks, Alaska*. pp. 1172–7. Washington, D.C.: Nat. Acad. Sci. – Nat. Acad. Press.

SMILEY, L. & ZUMBERGE, J.H. (1974). *Polar Deserts and Modern Man*. Tucson, Arizona: Univ. Arizona Press, 173 pp.

SMITH, D.J. (1987). Solifluction in the Southern Canadian Rockies. *The Canadian Geographer*, **31**, 4, 309–18.

SMITH, L.L. & CHEATHAM, J.B., JR. (1975). Plasticity of ice and sand ice systems. *Journal of Engineering for Industry*, **97**, 2, 579–84.

SMITH, M.W. (1975). Microclimatic influences on ground temperatures and permafrost distribution, Mackenzie Delta, Northwest Territories. *Canadian Journal of Earth Sciences*, **12**, 1421–38.

SMITH, M.W. (1976). *Permafrost in the Mackenzie Delta, Northwest Territories*. Geological Survey of Canada, Paper 75-28, 34 pp.

SMITH, M.W. (1977). *Computer simulation of microclimatic and ground thermal regimes in an arctic environment: Test results and program description*. Department of Indian Affairs and Northern Development, Ottawa, Canada, ALUR Report 75-76-72, 74 pp.

SMITH, M.W. (1985). Observations of soil freezing and frost heaving at Inuvik, Northwest Territories, Canada. *Canadian Journal of Earth Sciences*, **22**, 3, 283–90.

SMITH, M.W. (1986). The significance of climate change for the permafrost environment. In *Climate change impacts in the Canadian arctic*, ed. H.M. French. Proceedings of a Canadian Climate Program Workshop, March 3–5 1986, Geneva Park, Ontario 67–81.

SMITH, M.W. & BURN, C.R. (1987). Outward flux of vapour from frozen soils at Mayo, Yukon, Canada: results and interpretation. *Cold Regions Science and Technology*, **13**, 143–52.

SMITH, M.W. & RISEBOROUGH, D.W. (1983). Permafrost sensitivity to climatic change. *Proceedings of the Fourth International Conference on Permafrost, Fairbanks, Alaska*, pp. 1178–83. Washington, D.C.: Nat. Acad. Sci. – Nat. Acad. Press.

SMITH, M.W. & RISEBOROUGH, D.W. (1985). The sensitivity of thermal predictions to assumptions in soil properties. *Proceedings of the Fourth International Symposium on Ground Freezing, Sapporo, Japan*, vol. 1, pp. 17–23. Rotterdam: Balkema.

SPANNER, D.C. (1964). *Introduction to Thermodynamics*. Academic Press. 278 pp.

STUIVER, M., YANG, I.C. & DENTON, G.H. (1976). Permafrost oxygen isotope ratios and chronology of three cores from Antarctica. *Nature* **261**, 547–50.

SUSLOV, S.P. (1961). *Physical geography of Asiatic Russia*. Translated by N.D. Gershevsky, San Francisco: W.H. Freeman and Co.

SUTHERLAND, H.B. & GASKIN, P.N. (1973). Pore water and heaving pressures developed in partially frozen soils. *Proceedings of the Second International Conference on Permafrost, Yakutsk, USSR, North American Contribution*, pp. 409–19. Washington, D.C.: Nat. Acad. Sci.

TABER, S. (1918). Ice forming in clay soils will lift surface weights. *Engineering News-Record*, **80**, 6, 262–3.

TABER, S. (1929). Frost heaving. *Journal of Geology*, **37**, 5, 428–61.

TABER, S. (1930). The mechanics of frost heaving. *Journal of Geology*, **38**, 4, 303–17.

TAKAGI, S. (1971). *Numerical differentiation by spline functions and its application to analysing a lake temperature observation*. United States Army Cold Regions Research and Engineering Laboratory, Hanover, New Hampshire, Research Report 293, 18 pp.

TAYLOR, G.S. & LUTHIN, J.N. (1978). A model for coupled heat and moisture transfer during soil freezing. *Canadian Geotechnical Journal*, **15**, 548–55.

TERZAGHI, K. (1952). Permafrost. In *From theory to practice in soil mechanics: Selections from the writings of Karl Terzaghi*, pp. 246–95. New York and London: John Wiley.

TERZAGHI, K. & PECK, R. (1967). *Soil Mechanics in Engineering Practice*. New York: John Wiley and Sons. 729 pp.

THIE, J. (1974). Distribution and thawing of permafrost in the southern part of the discontinuous permafrost zone in Manitoba: *Arctic*, **27**, 189–200.

THOMPSON, E.G. & SAYLES, F.H. (1972). In-situ creep analysis of a room in frozen soil. *Soil Mech. Found. Divn., A.S.C.E.*, **98**, 899–916.

TICE, A.R., OLIPHANT, J.L., NAKANO, Y. & JENKINS, T.F. (1982). *Relationship between the ice and unfrozen water phases in frozen soil as determined by pulsed nuclear magnetic resonance and physical desorption data*. United States Army Cold Regions Research and Engineering Laboratory, Hanover, New Hampshire, Report 82-15. 8 pp.

TING, J.M. (1981). The creep of frozen sands: Qualitative and quantitative models. Unpublished Ph.D. Thesis, Department of Civil Engineering, M.I.T., 432 pp.

TOLSTIKHIN, N.I. & TOLSTIKHIN, O.N. (1974). *Groundwater and surface water in the permafrost region* (English translation). Canada, Inland Waters Directorate, Technical Bulletin 97 (1976), 25 pp.

TROLL, C. (1944). Strukturboden, Solifluktion und Frostklimate der Erde, *Geol. Rundschau*, **34**, 545–694. (Also as translation 43, Snow Ice and Permafrost Res. Estab. U.S. Army: *Structure soils, solifluction, and frostclimates of the earth*. 121 pp.)

TSYTOVICH, N.A. (1975). *The mechanics of frozen ground*. New York: Scripta, McGraw–Hill. 426 pp. (translation of: *Mekhanika Merzlykh Gruntov*, Moscow: Vysshaya Shkola Press, 1973).

TUFNELL, L. (1975). Hummocky microrelief in the Moor House area of the Northern Penines, England. *Biul. Perygl.*, **24**, 353–68.

TUMEL, N.V. & MUDROV, YU. V. (1973). Some patterns of formation of the horizon of discontinuous frost weathering. *Proceedings of the Second International Conference on Permafrost, Yakutsk, USSR, USSR Contribution*. pp. 207–9. Washington, D.C.: Nat. Acad. Sci.

UREY, H.C. (1947). The thermodynamic properties of isotopic substances. *J. Chem. Soc.*, **152**, 190–219.

VAN EVERDINGEN, R.O. (1974). Groundwater in permafrost regions of

Canada. In *Permafrost Hydrology*, pp. 83–93. Proceedings of a workshop seminar, Canadian National Committee for the Internatioanl Hydrologic Decade, Ottawa.

VAN EVERDINGEN, R.O. (1978). Frost mounds at Bear Rock, near Fort Norman, Northwest Territories, 1975–1976. *Canadian Journal of Earth Sciences*, **15**, 2, 263–76.

VAN EVERDINGEN, R.O. (1982). Frost blisters of the Bear Rock Spring area near Fort Norman, N.W.T. *Arctic*, **35**, 2, 243–65.

VAN VLIET-LANOE, B. (1982). Structures et microstructures associées à la formation de glace de ségrégation: Leurs consequences. *Proceedings of the Fourth Canadian Permafrost Conference, Calgary, Alberta*. pp. 116–22. Ottawa: N.R.C.C.

VAN VLIET-LANOE, B. (1985). From frost to gelifluction: A new approach based on micromorphology its applications to Arctic environment. *Inter-Nord*, **17**, 15–20.

VAN VLIET-LANOE, B., CLOUTARD, J.P. & PISSART, A. (1984). Structures caused by repeated freezing and thawing in various loamy sediments: A comparison of active fossil and experimental data. *Earth Surface Processes and Landforms.*, **9**, 553–65.

VIALOV, S.S., (ED.) (1965a). *The Strength and Creep of Frozen Soils and calculations for Ice–soil Retaining Structures*. Translation 76, United States Army, Cold Regions Research and Engineering Laboratory (from the Russian). Hanover, N.H.

VIALOV, S.S. (1965b). *Rheological Properties and Bearing Capacity of Frozen Soils*. Translation 74, United States Army, Cold Regions Research and Engineering Laboratory, (from the Russian). Hanover, N.H.

VIERECK, L. (1970). Forest succession and soil development adjacent to the Chena River in Interior Alaska. *Arctic and Alpine Research*, **2**, 1, 1–26.

VIGDORCHIK, M.E. (1980). *Arctic Pleistocene History and the Development of Submarine Permafrost*. Boulder, Colorado: Westview Press. 286 pp.

WALKER, H.J. & ARNBORG, L. (1963). Permafrost and ice-wedge effect on riverbank erosion. *Proceedings Permafrost International Conference*, pp. 164–71. Washington D.C. N.A.S.–N.R.C. Publication 1287.

WASHBURN, A.L. (1956). Classification of patterned ground and review of suggested origins. *Geol. Soc. Am. Bull.*, **67**, 823–65.

WASHBURN, A.L. (1979). *Geocryology*. Edward Arnold, 406 pp.

WASHBURN, A.L. (1980). Permafrost features as evidence of climatic change. *Earth Sciences Review*, **15**, 327–402.

WEAST, R.C. (1979). *Handbook of physics and chemistry*, Chem. Rubber Co.

WEAVER, J.S. & STEWART, J.M. (1982). *In situ* hydrates under the Beaufort Sea. *Proc. Fourth Can. Perm. Conf.* pp. 312–19.

WELLER, G. (1984). A monitoring strategy to detect carbon dioxide-induced climatic changes in the polar regions. In *The potential effects of carbon dioxide-induced climatic changes in Alaska*, ed. J.H. McBeath. School of Agriculture and Land Resources Management, University of Alaska-Fairbanks. Misc. Publ. 83-1, 23–30.

WILLIAMS, J.R. (1970). *Groundwater in permafrost regions of Alaska*. United States Geological Survey Professional Paper 696, 83 pp.

WILLIAMS, J.R. & VAN EVERDINGEN, R.O. (1973). Groundwater investigations in permafrost regions of North America: A review. *Proceedings of the Second International Conference on Permafrost, Yakutsk, USSR, North American Contribution*, pp. 435–46. Washington, D.C.: Nat. Acad. Sci.

WILLIAMS, P.J. (1957). Some investigations into solifluction features in Norway. *Geogr. Jour.*, **123**, 1, 42–58.

WILLIAMS, P.J. (1958). *The development and significance of stony earth circles.* Norsk. Vid. Akad., Oslo I Mat. – Naturv. Kl. 3, 14 pp.

WILLIAMS, P.J. (1963). Suction and its effects in unfrozen water of frozen soils. *Proceedings, Permafrost International Conference.* pp. 225–9. Washington D.C.: N.A.S./N.R.C. publication 1287.

WILLIAMS, P.J. (1964a). Experimental determination of apparent specific heats of frozen soil. *Geotechnique*, **140**, 2, 133–42.

WILLIAMS, P.J. (1964b). Unfrozen water content of frozen soils and soil moisture suction, *Geotechnique*, **14**, 3, 231–46.

WILLIAMS, P.J. (1966a). Suction and its effects in unfrozen water of frozen soils. *Proceedings, Permafrost International Conference.* pp. 225–9. N.A.S./N.R.C. publication 1287.

WILLIAMS, P.J. (1966b). Downslope soil movement at a sub-arctic location with regard to variations with depth. *Canadian Geotechnical Journal*, **3**, 191–203.

WILLIAMS, P.J. (1967). *Properties and behaviour of freezing soils.* Norwegian Geotechnical Institute Paper 72. 120 pp.

WILLIAMS, P.J. (1968). Ice distribution in permafrost profiles. *Canadian Journal of Earth Science*, **5**, 1381–6.

WILLIAMS, P.J. (1976). Volume change in frozen soils. *Laurits Bjerrum Mem. Vol.* Norw. Geot. Inst. Oslo. pp. 233–346.

WILLIAMS, P.J. (1982). *The surface of the earth: an introduction to geotechnical science.* London: Longman. 212 pp.

WILLIAMS, P.J. (1986). *Pipelines and Permafrost. Science in a Cold Climate.* Ottawa: Carleton University Press. 137 pp.

WILLIAMS, P.J. & WOOD, J.A. (1985). Internal stresses in frozen ground. *Canadian Geotechnical Journal*, **22**, 3, 413–6.

WOO, M-K. (1982). Upward flux of vapour from frozen materials in the high arctic. *Cold Regions Science and Technology*, **5**, 3, 269–74.

WOO, M-K. (1986). Permafrost hydrology in North America. *Atmosphere–Ocean*, **24**, 201–34.

WOO, M-K., HERON, R. & STEER, P. (1981). Catchment hydrology of a high Arctic lake. *Cold Regions Science and Technology*, **5**, 1, 29–41.

WOO, MING-KO. & STEER, P. (1983). Slope hydrology as influenced by thawing of the active layer, Resolute, N.W.T. *Canadian Journal of Earth Sciences*, **20**, 6, 978–86.

WOOD, J.A. & WILLIAMS, P.J. (1985a). Further experimental investigation of regelation flow with an ice sandwich permeameter. In *Freezing and thawing of soil–water systems.* ed. D.M. Anderson & P.J. Williams. pp. 85–94. Technical Council on Cold Regions Engineering Monograph, New York: American Society of Civil Engineers.

WOOD, J.A. & WILLIAMS, P.J. (1985b). Stress distribution in freezing soil. *Proceedings of Fourth International Symposium on Ground Freezing, Sapporo, Japan*, vol. 1, pp. 165–71.

WRIGHT, R.K. (1981). *The water balance of a lichen tundra underlain by permafrost.* McGill Subarctic Research Paper No. 33, 109 pp.

WU, T.H. (1984). Soil movements on permafrost-slopes near Fairbanks, Alaska. *Canadian Geotechnical Journal*, **21**, 699–709.

YONEYAMA, K., ISHIZAKI, T. & NISHIO, N. (1983). Water redistribution measurements in partially frozen soil by x-ray technique. *Proceedings of the*

Fourth International Conference on Permafrost, Fairbanks, Alaska. pp. 1445–50. Washington, D.C.: Nat. Acad. Sci. – Nat. Acad. Press.

YONG, R.N., BOONSINSUK, P. & YIN, C.W.P. (1985). Alteration of soil behaviour after cyclic freezing and thawing. *Fourth International Symposium on Ground Freezing, Sapporo, Japan*, pp. 187–95.

ZHOU, Y. & GUO DONGXIN. (1983). Some features of permafrost in China. *Proceedings of the Fourth International Conference on Permafrost, Fairbanks, Alaska*. pp. 1496–501. Washington, D.C.: Nat. Acad. Sci. – Nat. Acad. Press.

ZOLTAI, S.C. (1971). Southern limit of permafrost features in peat landforms, Manitoba and Saskatchewan. *Geological Association of Canada, Special Paper #9*, 305–310.

ZOLTAI, S.C. & TARNOCAI, C. (1975). Perenially frozen peatlands in the western Arctic and subarctic of Canada. *Canadian Journal of Earth Sciences*, **12**, 28–43.

ZHU, Y. & CARBEE, D.L. (1983). Creep behaviour of frozen silt under constant uniaxial stress. *Proceedings of the Fourth International Conference on Permafrost, Fairbanks, Alaska*. pp. 1507–12. Washington, D.C.: Nat. Acad. Sci. – Nat. Acad. Press.

ZHU, Y. & CARBEE, D.L. (1987). Creep and strength behaviour of frozen silt in uniaxial compression. *United States Army, Cold Regions Research and Engineering Laboratory*, Report 87-10. 67 pp.

Index